Contents

Introduction

New biological discoveries are made each day which add to our knowledge and understanding of the world and make it a safer and healthier place to live. With such discoveries also come the moral and ethical dilemmas on how to apply new technology in a way that does not threaten the long-term future of our world. The book has been written with these issues in mind. Factual information on topics such as genetic engineering, acid rain, the greenhouse effect and the human genome project is presented alongside an evaluation of their moral, ethical and economic implications.

Essential AS Biology has been written for the new advanced subsidiary (AS) specifications that lie mid-way between GCSE and Advanced level. The layout is clear and straightforward, with double-page spreads being used to separate topics up into manageable portions which can studied and absorbed easily. Within each spread, the information is organised into logical sections with extensive use of bullet-points to help understanding and to aid revision for examinations. The style and use of language is designed to make the text accessible and comprehension easier. Photographs and diagrams are used extensively to further aid clarity. The Summary Tests at the end of each spread test the reader's knowledge. The samples of AS examination questions at the end of each section are from recent papers set by Awarding Bodies and allow readers to apply their knowledge and to test their understanding. Further questions at the end of the book are of a longer type and may test information contained in several units.

Biology is, however, an integrated subject: the individual topics are not isolated but interconnected. For this reason considerable use is made of cross referencing to allow the reader to link the topics in a coherent and meaningful way. The use of double-page spreads allows flexibility in the use of the book. While certain fundamental topics are naturally covered in the early part of the book and each chapter contains related topics, there is nevertheless no requirement to study the topics in numerical order. Topics may be covered in the order that suits the student, teacher or the demands of the particular Awarding Body's specification. To aid this process, certain words are highlighted in **purple.** These words are terms that are not explained within that specific topic. Should the reader not understand them, they can turn to the glossary at the back for a short definition. The glossary is therefore not so much a comprehensive list of definitions as an explanatory collection of those terms that are used outside the context of the topic being studied.

The book has been written after a thorough analysis of all the Biology and Biology (Human) AS specifications of the five English and Welsh Awarding Bodies and therefore provides complete coverage of the units they contain. The content also includes the vast majority of the material required for Scottish Highers and the Northern Ireland Board examinations. The book will also provide valuable support to students studying vocational A-levels in Science or Health and Social Care as well as the International Baccalaureate. To enable candidates to select the material relevant to their own specification, there is a signpost at the start of each double-page spread to indicate which of the 6 major specifications the topic covers. This allows the reader to focus precisely on the material relevant for them. The reading of material from other specifications will, however, be useful, either as general interest or to provide valuable background information.

To allow students to test their knowledge and understanding, a website connected to the book has been established: **www.nelsonthornes.com/essentialbiology**.

This web site contains a variety of short answer questions to test knowledge and understanding of the test.

We trust that you will enjoy reading this book and find it interesting and informative. We hope that it will stimulate an interest in Biology that encourages you to pursue your study further. Above all, we hope it will contribute to your success in AS examinations.

Glenn and Susan Toole

Biological molecules

Water and inorganic ions

	EDEXCEL
	EDEXCEL (Human)
AQA.B	OCR

Although water is the most abundant liquid on Earth, it is certainly no ordinary molecule. Its unusual properties are due to its dipolar nature and the subsequent hydrogen bonding this allows.

1.1.1 The dipolar water molecule

A water molecule is made up of two atoms of hydrogen and one of oxygen as shown in figure 1.1. The atoms form a triangular shape. Although the molecule has no overall charge, the distribution of the negatively charged electrons is uneven because the oxygen atom draws them away from the hydrogen atoms. The oxygen atom therefore has a slight negative charge, while the hydrogen atoms have a slight positive one. In other words, the water molecule has both positive and negative poles and is therefore described as **dipolar**.

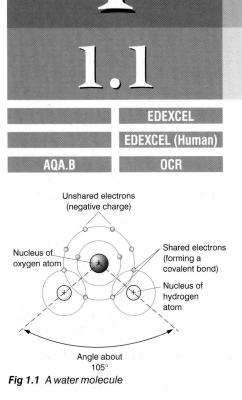

Unshared electrons (negative charge)

Nucleus of oxygen atom

Shared electrons (forming a covalent bond)

Nucleus of hydrogen atom

Angle about 105°

Fig 1.1 *A water molecule*

1.1.2 Water and hydrogen bonding

Different poles attract, and therefore the positive pole of one water molecule will be attracted to the negative pole of another water molecule. The attractive force between these opposite charges is called a **hydrogen bond** (Fig 1.2). Although each bond is fairly weak (about one-tenth as strong as a **covalent bond**), together they form important forces that cause the water molecules to stick together, giving water its unusual properties.

1.1.3 Specific heat capacity of water

As water molecules stick together, it takes more energy (heat) to separate them than would be needed if they did not bond to one another. For this reason the boiling point of water is higher than expected. Without its hydrogen bonding, water would be a gas (water vapour) at the temperatures commonly found on Earth, and life as we know it would not exist. For the same reason, it takes more energy to heat a given mass of water, i.e. water, has a high **specific heat capacity**. Water therefore acts against sharp temperature variations, making the aquatic environment a stable one as far as temperature is concerned.

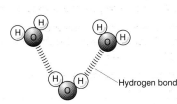

Hydrogen bond

Fig 1.2 *Water molcules, showing hydrogen bonding*

1.1.4 Latent heat of vaporisation of water

Hydrogen bonding between water molecules means it requires a lot of energy to evaporate one gram of water. This energy is called the **latent heat of vaporisation**. Evaporation of water such as sweat in mammals is therefore a very effective means of cooling.

1.1.5 Cohesion and surface tension in water

The tendency of molecules to stick together is known as **cohesion**. With its hydrogen bonding, water has large cohesion forces and these allow it to be pulled up through a tube such as a xylem vessel in plants. In the same way, water molecules at the surface of a body of water tend to be pulled back into the body of water rather than escaping from it. This force is called **surface tension** and means that the surface acts like a skin which is strong enough to support small organisms such as pond skaters.

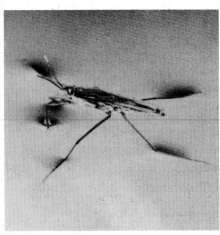

Pond skater walking on water

1.1.6 The density of water

Most substances are at their least dense when a gas and their most dense when a solid, with the liquid phase having an intermediate density. Water is different. Water is actually less dense as ice than when it is a liquid. This property is crucial to the survival of aquatic organisms as it means that ponds, lakes etc freeze from the top down rather than the bottom up. The ice formed at the top then acts as an insulating layer which delays the freezing of the water beneath it. Large bodies of water therefore almost never freeze completely, allowing their inhabitants to survive.

1.1.7 The importance of water to living organisms

Water is the main constituent of all organisms – up to 98% of a jellyfish is water and mammals are typically 65% water. It is also where life on Earth arose and is the environment in which many species still live. It is important for other reasons too.

Water in metabolism
- Water is used to break down many complex molecules by **hydrolysis**, e.g. proteins to amino acids.
- All chemical reactions take place in an aqueous medium.
- Water is a major raw material in photosynthesis.

Water as a solvent
Water readily dissolves other substances and is used for:
- transport, e.g. blood
- removal of wastes, e.g. ammonia, urea
- secretions, e.g. digestive juices, tears.

Water as a lubricant
Water's viscosity (stickiness) makes it a good lubricant. Examples include:
- mucus to aid external movement, e.g. of an earthworm, or internal movement, e.g. in the gut and vagina
- fluids to reduce friction, e.g. synovial fluid in joints, pleural fluid around the lungs and perivisceral fluid around internal organs.

Water giving support
Water is not easily compressed and therefore is used in:
- the **hydrostatic skeleton** of animals such as earthworms
- the amniotic fluid to support the foetus
- creating turgor pressure which supports the leaves of plants.

Other importance of water
- Its evaporation cools organisms and allows them to control their temperature.
- It is transparent and therefore aquatic plants can photosynthesise.

1.1.8 The roles of inorganic ions

Although needed only in tiny amounts, inorganic ions are essential to the normal functioning of organisms. The roles of ten of the most important ions are shown in table 1.1.

Table 1.1 *Roles of some important ions*

Ion	Functions
Nitrate NO_3^-	A component of amino acids, proteins, vitamins, coenzymes, nucleotides and chlorophyll. Some hormones contain nitrogen, e.g. auxins in plants and insulin in animals
Phosphate PO_4^{3-}	A component of nucleotides, ATP and some proteins. Used in the phosphorylation of sugars in respiration. A major constituent of bone and teeth. A component of cell membranes, in the form of **phospholipids**
Potassium K^+	Helps to maintain the electrical, osmotic and anion/cation balance across cell membranes. Assists active transport of certain materials across the cell membrane. Necessary for protein synthesis and is a co-factor in photosynthesis and respiration. A constituent of sap vacuoles in plants and so helps to maintain turgidity
Calcium Ca^{2+}	In plants, calcium pectate is a major component of the middle lamella of cell walls. It also aids the translocation of carbohydrates and amino acids. In animals, it is the main constituent of bones, teeth and shells. Needed for the clotting of blood and the contraction of muscle
Sodium Na^+	Helps to maintain the electrical, osmotic and anion/cation balance across cell membranes. Assists active transport of certain materials across the cell membrane. A constituent of the sap vacuole in plants and so helps maintain turgidity
Chlorine Cl^-	Helps to maintain the electrical, osmotic and anion/cation balance across cell membranes. Needed for the formation of hydrochloric acid in gastric juice. Assists in the transport of carbon dioxide by blood (chloride shift)
Magnesium Mg^{2+}	A constituent of chlorophyll. An activator for some enzymes, e.g. ATPase. A component of bones and teeth
Iron Fe^{2+} or Fe^{3+}	Found in electron carriers used in respiration and photosynthesis. Required to make chlorophyll. Forms part of the haemoglobin molecule
Sulphate SO_4^{2-}	Found in some proteins and certain coenzymes (e.g. acetyl coenzyme A)
Hydrogen-carbonate HCO_3^-	Helps to maintain the ionic balance of cells. Important in the transport of carbon dioxide in the blood

SUMMARY TEST 1.1

A water molecule is (1) because it has a positive and a negative pole as a result of the uneven distribution of (2) within it. This creates attractive forces called (3) between water molecules, causing them to stick together. This stickiness, otherwise called (4), of water means that its molecules are pulled inward at its surface. This force is called (5). Water is able to split large molecules into smaller ones by a process known as (6). Water is the raw material for the process of (7) in green plants and acts as a (8) in the synovial joints of animals. Three ions found in bone are (9), (10) and (11), whereas the three needed for maintaining an osmotic balance across cell membranes are (12), (13) and (14). The ion found in both vitamins and insulin is (15), whereas (16) is found in teeth but not in ATP or chlorophyll.

Carbohydrates – sugars

As the word suggests, carbohydrates are carbon molecules (carbo) combined with water (hydrate); their general formula is $C_x(H_2O)_y$. As with many organic molecules, they are made up of individual units called **monomers**, which can be combined to form larger ones called **polymers**. In carbohydrates the basic unit is a sugar or **saccharide**. A single unit is called a **monosaccharide**, which can be combined in pairs to form a **disaccharide**. Monosaccharides are usually combined in much larger numbers, to form **polysaccharides** (see unit 1.3).

1.2.1 Monosaccharides

Monosaccharides are sweet tasting, soluble substances which have the general formula $(CH_2O)_n$. While 'n' can be any number from 3 to 7, the three most common groups of monosaccharides are shown in table 1.2.

1.2.2 Structure of monosaccharides

Perhaps the best known monosaccaride is **glucose**. This molecule is a hexose sugar and has the formula $C_6H_{12}O_6$. However, the atoms of carbon, hydrogen and oxygen can be arranged in many different ways. Although the molecular arrangement is often shown as a straight chain for convenience, the atoms actually form a ring which can take a number of forms, as shown in figure 1.3. Different molecular structures are given different names, e.g. glucose, fructose, galactose, and further differences are shown by a letter before the molecule's name, e.g. α-glucose, β-glucose. Although some of these differences are small, they give the resulting molecules different properties.

Table 1.2 Types of monosaccharide

Formula	Name	Examples
$C_3H_6O_3$ (n = 3)	Triose	Glyceraldehyde
$C_5H_{10}O_5$ (n = 5)	Pentose	Ribose Deoxyribose
$C_6H_{12}O_6$ (n = 6)	Hexose	Glucose Fructose Galactose

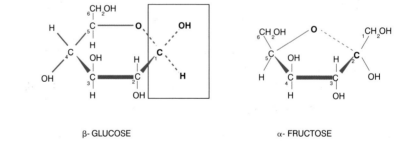

α- GLUCOSE   β- GLUCOSE   α- FRUCTOSE

Fig 1.3 Structure of different hexoses

1.2.3 Disaccharides

When combined in pairs, monosaccharides form a **disaccharide**. As table 1.3. shows, the two monosaccharides which combine can be the same or different. When they join, a molecule of water is removed and the reaction is therefore called a **condensation reaction**. The bond which is formed is called a **glycosidic bond**. As this bond is between carbon atom 1 of one monosaccharide and carbon atom 4 of the other, it is known as a 1,4 glycosidic bond. Figure 1.4 illustrates the formation of a disaccharide.

When water is added to a disaccharide under suitable conditions, it breaks the glycosidic bond into its constituent monosaccharides. This is called **hydrolysis** (breakdown by water). The breakdown is very slow however, unless it is catalysed by the appropriate enzyme.

Table 1.3 Types of disaccharide

glucose	+	glucose	=	maltose
glucose	+	fructose	=	sucrose
glucose	+	galactose	=	lactose

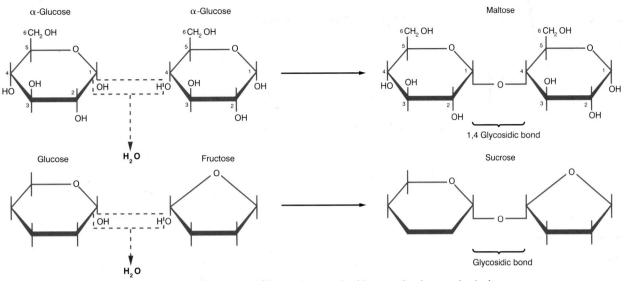

The removal of water (condensation) from the two hydroxyl groups (-OH) on carbons 1 and 4 of the respective glucose molecules forms a maltose molecule. Some carbon and hydrogen atoms have been omitted for simplicity. Sucrose is formed by a condensation reaction between one glucose and one fructose molecule. The process shown is much simplified.

Fig 1.4 *Formation of disaccharides*

1.2.4 Tests for reducing and non-reducing sugars

All monosaccharides and some disaccharides (e.g. maltose) are reducing sugars. The test for a reducing sugar is known as the **Benedict's test**. When a reducing sugar is heated with an alkaline solution of copper II sulphate (Benedict's reagent) it forms an insoluble precipitate of copper I oxide. The colour of the precipitate changes from green through yellow, orange and brown to deep red, depending on the quantity of reducing sugar present. Some disaccharides, such as sucrose, are non-reducing sugars. There is no direct test for a non-reducing sugar, but they can be identified by first hydrolysing them with a dilute acid and then detecting the resultant reducing sugars by the Benedict's test. The process is therefore:

- heat sample with Benedict's reagent. If there is no change (solution remains blue), then no reducing sugar is present
- heat the sample for 5 minutes with dilute hydrochloric acid to hydrolyse the non-reducing sugar and then neutralise with sodium hydrogen carbonate
- retest the resulting solution with Benedict's reagent, which will now turn yellow/brown/red due to the reducing sugars made from the non-reducing sugar.

1.2.5 Roles of monosaccharides and disaccharides

Generally, sugars function as respiratory substrates which are broken down to provide energy in the form of ATP for carrying out living processes. Table 1.4 lists the roles of the most important sugars.

Table 1.4 *Roles of sugars*

Name of carbohydrate	Function
Ribose / deoxyribose	Makes up part of nucleotides and as such gives structural support to the nucleic acids RNA and DNA. Constituent of hydrogen carriers such as NAD, NADP and FAD. Constituent of ATP
Glucose	Major respiratory substrate in plants and animals. Used in the synthesis of disaccharides and polysaccharides. Constituent of nectar. Major transporter of carbohydrate in mammals
Galactose	Respiratory substrate. Used in the synthesis of lactose
Fructose	Respiratory substrate. Constituent of nectar. Sweetens fruits, which attracts animals and aids seed dispersal
Sucrose	Respiratory substrate. Form in which most carbohydrate is transported in plants. Storage material in some plants
Lactose	Respiratory substrate. Mammalian milk contains 5% lactose and is therefore a major carbohydrate source for sucklings
Maltose	Respiratory substrate

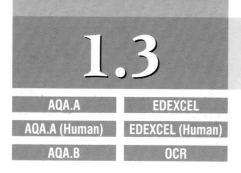

1.3 Carbohydrates – polysaccharides

AQA.A   EDEXCEL
AQA.A (Human)   EDEXCEL (Human)
AQA.B   OCR

The combining together of many monosaccharides forms a polymer known as a **polysaccharide**. The monosaccharides are joined by glycosidic bonds that are formed by condensation reactions. The resulting chain may vary in length, be branched and be folded in various different ways. All these features affect the properties of the polysaccharide that is formed. As polysaccharides are very large molecules **(macromolecules)**, they are insoluble – a feature which suits them for storage. When they are hydrolysed, polysaccharides break down into monosaccharides or disaccharides. Some polysaccharides such as cellulose are not used for storage, but give structural support to plant cells.

1.3.1 Starch

Starch is a polysaccharide which is found in many parts of a plant in the form of small granules, but in especially large amounts in seeds and storage organs such as potato tubers. It forms an important component of food and is the major energy source in most diets. Apart from the starch produced for eating, about 30 million tonnes are extracted from plants across the world for other purposes. These include wallpaper pastes, paper coatings, textiles, paints, cosmetics and medicines. Starch is a mixture of two substances – amylose and amylopectin.

Amylose is composed of between 200 and 5000 α-glucose units which are joined in a straight chain by 1,4 glycosidic bonds.

Amylopectin is made up of between 5000 and 100 000 β-glucose units joined to each other by 1,6 glycosidic bonds. A comparison of amylose and amylopectin is

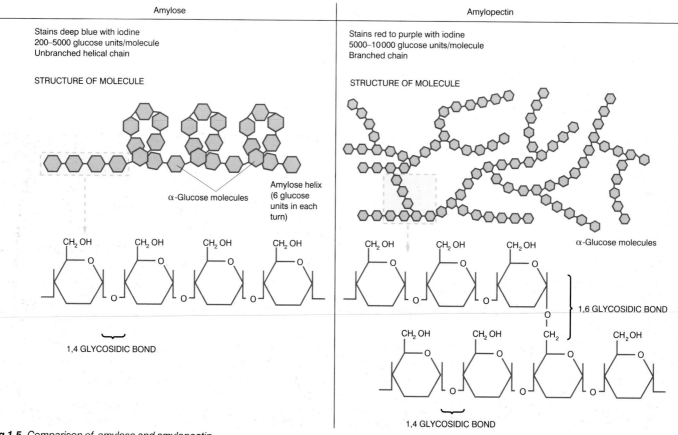

Fig 1.5 *Comparison of amylose and amylopectin*

given in figure 1.5. About 80% of starch is amylopectin and the remaining 20% is amylose. The main role of starch is for energy storage, something it is especially suited for because:

- it is insoluble and therefore does not have any osmotic influence within cells, i.e. it does not tend to draw water into the cells
- being insoluble, it does not easily diffuse out of cells
- as it is compact, a lot of it can be stored in a small place
- when hydrolysed, it forms glucose, which is both easily transported and readily used in respiration, to provide energy.

1.3.2 Glycogen

Glycogen is very similar in structure to amylopectin but has shorter chains and is more highly branched. It is sometimes called 'animal starch' because it is the major carbohydrate storage product of animals, in which it is stored mainly in the muscles and the liver, as small granules. Its structure suits it for storage for the same reasons as given for starch (see section 1.3.1), except that because it is made up of smaller chains it is even more readily hydrolysed to α-glucose.

1.3.3 Cellulose

Cellulose differs from starch and glycogen in one major respect – it is made of monomers of β-glucose rather than α-glucose. This seemingly small variation produces fundamental differences in the structure and function of this polysaccharide. Rather than forming a coiled chain like starch, cellulose has straight, unbranched chains. These run parallel to one another, allowing hydrogen bonding between them to form cross linkages, adding to its stability and making it a valuable structural material. As a result, the chains can be bundled together in groups of up to 2000 to form a **microfibril** (Fig 1.6).

Cellulose typically makes up 50% of a plant cell wall and its stability makes it difficult to digest. It is therefore not a useful source of food for animals, which rarely produce cellulose-digesting enzymes. Some animals get round this by forming mutualistic relationships with microorganisms which digest cellulose (see section 9.2.2). The structural strength of cellulose has been made use of by humans. Cotton and rayon used in fabrics are largely cellulose. Cellophane used in packaging and celluloid, used in photographic films, are also derived from cellulose. Paper is perhaps the best known cellulose product.

TEST FOR STARCH

Starch is easily detected by its ability to turn iodine in potassium iodide solution from a yellow colour to blue-black. The colouration is due to the iodine molecules becoming fixed in the centre of the helix of each starch molecule. It is important that this test is carried out at room temperature (or below), as high temperatures cause the starch helix to unwind, releasing the iodine, which assumes its usual yellow colouration

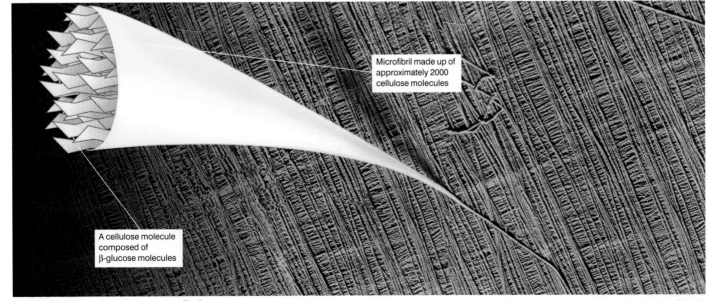

Microfibril made up of approximately 2000 cellulose molecules

A cellulose molecule composed of β-glucose molecules

Fig 1.6 *Structure of a cellulose microfibril*

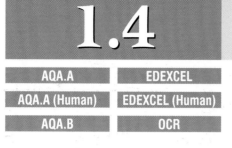

Lipids

Lipids are a varied and diverse group of substances which share the following characteristics:
- they contain carbon, hydrogen and oxygen
- the proportion of oxygen is smaller than in carbohydrates
- they are insoluble in water
- they are soluble in organic solvents such as alcohols and acetone.

The main groups of lipids are **triglycerides (fats and oils)**, **phospholipids** and **waxes**. Other forms of lipids include steroids and cholesterol.

1.4.1 Triglycerides (fats and oils)

There is no fundamental chemical difference between a fat and an oil. Fats are solid at room temperature (10–20°C), whereas oils are liquid. Triglycerides are so called because they have three (tri) fatty acids combined with glycerol (glyceride). Each fatty acid forms an ester bond with glycerol in a condensation reaction (see figure 1.7). Hydrolysis of a triglyceride therefore produces glycerol and three fatty acids.

The three triglycerides may all be the same, thereby forming a simple triglyceride, or they may be different, in which case a mixed triglyceride is produced. In either case it is a condensation reaction.

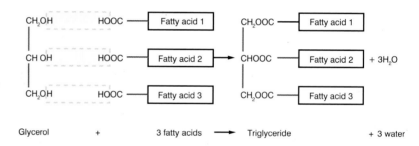

Fig 1.7 *Formation of a triglyceride*

1.4.2 Fatty acids

As the glycerol molecule in all triglycerides is the same, the differences in the properties of different fats and oils come from variations in the fatty acids. There are over 70 fatty acids and all have a carboxyl (-COOH) group with a hydrocarbon chain attached. This chain may possess no double bonds and is then called **saturated**, because all the carbon atoms are linked to the maximum possible number of hydrogen atoms, i.e. they are saturated with hydrogen atoms. If the chain has some double bonds it is called **unsaturated**. If there is a single double bond it is **mono-unsaturated**; if more than one double bond is present, it is **polyunsaturated**.

1.4.3 Phospholipids

Phospholipids are similar to lipids except that one of the fatty acid molecules is replaced by a phosphate molecule (Fig 1.8). Whereas fatty acid molecules repel water (are **hydrophobic**), phosphate molecules attract water (are **hydrophilic**).

Fig 1.8 *Structure of a phospholipid*

Phospholipids are important components of cell membranes. Both the inside of a cell and the environment outside are watery, and the phospholipids in cell membranes form a double layer, with the hydrophilic heads of the molecules pointing into either the watery environment outside the membrane or the watery medium inside the cell. The hydrophobic tails point into the middle of the membrane (Fig 1.9). This **bilayer** arrangement makes cell membranes fluid and easily traversed by lipid-soluble substances.

1.4.4 Waxes

Waxes are similar to triglycerides, except that glycerol is replaced by a longer chain alcohol, making them more complex structures. They are more solid than fats at room temperature and their major role is in producing a protective waterproofing cuticle on the surfaces of leaves and insects.

1.4.5 Roles of lipids

Due to their variety of types and forms, lipids perform many different functions in living organisms. These include:

- **An energy source** – Lipids provide more than twice as much energy as carbohydrate when they are oxidised. This makes them excellent stores of energy, especially in animals and plant seeds, both of which have to move, or be moved, from place to place and therefore need to keep their mass to a minimum. Being insoluble in water, lipids are not easily leached from cells.
- **Insulation** – Fats are slow conductors of heat and so are often stored beneath the body surface in **endothermic** animals, which helps retain body heat. This sub-cutaneous fat is especially important in aquatic organisms, such as whales and seals, where hair is ineffective.
- **Protection** – Fat is often stored around delicate organs such as the kidney, where it acts as packing material to protect the organ from physical damage.
- **Waterproofing** – Water conservation is important to all terrestrial organisms, but especially those in hot and dry regions of the world. Plants and insects both have waxy cuticles that conserve water, while mammals produce an oily secretion from the sebaceous glands in the skin. The hydrophobic fatty acids repel water and so lipids are especially efficient at preventing water loss. In addition, the oxidation of lipids in respiration provides more water than the oxidation of an equivalent amount of carbohydrate – a considerable bonus where water is in short supply.
- **Buoyancy** – Lipids are less dense than water and so aquatic mammals which have fat for insulation enjoy the added advantage of being more buoyant. This is important, as aquatic mammals breathe air. Aquatic birds and insects use oil droplets to aid buoyancy.
- **Cell membranes** – Phospholipids are important in cell membranes, contributing to their flexibility and the transfer of lipid-soluble substances across them.

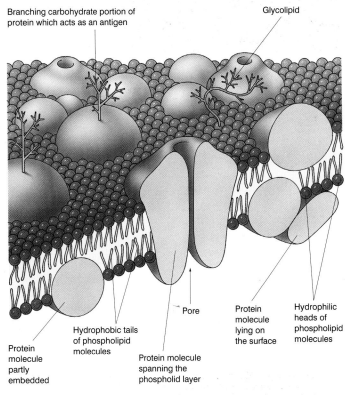

Fig 1.9 *Arrangement of the phospholipid bilayer in the cell membrane*

TEST FOR LIPIDS

The test for lipids is known as the **emulsion test** and is carried out as follows:

- Take a completely dry and grease-free test tube.
- Add about 2cm³ of the sample being tested and add 5cm³ of ethanol.
- Shake the tube thoroughly to dissolve any lipid in the sample.
- Add 5cm³ of water and shake gently.
- A cloudy-white colour indicates the presence of a lipid.
- As a control, repeat the procedures using water instead of the sample; the final solution should remain clear.

The cloudy colour is due to any lipid in the sample being finely dispersed in the water to form an emulsion. Light passing through this emulsion is refracted as it passes from oil droplets to water droplets, making it appear cloudy

SUMMARY TEST 1.4

Fats and oils make up a group of lipids called **(1)** which when hydrolysed form **(2)** and fatty acids. A fatty acid with more than one double bond is called **(3)**. In a phospholipid the number of fatty acid molecules is **(4)**; these are called **(5)** because they repel water. Lipids are important as an **(6)** in seeds and as **(7)** in endothermic animals. Around the kidneys they perform the role of **(8)** whereas in plant and insect cuticles they are used for **(9)**.

Amino acids and polypeptides

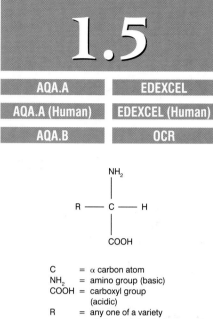

C = α carbon atom
NH₂ = amino group (basic)
COOH = carboxyl group (acidic)
R = any one of a variety of chemical structures
H = hydrogen atom

Fig 1.10 *General structure of an amino acid*

Amino acids are the monomers which make up proteins. Of the 100 or more amino acids identified, 20 are found naturally in proteins.

1.5.1 Amino acid structure

The name amino acid comes from the presence of an amino group ($-NH_2$) and an acid carboxyl group ($-COOH$) in these molecules. The general formula for an amino acid is shown in figure 1.10. As the carboxyl group is acid and the amino group is basic, an amino acid is both an acid and a base – it is **amphoteric**. Amphoteric compounds act as **buffer** solutions and therefore resist any tendency to alter their pH. This property is important in helping to keep a stable pH, which is necessary for the functioning of enzymes. Green plants make amino acids by combining nitrates absorbed from the soil with the carbohydrates made by photosynthesis. They are therefore able to manufacture all the amino acids they need. Animals, however, can manufacture only a dozen amino acids and the remaining eight, known as **essential amino acids**, must be obtained directly from the food eaten (section 14.7.3).

1.5.2 Dipeptides

Monomers such as amino acids can be combined together, with the loss of a water molecule, in what is known as a **condensation reaction** (see section 1.2.3). The water in this case is derived from the -OH of the carboxyl group and the H from the amino group. The resulting bond between two amino acids is known as a **peptide bond** and the new molecule is called a **dipeptide**. The formation of a dipeptide is illustrated in figure 1.11.

1.5.3 Polypeptides

Through a series of condensation reactions, many amino acid monomers can be joined together to form a polymer in a process called **polymerisation**. The resulting chain of many hundred amino acids is called a **polypeptide**. The amino acids in the chain form hydrogen bonds between one another which create a three-dimensional shape. Typically this three-dimensional shape is either a rod-like formation called an **α-helix** or a folded sheet known as a **β-pleated sheet** (Fig 1.12).

1.5.4 Chromatography

Chromatography is the name given to a number of techniques used in the chemical analysis of substances. It involves the separation of molecules such as amino acids so that each one can be identified or isolated for further investigation. This is done by moving the mixture **(mobile phase)**, usually a gas or liquid, over a **stationary phase**, usually a solid. **Paper chromatography** (Fig 1.13) is used to separate mixtures such as amino acids, sugars or photosynthetic pigments. Drops of the mixture are spotted onto one corner of a sheet of paper, one at a time, allowing them to dry each time. In this way a concentrated spot of the mixture is formed. The end of the paper is then dipped into a suitable solvent which moves up the paper by **capillarity**, carrying the molecules with it. Each molecule is carried a different distance according to its **relative molecular mass** and its solubility in the solvent. Where molecules such as amino acids are colourless, they can be revealed by treatment with chemicals which dye them different colours.

Two amino acids – R₁ and R₂ – represent any of the 20 or so groups commonly found in naturally occurring amino acids

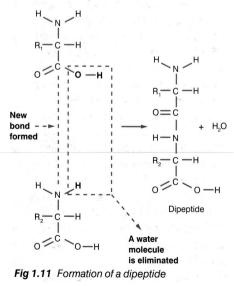

New bond formed

+ H₂O

Dipeptide

A water molecule is eliminated

Fig 1.11 *Formation of a dipeptide*

α-Helix　　　　　　　　　　　　　**β-Pleated sheet**

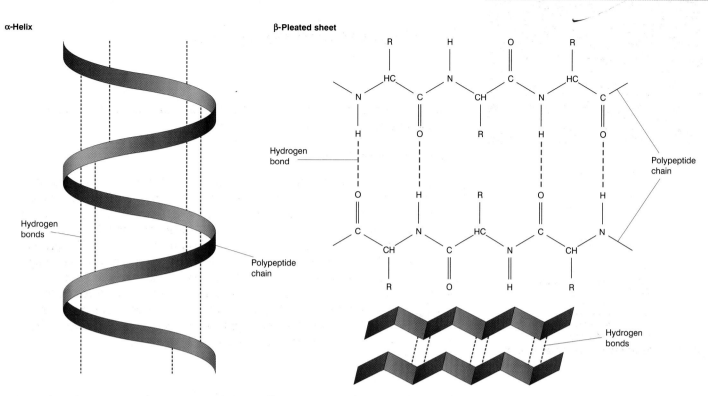

Fig 1.12 *Structure of the α-helix and the β-pleated sheet*

Each separated part may itself be a mixture of molecules. These can be further separated by turning the paper through 90° and repeating the process at right angles to the original using a different solvent. This process is known as **two-way chromatography**. The identification of a particular substance can be made on the basis of its colour and its position, relative to the distance moved by the solvent. Each substance has its own **Rf value** (retardation factor) for a given solvent. This value is calculated as:

$$Rf = \frac{\text{Distance travelled by the substance}}{\text{Distance travelled by the solvent front}}$$

Other forms of chromatography such as thin layer chromatography, column chromatography and gas chromatography are based on the same principle.

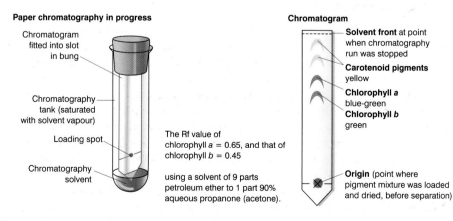

Fig 1.13 *Paper chromatography*

Protein structure

AQA.A	EDEXCEL
AQA.A (Human)	EDEXCEL (Human)
AQA.B	OCR

Proteins are large molecules with relative molecular masses from many thousands up to 40 million. While the carbohydrates and lipids of all organisms are relatively few and very similar, their proteins are numerous (around 10 000 types in humans) and differ from species to species. How proteins function is related to their structure.

1.6.1 Primary structure of proteins

The primary structure of a protein is the sequence of amino acids that makes up each of its polypeptide chains (Fig 1.14). This sequence of amino acids sets the pattern of additional bonding, such as hydrogen bonding, between the amino acids and therefore how the chain is folded. This in turn determines the overall three-dimensional shape of the molecule.

1.6.2 Secondary structure of proteins

Hydrogen bonding can occur between hydrogen atoms on the NH groups of one peptide bond and oxygen atoms on the CO group of another peptide bond within a single polypeptide molecule (Fig 1.15). Although each individual bond is weak, the sheer number of them means they play a considerable role in the shape and stability of a polypeptide molecule. The polypeptide is most often coiled into an **α-helix** or folded into a **β-pleated sheet** (Fig 1.12, section 1.5.3).

(a) The primary structure of a protein is the sequence of amino acids found in its polypeptide chains. This sequence determines its properties and shape. Following the elucidation of the amino acid sequence of the hormone insulin, by Frederick Sanger in 1954, the primary structure of many other proteins is now known.

(b) The secondary structure is the shape which the polypeptide chain forms as a result of hydrogen bonding. This is most often a spiral known as the α-helix, although other configurations occur.

(c) The tertiary structure is due to the bending and twisting of the polypeptide helix into a compact structure. All three types of bond, disulphide, ionic and hydrogen, contribute to the maintenance of the tertiary structure.

(d) The quaternary structure arises from the combination of a number of different polypeptide chains, and associated non-protein (prosthetic) groups, into a large, complex protein molecule.

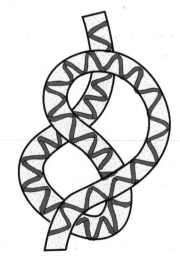

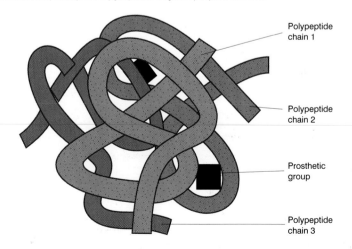

Polypeptide chain 1

Polypeptide chain 2

Prosthetic group

Polypeptide chain 3

Fig 1.14 Structure of proteins

1.6.3 Tertiary structure of proteins

The helices and/or sheets of the secondary protein structure can be twisted and folded even more, to give the complex, and often unique, three-dimensional structure of each protein (Fig 1.14). This is known as the **tertiary structure** and is the result of four possible types of bonds which can arise between the R groups (Fig 1.10) of each amino acid:

- **Disulphide bridges** are found between sulphur atoms in the molecules of the amino acid, cysteine. They are covalent bonds and, as such, form very strong links which make the tertiary protein structure very stable.
- **Ionic bonds** occur between any carboxyl and amino groups that have not been involved in forming peptide bonds. These groups ionize to give NH_3^+ and COO^- groups which then form bonds, due to their mutual attraction. These bonds are weaker than disulphide bridges and can be broken by changes in pH.
- **Hydrogen bonds** result from attraction between the electronegative oxygen atoms on the CO groups and the electropositive H atoms on either the OH or NH groups. Although they are individually weaker than ionic bonds, their large number makes them an important factor in maintaining the tertiary structure of a protein.
- **Hydrophobic interactions** are due to certain non-polar R groups in amino acids that have side groups which repel water. As a result they may fold or twist the polypeptide chain as they take up a position towards the centre of the protein, further away from the watery medium outside.

1.6.4 Quaternary structure of proteins

Large proteins often form complex molecules that contain a number of individual polypeptide chains which are linked in various ways. There may also be non-protein groups associated with the molecules (Fig 1.14). Examples of quaternary structure are illustrated by the blood protein haemoglobin and the hormone insulin (unit 1.7).

A simplified representation of a polypeptide chain to show three types of bonding responsible for shaping the chain. In practice the polypeptide chains are longer, contain more of these three types of bond and have a three-dimensional shape.

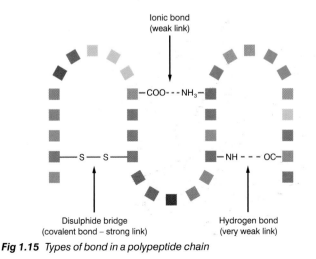

Fig 1.15 *Types of bond in a polypeptide chain*

SUMMARY TEST 1.6

The primary structure of proteins is determined by the sequence of (**1**) which make up the (**2**) chain. The secondary structure results from coiling or folding of the chain due to (**3**) formed between the NH and the (**4**) group of the (**5**) bonds. Three additional bonds cause further twisting and folding of the chain. The first of these bonds arise between (**6**) atoms in cysteine molecules and are called (**7**). The second type are called (**8**) and result from electrostatic forces between carboxyl and amino groups of amino acids. Thirdly there are forces due to amino acid side groups which repel water and these are called (**9**). The strongest of these bonds are the (**10**), followed by (**11**), while (**12**) are the weakest of the four. The quaternary structure of proteins results from a number of chains combining, sometimes also incorporating non-protein groups known as (**13**) groups.

Protein function

Proteins perform many different roles in living organisms. In one form or another they are essential for the efficient functioning of every characteristic of life. Their roles depend on their basic molecular configurations, which are of two basic types:

- **Fibrous proteins** such as collagen and keratin have structural functions.
- **Globular proteins** such as enzymes, haemoglobin and insulin carry out metabolic functions.

1.7.1 Fibrous proteins – e.g. collagen

Fibrous proteins form long chains which run parallel to one another. These chains are linked by cross bridges and so form very stable molecules. One example is **collagen**, a protein found in tissues requiring physical strength, e.g. tendons. It has a primary structure which is largely a repeat of the amino acid sequence glycine-proline-alanine and forms a long unbranched chain. Three of these chains are coiled around one another to give a triple helix (Fig 1.16). This provides a structure that is flexible but cannot be stretched, suiting it to its role in tendons, which attach muscles to bones. Another fibrous protein is keratin, which is found in horn, nails and hair.

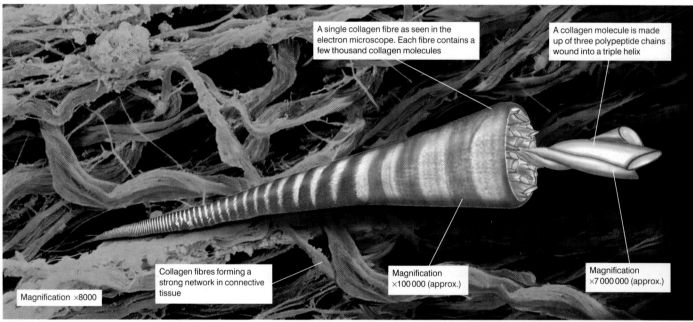

A single collagen fibre as seen in the electron microscope. Each fibre contains a few thousand collagen molecules

A collagen molecule is made up of three polypeptide chains wound into a triple helix

Collagen fibres forming a strong network in connective tissue

Magnification ×100 000 (approx.)

Magnification ×7 000 000 (approx.)

Magnification ×8000

Fig 1.16 *Fine structure of the fibrous protein, collagen*

1.7.2 Globular proteins – e.g. insulin

The sequence of amino acids in globular proteins is far more varied, and they form a more compact structure than a fibrous protein. If the polypeptide chains of a fibrous protein are thought of as string twisted into a rope, then a globular protein is like the same string rolled into a ball. One example of a globular protein is haemoglobin (Fig 1.16 and section 8.2.1). It has two α-polypeptide chains each made up of 141 amino acids, and two β-polypeptide chains both of 146 amino acids. Associated with each of these four chains is a single iron-containing haem group (Fig 1.17). The hormone insulin is another example of a globular protein. It is made up of 51 amino acids in two polypeptide chains which are linked by two disulphide bridges (Fig 1.18). Globular proteins form highly specific shapes which are essential for their functions as hormones and enzymes (unit 2.1).

Four polypeptide chains make up the haemoglobin molecule. Each molecule contains 574 amino acids

β   β

α   α

Each chain is attached to a haem group that can combine with oxygen

Fig 1.17 *Quaternary structure of a haemoglobin molecule*

1.7.3 Difference between fibrous and globular proteins

Table 1.5 *Comparison of fibrous and globular proteins*

Fibrous proteins	Globular proteins
Repetitive regular sequences of amino acids	Irregular amino acid sequences
Actual sequences may vary slightly between two examples of the same protein	Sequence highly specific and never varies between two examples of the same protein
Polypeptide chains form long parallel strands	Polypeptide chains folded into a spherical shape
Length of chain may vary in two examples of the same protein	Length always identical in two examples of the same protein
Stable structure	Relatively unstable structure
Insoluble	Soluble – forms **colloidal** suspensions
Support and structural functions	Metabolic functions
Examples include collagen and keratin	Examples include all enzymes, some hormones (e.g. insulin) and haemoglobin

1.7.4 Other important proteins and their functions

Table 1.6 *Protein functions*

Protein	Function of protein
Trypsin/pepsin	Digestion of proteins/polypeptides
Myoglobin	Stores oxygen in muscle
Actin/myosin	Needed for contraction of muscle
Antibodies	Defend against bacterial invasion
Gluten	Storage protein in seeds
Chromatin	Gives structural support to chromosomes

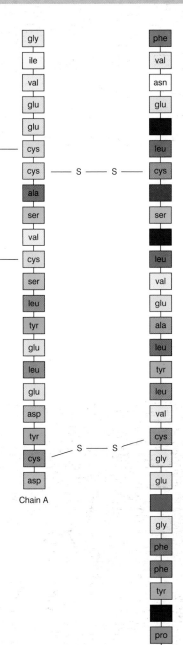

Fig 1.18 *The insulin molecule*

SUMMARY TEST 1.7

Proteins are of two basic types: fibrous proteins such as collagen and **(1)** and globular proteins such as the blood pigment **(2)**. Proteins such as actin and myosin in muscle have a structural function and are therefore examples of a **(3)** protein. Collagen is a fibrous protein with the repeating amino acid sequence glycine-**(4)**-alanine. It is found in **(5)** which attach muscle to bone, where its properties of **(6)** and **(7)** suit it to its role.

1 A tripeptide is made up of three amino acids. The diagram shows the molecular structure of a tripeptide.

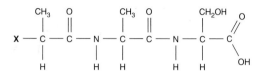

a (i) Give the formula of the chemical group at position **X** on the molecule. *(1 mark)*

(ii) Give **one** piece of evidence from the diagram that this molecule is made up from three amino acids. *(1 mark)*

b This tripeptide was broken down into its amino acids. These were separated and identified using chromatography. The diagram shows the resulting chromatogram.

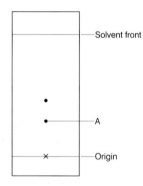

(ii) Mark the diagram with a line to show where the solvent should come to when the apparatus is set up. *(1 mark)*

(iii) The tripeptide was completely broken down into its amino acids but there are only two spots on the chromatogram. Explain why. *(1 mark)*
(Total 4 marks)
AQA June 2001, B/HB (A) BYA1, No.5

2 a Starch is an important storage substance in plants. Give **two** features of starch molecules and explain how each enables starch to act as an efficient storage substance. *(2 marks)*

Glucose syrup is used in the production of many human foods. It is produced from starch in a series of enzyme-controlled reactions.

b One way of monitoring the progress of these reactions is to measure the amount of reducing sugar produced.

(i) Describe a chemical test that would enable you to show that glucose syrup contained reducing sugar.

(ii) Suggest how you could use this test to compare the amount of reducing sugar in two solutions. *(4 marks)*

c The progress of these reactions can also be monitored by finding the dextrose equivalent (DE).

Dextrose equivalent can be calculated from the formula:

$$DE = \frac{\text{number of glycosidic bonds hydrolysed} \times 100}{\text{number of glycosidic bonds present in starch}}$$

Explain why pure glucose obtained from starch has a dextrose equivalent of 100. *(1 mark)*
(Total 7 marks)
AQA Jan 2001, B/HB (A) BYA1, No.3

3 A solution thought to contain either a reducing sugar or a non-reducing sugar was tested with Benedict's reagent.

a Describe how the presence of a reducing sugar is detected using Benedict's reagent. *(2 marks)*

b If the test was negative for reducing sugars, describe what steps you would need to carry out before you could show that a non-reducing sugar was present. *(3 marks)*

c Describe how Benedict's reagent could be used to compare the concentrations of reducing sugar present in two solutions. *(3 marks)*
(Total 8 marks)
Edexcel 6101/01 June 2001, B B(H) AS/A, No.6

4 The table below refers to some disaccharides, their constituent monomers and their roles in living organisms.

Disaccharide	Constituent monomers	ONE role in living organisms
Lactose		Carbohydrate source in mammalian milk
	Glucose + glucose	
		Form in which sugars are transported in plants

Complete the table by writing in the appropriate word or words in the empty boxes. *(Total 5 marks)*
Edexcel 6101/01 Jan 2001, B B(H) AS, No.2

5 a Explain what is meant by the **primary structure** of a protein molecule. *(2 marks)*

b Explain the role of hydrogen bonding in maintaining the structure of a globular protein such as insulin. *(3 marks)*

c Describe how the structure of a fibrous protein, such as collagen, differs from the structure of a globular protein. *(3 marks)*
(Total 8 marks)
Edexcel 6101/01 Jan 2001, B B(H) AS, No.4

6 a Extracts containing the pigments from carrot leaf and carrot root were compared by chromatography. The diagram shows the chromatogram that was produced.

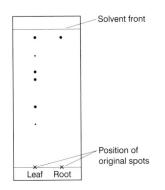

(i) Calculate the Rf value of the pigment that was found in both the leaf and the root. Show your working. *(2 marks)*

(ii) The investigators thought that some of the spots might contain more than one pigment. Describe how they could find out if this was true. *(2 marks)*

b Describe a biochemical test that you could use to test an extract of carrot root for the presence of non-reducing sugar. *(3 marks)*

(Total 7 marks)

AQA Jan 2001, B (B) BYB1, No.1

7 a Some cereal grains were soaked in water for 24 hours and then cut in half.

The diagram shows the cut surface of one grain.

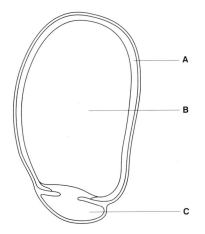

Three biochemical tests were carried out on samples from the parts of the grain labelled **A**, **B** and **C**. The results are shown in the table.

Part of grain	Tested with iodine solution	Biuret test	Emulsion test	Substances shown to be present
A	Brownish yellow	Lilac/mauve	Colourless	
B	Blue-black	Pale blue	Colourless	
C	Brownish yellow	Lilac/mauve	Cloudy white	

Complete the table to show which substances were present in each of the parts, **A**, **B** and **C**. *(3 marks)*

b Cut grains were placed with the cut surface downwards on a starch agar medium in a Petri dish. After 24 hours the grains were removed and the dish was flooded with iodine solution. The results are shown in the diagram.

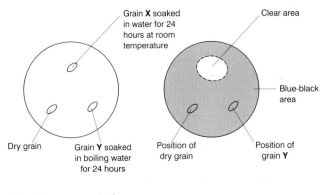

Explain the result for

(i) grain **X**; *(2 marks)*

(ii) grain **Y**. *(1 mark)*

(Total 6 marks)

AQA June 2001, B (B) BYB1, No.2

8 a State **one** role for each of the following in living organisms.

(i) calcium *(1 mark)*

(ii) magnesium *(1 mark)*

(iii) phosphate *(1 mark)*

(iv) sodium *(1 mark)*

(v) water *(1 mark)*

A solution is thought to contain both sucrose and glucose. A student carried out a test and confirmed that a small amount of glucose was present in the solution.

b Describe how the student could determine that the solution also contains sucrose. *(4 marks)*

c Describe the molecular structure of starch (amylose and amylopectin).

(In this question, 1 mark is available for the quality of written communication) *(8 marks)*

(Total 17 marks)

OCR 2801 June 2001, B (BF), No.3

2.1

Enzyme structure and mode of action

AQA.A	EDEXCEL
AQA.A (Human)	EDEXCEL (Human)
AQA.B	OCR

Enzymes are globular proteins which act as biological catalysts. A catalyst alters the rate of a chemical reaction without itself undergoing permanent change; it can therefore be used repeatedly and so is effective in tiny amounts. Enzymes do not make a reaction happen, they simply alter the speed of ones which already occur.

2.1.1 Enzyme structure

As globular proteins, enzymes have a specific three-dimensional shape which is determined by their quaternary structure (see section 1.6.4). Despite their large overall size, enzyme molecules have only a small region which is functional. This is known as the **active site**. Only a few amino acids of the enzyme molecule make up this active site; the remainder are used to maintain its shape.

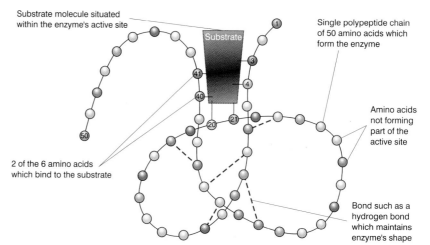

Substrate molecule situated within the enzyme's active site

Single polypeptide chain of 50 amino acids which form the enzyme

Amino acids not forming part of the active site

2 of the 6 amino acids which bind to the substrate

Bond such as a hydrogen bond which maintains enzyme's shape

Fig 2.1 Enzyme–substrate complex showing the six out of 50 enzyme amino acids which form the active site

2.1.2 Enzymes and activation energy

Consider a typical chemical reaction:

Sucrose + water   →   glucose + fructose
(substrates)   (products)

For such a reaction to occur naturally, the energy of the products (glucose + fructose) must be less than that of the substrate (sucrose). Such reactions, however, need an initial boost of energy to get them kick-started. This is known as the **activation energy**. It can be thought of as a stone lying on a hillside. If it is disturbed from its resting position, e.g. by being pushed, it will move downhill, rather than uphill, because this lowers its potential energy. Once set going, however, it gathers its own momentum and reaches the bottom with no further input of energy. This comparison shows how an initial input of energy (activation energy) can cause a reaction to continue on its own. In other words, there is an energy hill or barrier which must be overcome before the reaction can proceed. What enzymes do is to lower this activation energy level, so that the reaction can happen more easily (Fig 2.2). For example, they allow many reactions to take place at a lower temperature than normal.

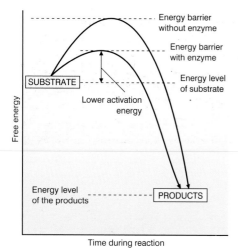

Energy barrier without enzyme

Energy barrier with enzyme

Energy level of substrate

Lower activation energy

SUBSTRATE

Free energy

Energy level of the products

PRODUCTS

Time during reaction

Fig 2.2 How enzymes lower the activation energy

2.1.3 How enzymes work

In one sense, enzymes operate in the same way that a key operates a lock: each key has a very specific shape which, on the whole, only fits and operates one lock. In the same way, a substrate will only fit the active site of one particular enzyme. The shape of the substrate (key) exactly fits the active site of the enzyme (lock). This is known as the **lock and key theory** and explains, in a simple way, what exactly happens (Fig 2.3). In practice, the process is more refined: it is suggested that, unlike a rigid lock, the enzyme actually changes its form slightly to fit the shape of the substrate. In other words, it is flexible and moulds itself around the substrate just as a glove moulds itself to the shape of someone's hand. The enzyme has a certain basic shape just as a glove has, but this becomes slightly different as it alters in the presence of the substrate. As it alters its shape, the enzyme puts a strain on the substrate molecule, and thereby lowers its activation energy. This whole process is called the **induced fit theory** of enzyme action.

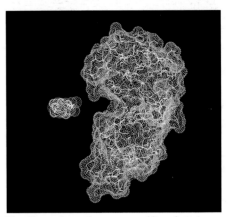

The ribonuclease A enzyme and its substrate

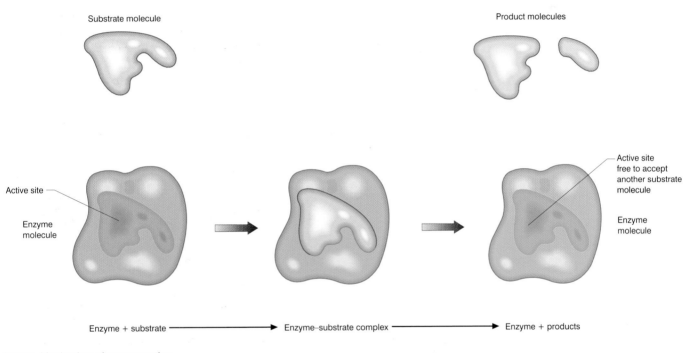

Fig 2.3 *Mechanism of enzyme action*

SUMMARY TEST 2.1

Enzymes act as biological **(1)**. They are **(2)** proteins that have a specific shape within which there is a functional portion known as the **(3)**. Enzymes lower the **(4)** of a reaction, allowing it to proceed at a lower temperature than it would normally. In an enzyme-controlled reaction, the general term for the substance on which the enzyme acts is **(5)** and the substances formed at the end of the reaction are known as the **(6)**. The enzyme molecule and the substance it acts on fit together very precisely, giving rise to the name **(7)** theory of enzyme action. In practice, the enzyme is thought to change shape slightly and so mould itself to the shape of the substance it acts on. This is called the **(8)** theory of enzyme action.

Enzyme properties

For an enzyme to work it must
- come into physical contact with its substrate
- have an active site which fits the substrate.

Many factors influence the rate at which an enzyme works, but almost all do so by affecting one or other of the above conditions.

2.2.1 Measuring enzyme-catalysed reactions

The measurement of enzyme reactions is usually made in one of two ways:
- time-course
- rate of reaction.

The **time-course** of an enzyme reaction can be plotted by measuring
- **the formation of the products** of the reaction, e.g. the volume of oxygen produced when catalase acts on hydrogen peroxide (Fig 2.4)
- **the disappearance of the substrate**, e.g. the reduction in the concentration of starch when it is acted on by amylase (Fig 2.5).

The rate of reaction is measured by the amount of substrate which is converted to product in a given period of time. Typically, this period of time is 1 minute.

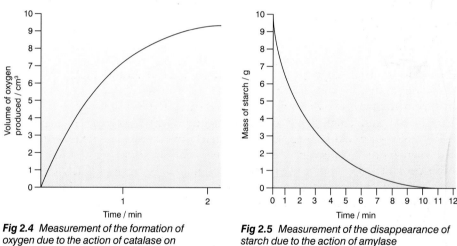

Fig 2.4 *Measurement of the formation of oxygen due to the action of catalase on hydrogen peroxide*

Fig 2.5 *Measurement of the disappearance of starch due to the action of amylase*

2.2.2 Enzyme concentration

Being catalysts, enzymes are not used up in reactions, so they can operate repeatedly, making small amounts highly effective. In some cases, a single enzyme molecule can act on millions of substrate molecules in 1 minute. As long as there is an excess of substrate, an increase in the amount of enzyme leads to a proportionate increase in the rate of reaction. This is because with high substrate concentrations it is more likely that a substrate molecule will collide with an enzyme molecule. However, if the amount of substrate is limited, an increase in the enzyme concentration will not affect the rate of reaction, because there is already enough enzyme to cope with the amount of substrate. The rate of reaction will then stabilise at a constant level (Fig 2.6).

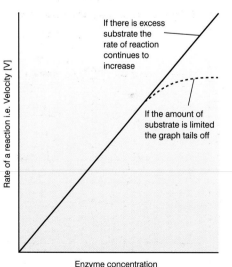

If there is excess substrate the rate of reaction continues to increase

If the amount of substrate is limited the graph tails off

Fig. 2.6 *Effect of enzyme concentration on the rate of an enzyme-controlled reaction*

2.2.3 Substrate concentration

If the amount of enzyme is fixed at a constant level and substrate is slowly added, the rate of reaction increases in proportion to the amount of substrate which is added. This is because the enzyme molecules have only a limited number of substrate molecules to collide with, and therefore the active sites of the enzyme are not working to full capacity. As more substrate is added, the active sites gradually become fully utilised, until the point where all of them are working as fast as they can. After that, the addition of more substrate will have no effect on the rate of reaction (Fig 2.7).

2.2.4 Temperature

A rise in temperature increases the kinetic energy of molecules, which therefore move around more rapidly and collide with one another more often. In an enzyme-catalysed reaction, this means that the enzyme and substrate molecules come together more often in a given time, so that the rate of reaction is increased. Shown on a graph, this gives a rising curve. However, the temperature rise also increases the energy of the atoms which make up the enzyme molecules. Its atoms begin to vibrate and cause the hydrogen and other bonds which hold it in shape to break. Gradually, the shape of the active sites is disrupted, until they no long fit their substrate molecules. At this point, usually around 60 °C, the enzyme stops working and is said to be **denatured**. Shown on a graph, the rate of this reaction follows a falling curve. The actual effect of temperature on the rate of an enzyme reaction is a combination of these two factors (Fig 2.8). The optimum working temperature differs from enzyme to enzyme. Some work best at around 10°C, and others continue to work well at 80°C.

2.2.5 pH

The pH of a solution is a measure of its hydrogen ion concentration. Each enzyme has an optimum pH at which it works best (Fig 2.9). This is because the exact arrangement of the active site of an enzyme is partly fixed by hydrogen bonds between NH_2 and COOH groups of the polypeptides which make up the enzyme. Even small changes in pH affect this hydrogen bonding, causing changes of shape in the active site which reduce the effectiveness of the enzyme. Solutions, known as **buffer solutions**, can be used to prevent fluctuations in pH. A buffer solution is a mixture of at least two chemicals which counteract the effect of acids and alkalis. A buffer solution does not therefore change its pH when a small amount of acid or alkali is added.

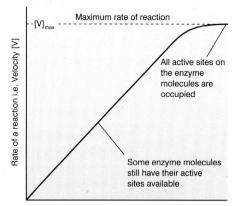

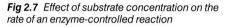

Fig 2.7 Effect of substrate concentration on the rate of an enzyme-controlled reaction

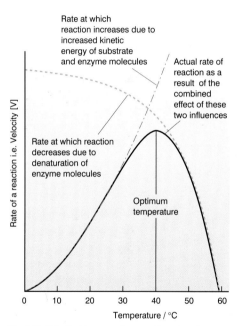

Fig 2.8 Effect of temperature on the rate of an enzyme-controlled reaction

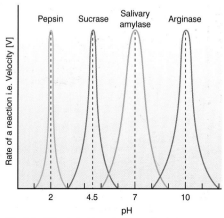

Fig 2.9 Effect of pH on the rate of an enzyme-controlled reaction

---

SUMMARY TEST 2.2

The two ways in which an enzyme reaction can be measured are **(1)** and time-course. Time-course reactions are usually plotted by measuring either the **(2)** or the **(3)**. If the temperature is increased, the rate of an enzyme reaction will **(4)** up to a point at which its molecular structure is disrupted. At this point the enzyme is said to be **(5)**. With a fixed amount of enzyme, the addition of more substrate will cause the rate of reaction to increase until all enzyme molecules are being used. At this point the rate of reaction levels off because the enzyme is **(6)** the reaction. An increase in the amount of enzyme will cause a proportional increase in the rate of reaction, provided that there is excess **(7)**. Enzymes work in a narrow range of pH outside of which the **(8)** bonds between the NH and **(9)** groups are broken. A solution that prevents changes in pH is called a **(10)**.

2.3 Enzyme inhibition and control of metabolic pathways

AQA.A	EDEXCEL
AQA.A (Human)	EDEXCEL (Human)
AQA.B	OCR

Enzyme inhibitors are substances which directly or indirectly interfere with the functioning of the active site of an enzyme and so reduce its activity. Sometimes the inhibitor binds itself so strongly to the active site that it cannot be removed and so permanently prevents the enzyme functioning. These are so called **non-reversible** or **permanent inhibitors**; they include heavy metal ions such as mercury and silver. Most inhibitors only make temporary attachments to the active site. These are called **reversible inhibitors** and are of two types:
- **competitive** (active site directed)
- **non-competitive** (non-active site directed).

2.3.1 Competitive (active site directed) inhibitors

Competitive inhibitors have a molecular shape which allows them to occupy the active site of an enzyme. They therefore compete with the substrate for the available active sites (Fig 2.10). It is the difference between the concentrations of the inhibitor and the substrate which determines the effect this has on enzyme activity: if the substrate concentration is increased, the effect of the inhibitor is weaker. The inhibitor is not permanently bound to the active site and so, when it leaves, another molecule can take its place. This could be a substrate or inhibitor molecule, depending on how much of each type is present. Sooner or later, all the substrate molecules will find an active site, but the greater the concentration of inhibitor, the longer this will take. Examples of competitive inhibitors include malonic acid, which inhibits succinic dehydrogenase in the Krebs cycle, and hirudin, which is used by leeches to inhibit thrombin and so prevent clotting when they take a blood meal.

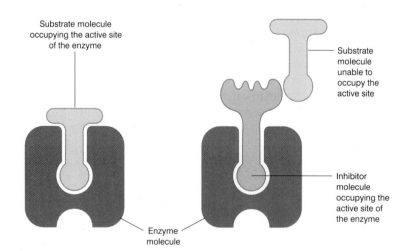

Fig 2.10 Competitive inhibition

2.3.2 Non-competitive (non-active site directed) inhibitors

Non-competitive inhibitors attach themselves to the enzyme at a site which is not the active site. This is known as the **allosteric site** (allosteric = 'at another place'). By attaching, the inhibitor alters the shape of the active site in such a way that the substrate cannot occupy it, and so the enzyme cannot function (Fig 2.11). As the substrate and the inhibitor are not competing for the same site, an increase in substrate concentration does not decrease the effect of the inhibitor (Fig 2.12). An example of a non-competitive inhibitor is cyanide, which inhibits the respiratory enzyme, cytochrome oxidase.

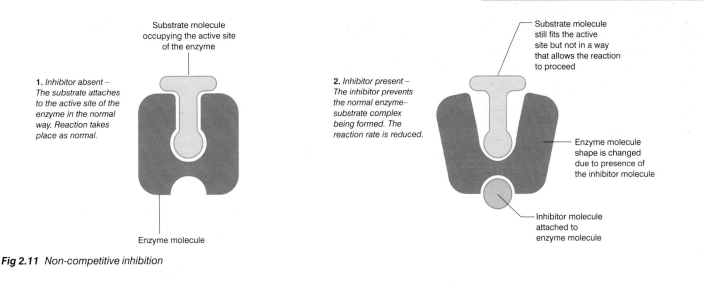

Fig 2.11 *Non-competitive inhibition*

2.3.3 **Control of metabolic pathways**

In the tiny space inside a single cell, many hundreds of different reactions take place. The process is not at all haphazard, it is highly structured. The enzymes which control a process are often attached to the inner membrane of a cell organelle in a very precise sequence. This increases the chance of each enzyme coming into contact with its substrate, and leads to greater efficiency. Inside each organelle, there are optimum conditions for the functioning of these enzymes, e.g. the pH may vary from organelle to organelle. To keep a steady level of a particular metabolite in a cell, the metabolite often acts as an inhibitor of an enzyme at the start of a reaction. In the example below, the end product inhibits enzyme A. The more end product there is, the greater the inhibition and the less end product is produced. A reduction in the amount of end product lessens the inhibition, so bringing the level of the metabolite back to normal. This is known as **end-product inhibition** and is a form of **homeostasis**. The enzyme which is inhibited therefore controls the rate of the reaction and is known as a **regulatory enzyme**. The type of inhibition involved is usually non-competitive.

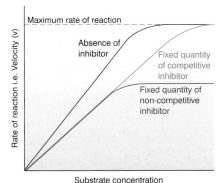

Fig 2.12 *Comparison of competitive and non-competitive inhibition on the rate of an enzyme-catalysed reaction, at different substrate concentrations*

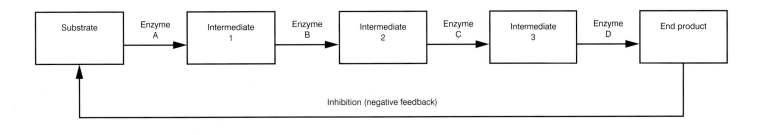

SUMMARY TEST 2.3

Inhibitors reduce the activity of enzymes. Where the inhibitor cannot be removed from the active site of the enzyme it is known as a **(1)** inhibitor. Inhibitors that can be removed from the active site are of two types. Competitive inhibitors are also known as **(2)** and compete with the substrate for active sites. Therefore if more substrate is added the effect of the inhibition is **(3)**. Non-competitive inhibitors become fixed to the enzyme at a point other than the active site. This point is called the **(4)**. If more substrate is added the inhibition is **(5)**.

Commercial uses of enzymes

The manufacture of enzymes by organisms, especially microorganisms, has been used by humans for thousands of years, in the production of foods such as breads, cheeses, yoghurts, wines, beers and vinegar. Our ancestors were unaware that the production of these foods was the result of microbial activity. Then, in 1857, Louis Pasteur showed that microorganisms were responsible for wine production by fermentation. Since his discovery, mankind has exploited the advantages of using microbes to produce enzymes for commercial purposes.

2.4.1 Why use enzymes for commercial purposes?

Enzymes are produced by cells either for internal use (**intracellular enzymes**) or for external use (**extracellular enzymes**). These enzymes have advantages over inorganic catalysts in commercial processes because they:
* are highly effective in tiny quantities
* function at lower temperatures, often needing little or any external heat
* operate at atmospheric pressure, reducing energy needs
* are highly specific and therefore produce a very pure product – essential for the food and drinks industry
* are biodegradable and therefore environmentally acceptable.

Enzymes have a wide range of commercial uses in the food, fuel and pharmaceutical industries, as well as in waste disposal and agriculture.

2.4.2 Pectinases in the food industry

Pectin produced by plants holds together the walls of neighbouring cells. As fruits ripen, pectin is modified by enzymes and the fruit becomes softer (and so more attractive to the animals that disperse it). When fruits are pressed commercially to extract their juice, these modified pectins make the juice cloudy and thicker, and adversely affect its flavour. Fruit juice manufacturers therefore add commercially produced **pectinase** to break down the pectin and overcome these problems. Cellulases and amylases may also be used to break down starch and cellulose respectively, which clarifies the juice even more. These and other examples of enzymes used in the food industry as listed in table 2.1.

2.4.3 Proteases in biological detergents

The detergent industry is one of the main consumers of commercially manufactured enzymes. Many stains on clothes are of a protein nature – blood, food, grass and sweat. Proteases are used to break down peptide bonds, reducing the protein to soluble polypeptides and amino acids, which are then removed in the washing water. The proteases are produced by varieties of the bacteria, *Bacillus subtilis* and *Bacillus licheniformis*, which have been genetically modified to make the enzyme **thermostable**, i.e. not denatured at higher temperatures and therefore able to work even in a hot wash. They also function well despite the high pHs created by the phosphates used in detergents.

Biological detergents are more effective if the clothes are pre-soaked with detergent before washing. This allows the enzymes to work on stains. The colder the water that the clothes are soaked in, the longer enzymes take to work because the molecules move more slowly. One hour at 40°C is as effective as 5 hours at 10°C.

FACTORS TO BE CONSIDERED WHEN CHOOSING ENZYMES TO BE USED IN INDUSTRIAL PROCESSES

* pH tolerance
* temperature tolerance
* specificity
* activators needed
* inhibition problems
* potency
* availability
* production costs
* technical support required

Table 2.1 *Some enzymes produced by microorganisms and used in the food industry*

Enzyme	Application
α-Amylase	Breakdown of starch in beer production
	Improving flour
	Preparation of glucose syrup
	Thickening of canned sauces
Glucose isomerase	Sweetener for soft drinks
	Cake fillings
Lactase	Lactose removal from whey
	Sweetener for milk drinks
Lipase	Flavour development in cheese
Pectinase	Clearing of wines, fruit juices and cider
Protease	Meat tenderisers
Sucrase	Confectionery production

2.4.4 Enzymes as analytical agents

Enzymes make excellent analytical agents because they are:
- very specific in the reactions they catalyse, and can therefore be used to identify, very precisely, one type of molecule amongst a mixture of many
- highly sensitive, and so are able to detect molecules even when they are present in minute amounts.

For these reasons, enzymes are used as **biosensors**. Biosensors enable rapid and accurate measurement of the concentration of a particular chemical. They are used extensively in medicine, forensic science, agriculture and industry. One example is the use of **glucose oxidase** in assaying the amount of glucose, even in the presence of other sugars. Glucose oxidase catalyses the conversion of glucose to hydrogen peroxide:

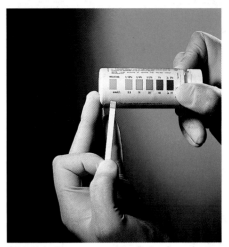

Clinistix being compared against colour chart

$$\text{glucose} + \text{oxygen} \xrightarrow{\text{glucose oxidase}} \text{gluconic acid} + \text{hydrogen peroxide}$$

The hydrogen peroxide can then be detected through its ability to convert a colourless chemical (a hydrogen donor) to a coloured one:

$$\text{colourless hydrogen donor} + \text{hydrogen peroxide} \xrightarrow{\text{peroxidase}} \text{water} + \text{coloured compound}$$

The glucose oxidase, peroxidase and colourless hydrogen donor are immobilised (see section 2.5.5) on a fibre pad to form 'Clinistix'. One use of these is to measure the blood sugar level in people with diabetes, enabling them to adjust the dose of insulin they require at any injection. The actual glucose level can be measured by comparing the density of the colour on the pad against a table which shows the colour produced by a known concentration of glucose. Many enzyme biosensors work in conjunction with a **transducer** which produces an electrical signal: the intensity of the signal gives a measure of the quantity of substrate.

SUMMARY TEST 2.4

Enzymes produced and used within cells are called **(1)** enzymes; those used outside cells are known as **(2)** enzymes. The advantage over inorganic catalysts using enzymes commercially are that they function at lower temperatures and atmospheric pressure, thereby using less **(3)**. Effective in small amounts and highly specific, they also have less effect on the environment because they are **(4)**. Commercial uses of enzymes include the clarification of fruit juices by **(5)**, **(6)** and **(7)** and the development of flavour in cheeses by **(8)**. Proteases are used in detergents to break down the **(9)** bonds of proteins. These proteases are **(10)** to allow them to operate at high temperatures without being **(11)**. Enzymes used as analytical agents act as **(12)** to give an accurate measure of a particular chemical. For example, the amount of glucose in a sample can be measured using the enzyme **(13)**, which catalyses the conversion of glucose to gluconic acid and **(14)**, which is then detected because it can change a colourless compound to a coloured one. One use of this technique is in measuring the blood sugar levels of people suffering from **(15)**. Some enzymes work in a similar way with substances called **(16)** that produce an electrical charge.

Enzyme technology

Enzymes are widely used in industrial processes for the reasons outlined in section 2.4.1. The study and use of these processes is called **enzyme technology**. The large-scale commercial production of enzymes takes place in five stages:

- **Growth of the microorganism** to build up adequate stock for mass production.
- **Isolation of the enzyme** from the microorganism that produces it.
- **Purification of the enzyme** to remove any contaminants.
- **Stabilisation of the enzyme** to ensure it remains effective under the conditions in which it operates.
- **Immobilisation of the enzyme** to allow continuous production.

2.5.1 Growth of microorganisms

As enzymes are found in all living cells, it is theoretically possible to obtain them from any tissue of any organism. In practice, microorganisms are the best commercial source because they:

- are easily grown in a laboratory where the conditions for maximum yield can be carefully controlled
- have rapid growth rates and so build up numbers quickly
- feed on a wide variety of substances, many of which are cheap, e.g. waste material
- produce enzymes more effectively for a given mass than any other organism
- are varied and so types can be found to suit a wide range of different production techniques
- can easily be genetically manipulated to produce large quantities of a desired enzyme.

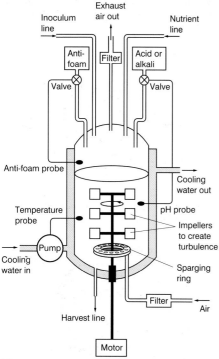

Fig 2.13 Microorganisms are usually grown in fermenters such as the one illustrated above

The **culture media** on which the microorganisms are grown must contain the correct balance of nutrients in suitable concentrations. The proportion of each nutrient needed varies from microorganism to microorganism, but normally includes the following:

- **Water** for a variety of metabolic functions (section 1.1.7).
- **Major nutrients** which provide carbon, nitrogen, sulphur and phosphorus in relatively large quantities.
- **Macronutrients** such as potassium, calcium, magnesium and iron needed in small amounts.
- **Micronutrients** such as manganese, cobalt, zinc, copper and molybdenum required in trace quantities.
- **Growth factors**, which are complex molecules that certain microorganisms cannot manufacture for themselves. They include vitamins, amino acids, purines and pyrimidines.

To avoid contamination, especially by other microorganisms, growth must take place in **aseptic conditions**. All equipment, instruments and the culture medium must be sterilised, usually by steam or dry heat. Suitable disinfectants such as hypochlorite may be used on equipment, and the air entering the apparatus is either filtered free of microorganisms or irradiated with ultra-violet light to kill them.

Fermenter in use

2.5.2 Isolation of the enzyme

Once the microorganisms have been grown, the required enzyme has to be isolated from the culture. If the product is an extracellular enzyme, the process can begin immediately, but if it is an intracellular enzyme, the cells must first be disrupted by centrifugation or **ultrafiltration**. The process of separating the cells or cell debris from the enzyme is called **downstream processing** and involves the following:

- **Settlement** – The cells are precipitated out, a process which can be accelerated by the addition of **flocculating agents**.
- **Centrifugation** – The culture is spun at high speed and the cells and debris are forced out of suspension.
- **Ultrafiltration** – An alternative to centrifugation, in which the culture is forced through filters with a pore size of less than 0.5µm, trapping the cell debris and allowing only liquid through.

2.5.3 Purification of the enzyme

To meet the high standards required by industry, the enzyme extract must be further purified. This is achieved in a number of ways:

- **Precipitation** – The enzyme is precipitated out of solution using an appropriate chemical such as alcohol or ammonium sulphate.
- **Electrophoresis** – The enzyme is separated out according to its electrical charge (section 12.3.1).
- **Chromatography** – The enzyme is separated from other molecules according to its relative molecular mass and its solubility in the solvent used (section 1.5.4).

2.5.4 Stabilisation of the enzyme

Many industrial processes require conditions which would denature most enzymes, e.g. extremes of temperature and pH or the presence of harmful chemicals. Enzymes which can withstand such hostile conditions have been developed by transferring the genes from naturally occurring microorganisms that live in extreme environments into varieties which grow rapidly under laboratory conditions. For example, the genes from bacteria which live in hot springs at 75°C have been transferred to the easily grown *Bacillus subtilis*, to produce **thermostable** enzymes.

2.5.5 Immobilisation of the enzymes

In industrial processes, there are advantages in fixing the enzyme in one position and continuously moving the substrate over it, collecting the product at the other end. The advantages of such enzyme immobilisation are:

- the enzyme can be used repeatedly, making it economic, especially if it is expensive to produce
- the material in which the enzyme is held makes it more stable by protecting it from pH and temperature changes
- because the enzyme is held in place, it does not contaminate the product
- a number of different enzymes can be fixed in a precise order, allowing a sequence of reactions to take place one after the other, and giving greater control over the process
- the process can be continuous, with substrate being added at one end of a column of immobilised enzymes and the product being removed at the other.

The four main types of enzyme immobilisation are illustrated in figure 2.14.

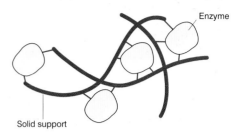

Binding *– Enzyme is bonded to supporting material such as nylon or cellulose, either covalently or by means of a binding chemical.*

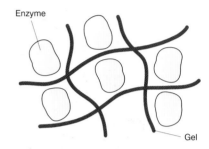

Entrapment *– Enzyme is trapped inside a gel such as silica gel or collagen.*

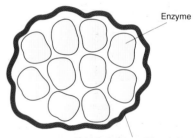

Encapsulation *– Enzyme is held inside a partially permeable membrane, e.g. nylon.*

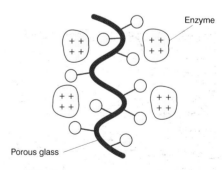

Adsorption *– Enzyme is adsorbed onto the surface of an insoluble matrix such as collagen or fibreglass.*

Fig 2.14 *Methods of immobilising enzymes*

1 a Diagram **A** shows an enzyme, and **B** is the substrate of this enzyme.

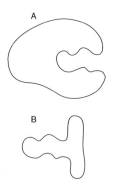

By drawing on this diagram, show how a competitive inhibitor would affect the activity of the enzyme.

(2 marks)

b The graph shows the effect of changing substrate concentration on the rate of an enzyme controlled reaction.

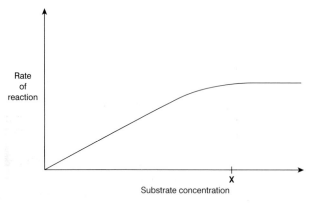

Explain why increasing substrate concentration above the value shown at **X** fails to increase the rate of reaction further.

(2 marks)

c Explain how adding excess substrate could overcome the effect of a competitive inhibitor.

(2 marks)

(Total 6 marks)

AQA (specimen), B/HB (A) BYA1, No.3

2 Urease is an enzyme which catalyses the breakdown of urea to ammonia and carbon dioxide.

An experiment was carried out into the effect of pH on the activity of urease. $10\,cm^3$ of pH 3 buffer solution was mixed with $1\,cm^3$ of urease solution. This mixture was then added to $10\,cm^3$ of urea solution and the concentration of ammonia in the mixture was measured after 60 minutes. This procedure was repeated using buffer solutions of pH 4, 5, 6, 7, 8 and 9.

The results are shown in the following graph.

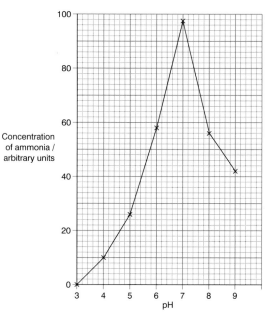

a What do these results suggest is the optimum pH for urease activity? *(1 mark)*

b Suggest how the experiment could be modified to determine the optimum pH more accurately. *(1 mark)*

c Explain why no ammonia was produced at pH 3. *(3 marks)*

d Explain why less ammonia is produced at pH 9 than at pH 8. *(2 marks)*

e Describe how this experiment could be modified to determine the effect of enzyme concentration on the activity of urease. *(4 marks)*

(Total 11 marks)

Edexcel 6101/01 Jan 2001, B B(H) AS, No.7

3

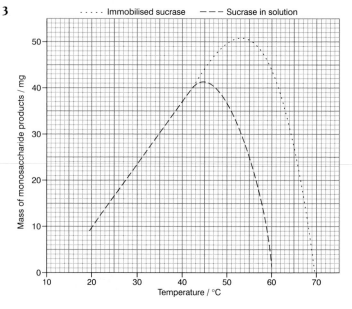

Sucrase is an enzyme that catalyses the hydrolysis of sucrose. An investigation was carried out to compare the activity of sucrase in solution with immobilised sucrase, over a range of temperatures.

The enzyme in solution was incubated with a solution of sucrose for 5 minutes at different temperatures. The mass of monosaccharide produced at each temperature was determined. This was repeated using immobilised sucrase.

The results of the investigation are shown in the graph.

a Name the **two** monosaccharides produced from the hydrolysis of sucrose. *(2 marks)*

b What evidence from the graph suggests that the concentrations of sucrase in solution and immobilised sucrase were equivalent? *(1 mark)*

c Compare the effect of temperature on the activity of sucrase in solution with that on immobilised sucrase. *(3 marks)*

d Suggest why temperatures above 45°C have different effects on immobilised sucrase and sucrase in solution. *(2 marks)*

e Describe how this investigation could be adapted to compare the activity of sucrase in solution with that of immobilised sucrase over a range of pH values. *(4 marks)*
(Total 12 marks)
Edexcel 6101/01 June 2001, B B(H) AS/A, No.8

4 a Starch molecules do not break down in boiling water. In the body starch is digested by amylase.

(i) Name the product of the digestion of starch by amylase. *(1 mark)*

(ii) Explain how amylase makes it possible for starch to be digested at body temperature. *(3 marks)*

b In an investigation, a sample of amylase was placed in a water bath at 60°C. Each minute, a small amount of the amylase was removed and mixed with starch solution at 35°C. The rate of activity of the amylase was measured. The results are shown in the graph.

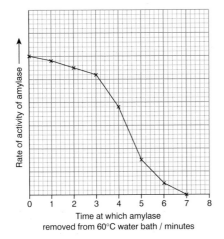

Rate of activity of amylase (y-axis)
Time at which amylase removed from 60°C water bath / minutes (x-axis)

Use your knowledge of enzymes to explain why the activity of the amylase decreased. *(4 marks)*
(Total 8 marks)
AQA Jan 2001, B (B) BYB1, No.4

5 Amylase is an enzyme which catalyses the hydrolysis of starch to maltose.

a (i) Name the bond which must be broken by this enzyme. *(1 mark)*

(ii) Name the reagent that you would use to carry out a test for starch. *(1 mark)*

(iii) State the colour you would expect to see if you carried out the test before and after the action of the enzyme. *(2 marks)*

In an investigation into the action of amylase, equal volumes of enzyme solution and starch solution were mixed. The quantity of maltose produced was measured during the course of two separate experiments, one carried out at 18°C and the other at 23°C. The results are shown in the figure.

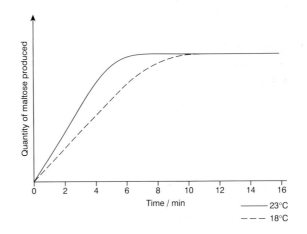

Quantity of maltose produced (y-axis)
Time / min (x-axis)
——— 23°C
– – – 18°C

b (i) Explain why the quantity of maltose produced eventually became constant. *(1 mark)*

(ii) Explain why, after 4 minutes, more maltose had been produced at 23°C than at 18°C. *(2 marks)*

The experiment was repeated with the same volume and concentration of amylase, but with a higher concentration of starch solution.

c Sketch, on the figure, the curve you would expect to obtain if the experiment were repeated at 23°C with an increased starch concentration. *(2 marks)*
(Total 9 marks)
OCR 2801 Jan 2001, B (BF), No.4

Nucleic acids and protein synthesis

Nucleotides and ribonucleic acid (RNA)

Adenosine monophosphate (adenylic acid)

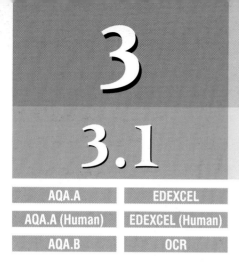

Fig 3.1 *Structure of a nucleotide*

NAME OF MOLECULE	REPRESENTATIVE SHAPE
Phosphate	
Pentose sugar	
Adenine (a purine)	Adenine
Guanine (a purine)	Guanine
Cytosine (a pyrimidine)	Cytosine
Thymine (a pyrimidine)	Thymine
Uracil (a pyrimidine)	Uracil

Fig 3.2 *Molecules found in nucleotides*

Nucleotides are the basic units which make two of the most important chemicals in all organisms. These are the **nucleic acids**, of which the best known are:
- **ribonucleic acid (RNA)**
- **deoxyribonucleic acid (DNA)**.

3.1.1 Nucleotide structure

Individual nucleotides are made up of three components:
- **a pentose sugar**, of which there are two types: **ribose** and **deoxyribose**
- **a phosphate group**
- **an organic base**, of which there are five different forms found in nucleic acids.

The five organic bases are divided into two groups:
- **Pyrimidines**, which are made up of a single six-sided ring, include **cytosine**, **thymine** and **uracil**.
- **Purines**, which are made up of a six-sided ring joined to a five-sided one. The two examples found in nucleic acids are **adenine** and **guanine**.

The pentose sugar, phosphate and organic base are combined, as a result of **condensation reaction**, to give a **mononucleotide** (Fig 3.1). Two mononucleotides may, in turn, be combined together as a result of a condensation reaction between the pentose sugar of one mononucleotide and the phosphate group of another. The new structure is called a **dinucleotide**. Continued linking of mononucleotides in this way forms a **polynucleotide**, such as ribonucleic acid (RNA). While RNA and DNA perform essential fuctions in protein synthesis and heredity, they are by no means the only biologically important molecules containing nucleotides. A number of the others are listed in table 3.1.

Table 3.1 *The functions of biologically important molecules containing nucleotides*

Molecule	Abbreviation	Function
Deoxyribonucleic acid	DNA	Contains the genetic information of cells
Ribonucleic acid	RNA	All three types play a vital role in protein synthesis
Adenosine monophosphate Adenosine diphosphate Adenosine triphosphate	AMP ADP ATP	Coenzymes important in making energy available to cells for metabolic activities, osmotic work, muscular contractions, etc
Nicotinamide adenine dinucleotide Flavine adenine dinucleotide	NAD FAD	Electron (hydrogen) carriers important in respiration in transferring hydrogen atoms from the Krebs cycle along the respiratory chain
Nicotinamide adenine dinucleotide phosphate	NADP	Electron (hydrogen) carrier important in photosynthesis for accepting electrons from the chlorophyll molecule and making them available for the photolysis of water
Coenzyme A	CoA	Coenzyme important in respiration in combining with pyruvate to form acetyl coenzyme A and transferring the acetyl group into the Krebs cycle

3.1.2 Ribonucleic acid (RNA) structure

Ribonucleic acid is a polymer made up of repeating mononucleotide sub-units. It forms a single strand in which the pentose sugar is always **ribose** and the organic bases are adenine, guanine, cytosine and uracil (Fig 3.3). There are three types of RNA, all of which are important in protein synthesis:
- **ribosomal RNA (rRNA)**
- **transfer RNA (tRNA)**
- **messenger RNA (mRNA)**.

3.1.3 Ribosomal RNA (rRNA)

Ribosomal RNA is a large, complex molecule which is a major component of ribosomes, making up over half of their mass. It has a sequence of organic bases which is very similar in all organisms.

3.1.4 Transfer RNA (tRNA)

Transfer RNA is a relatively small molecule which is made up of around 80 nucleotides. It is manufactured by DNA and makes up 10–15% of the total RNA in a cell. Although there are a number of types of tRNA, they are very similar, each having a single-strand chain folded into a clover-leaf shape, with one end of the chain extending beyond the other. This extended chain always has the organic base sequence cytosine-cytosine-adenine; this is the part of the tRNA molecule to which amino acids can easily attach. There are at least 20 types of tRNA, each able to carry a different amino acid. At the opposite end of the tRNA molecule is a sequence of three other organic bases, known as the **anticodon**. For each amino acid there is a different sequence of organic bases on the anticodon. During protein synthesis, this anticodon pairs with the complementary three organic bases which make up the triplet of bases on messenger RNA, known as the **codon**. The tRNA structure (Fig. 3.4), with its end chain for attaching amino acids and its anticodon for pairing with the codon of the mRNA, is structually suited to its role of lining up amino acids on the mRNA template during protein synthesis.

3.1.5 Messenger RNA (mRNA)

Consisting of thousands of mononcleotides, messenger RNA is a long strand which is arranged in a single helix. Because it is manufactured when DNA forms a mirror-copy of part of one of its two strands, there is a great variety of different types of mRNA. Once formed, mRNA leaves the nucleus via pores in the nuclear membrane and enters the cytoplasm, where it associates with the ribosomes. There it acts as a template on which proteins are built. Its structure is suited to this function, because it possesses the correct sequence of many triplets of organic bases which code for specific polypeptides. It is also easily broken down, and so exists only for as long as it is needed to manufacture a given protein.

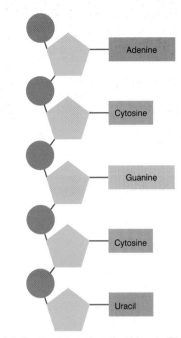

Fig 3.3 *Section of a polynucleotide, e.g. RNA*

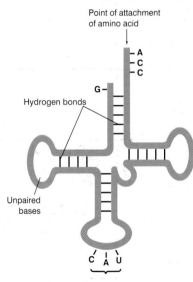

Fig 3.4 *Structure of transfer RNA*

SUMMARY TEST 3.1

Nucleotides are organic compounds that contain the elements carbon, hydrogen, oxygen, **(1)** and **(2)**. A mononucleotide contains **(3)** sugar which has **(4)** carbon atoms and has two forms, **(5)** and **(6)**. It also contains one of five organic bases, that fall into two groups. Those with a six-carbon ring only are called **(7)** and exist in three forms, thymine, **(8)** and **(9)**. The second group, called **(10)**, have a six-sided ring joined to a **(11)**-sided ring; there are two such molecules, **(12)** and **(13)**. Ribonucleic acid, which never has the organic base **(14)**, exists in three forms. The form that has the same sequence of organic bases in all living organisms is called **(15)**, the form that has a sequence of three bases called an anticodon is **(16)** and the remaining form upon which proteins are formed is **(17)**.

3.2

Deoxyribonucleic acid (DNA)

AQA.A	EDEXCEL
AQA (Human)	EDEXCEL (Human)
AQA.B	OCR

Deoxyribonucleic acid is made up of two nucleotide polymer strands. In DNA, the pentose sugar is **deoxyribose** and the organic bases are adenine, guanine, cytosine and thymine. Each of the two polynucleotide strands is extremely long, and they are wound around one another to form a double helix. The differences between RNA and DNA are listed in table 3.2 opposite.

3.2.1 DNA structure

In 1953, James Watson and Francis Crick worked out the structure of DNA and thus opened the door for many of the major developments in biology over the next half-century. In its simplified form, DNA can be thought of as a ladder, in which the phosphate and deoxyribose molecules alternate to form the uprights and the organic bases pair together to form the rungs (Fig 3.5). The organic bases are of two types: the purines (adenine and guanine) are longer molecules than the pyrimidines (cytosine and thymine). It follows that, if the rungs of the DNA ladder are to be the same length, the base pairs must always be made up of one purine and one pyrimidine. In fact, the pairings are even more precise than this:

- Adenine always pairs with thymine by means of two hydrogen bonds.
- Guanine always pairs with cytosine by means of three hydrogen bonds.

It follows that the quantity of adenine and thymine in DNA is always the same, as is that of guanine and cytosine. However, the ratio of adenine and thymine to guanine and cytosine varies from species to species.

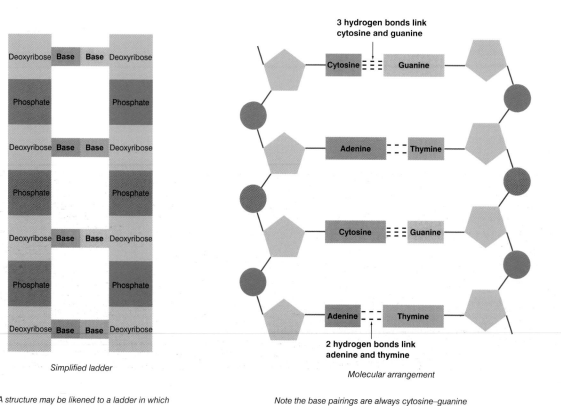

Simplified ladder

DNA structure may be likened to a ladder in which alternating phosphate and deoxyribose molecules make up the 'uprights' and pairs of organic bases comprise the 'rungs'.

Molecular arrangement

Note the base pairings are always cytosine–guanine and adenine–thymine. This ensures a standard 'rung' length. Note also that the 'uprights' run in the opposite direction to each other (i.e. are antiparallel).

Fig 3.5 *Basic structure of DNA*

To appreciate the true structure of DNA, however, you have to imagine this ladder twisted, so that the uprights wind around one another to form a **double helix**. These uprights run in the opposite direction to each other and are therefore said to be **antiparallel**. For each complete turn of this helix, there are 10 base pairs (Fig 3.6). In total, there are around 6 billion base pairs in the DNA of a typical mammalian cell. This vast number means that there is an almost infinite number of sequences of bases along the length of a DNA molecule, and it is this variety which provides the immense genetic diversity within living organisms.

The DNA molecule is adapted to carry out its functions in a number of ways:

- It is very stable and can pass from generation to generation without change.
- Its two strands are, however, joined only with hydrogen bonds, allowing them to separate during replication (unit 3.3) and form mRNA during protein synthesis (unit 3.5).
- It is an extremely large molecule and it therefore carries an immense amount of genetic information.

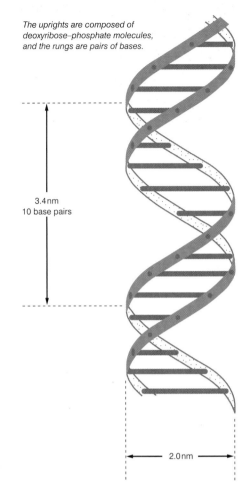

The uprights are composed of deoxyribose–phosphate molecules, and the rungs are pairs of bases.

3.4 nm
10 base pairs

2.0 nm

Fig 3.6 *The DNA double helix structure*

Table 3.2 *Differences between RNA and DNA*

RNA	DNA
Single polynucleotide chain	Double polynucleotide chain
Smaller molecular mass (20 000–2 000 000)	Larger molecular mass (100 000–150 000 000)
May have a single or double helix	Always a double helix
Pentose sugar is ribose	Pentose sugar is deoxyribose
Organic bases present are adenine, guanine, cytosine and uracil	Organic bases present are adenine, guanine, cytosine and thymine
Ratio of adenine and uracil to cytosine and guanine varies	Ratio of adenine and thymine to cytosine and guanine is one
Manufactured in the nucleus but found throughout the cell	Found almost entirely in the nucleus
Amount varies from cell to cell (and within a cell according to metabolic activity)	Amount is constant for all cells of a species (except gametes and spores)
Chemically less stable	Chemically very stable
May be temporary – existing for short periods only	Permanent
Three basic forms: messenger, transfer and ribosomal RNA	Only one basic form, but with an almost infinite variety within that form

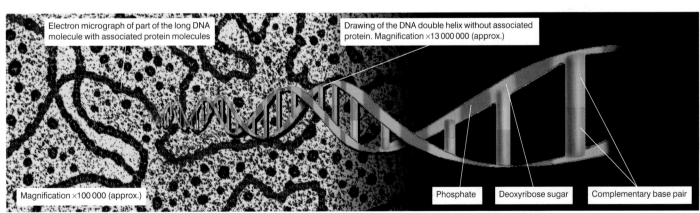

Electron micrograph of part of the long DNA molecule with associated protein molecules

Drawing of the DNA double helix without associated protein. Magnification ×13 000 000 (approx.)

Magnification ×100 000 (approx.)

Phosphate   Deoxyribose sugar   Complementary base pair

Fig 3.7 *Deoxyribonucleic acid*

3.3

DNA replication

AQA.A   EDEXCEL
AQA.A (Human)   EDEXCEL (Human)
AQA.B   OCR

We know that DNA is the hereditary material responsible for passing genetic information from cell to cell and generation to generation (section 3.4.1). We have only to look at identical twins to see just how perfectly the 6 million base pairs of DNA in the **human genome** can be copied. The process of DNA replication is clearly very precise. How then is it achieved?

3.3.1 Semi-conservative replication

The Watson–Crick model of DNA structure (unit 3.2) allows for a logical explanation of how DNA produces exact copies of itself. Basically, the **hydrogen bonds** linking the base pairs of DNA break and the double helix separates progressively, from one end, into its two strands. Each exposed strand then acts as a template to which complementary nucleotides are attracted. These nucleotides zip together to form the 'missing' strand on each half, and two identical DNA molecules result (Fig 3.8). Each of the new DNA molecules possesses one of the original DNA strands – i.e. half the original DNA has been saved and built into each of the new DNA molecules. The process is therefore termed **semi-conservative replication**. It takes place during interphase in the cell cycle (unit 5.1).

1. A representative portion of DNA, which is about to undergo replication.

2. Helicase enzymes cause the two strands of the DNA to separate. **Binding proteins** then keep the two strands apart.

3. The helicases complete the splitting of the strand. Meanwhile, free nucleotides are attracted to their complementary bases.

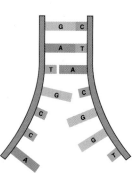

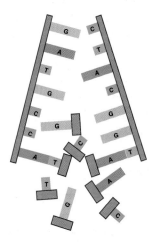

4. Once the nucleotides are lined up, they are joined together by DNA polymerase (bottom three nucleotides). The remaining unpaired bases continue to attract their complementary nucleotides.

5. Finally, all the nucleotides are joined to form a complete polynucleotide chain. In this way, two identical strands of DNA are formed. As each strand retains half of the original DNA material, this method of replication is called the semi-conservative method.

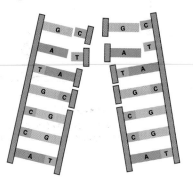

Fig 3.8 The semi-conservative replication of DNA

The process as described is a simplified one. In practice, the mechanism is more complex because:

- An enzyme called **topoisomerase** is used to break one strand of the parent DNA, allowing it to untwist, before replication.
- The DNA helix splits at a number of points along its length, each called a **replication fork**.
- The new DNA strands are built up simultaneously in each of these sections, which are then joined together using the enzyme **DNA ligase**. This makes the process much quicker – an important factor, given the considerable length of a single DNA molecule.
- Within each section, only one of the new strands is constructed by linking nucleotides one by one in a continuous process. The other strand is constructed by making up a series of short lengths which are then joined together using **DNA ligase**.
- Errors during replication could have disastrous effects, so an enzyme called **proofreading endonuclease** is associated with DNA polymerase. This enzyme cuts off any bases which are incorrectly paired.

3.3.2 Evidence for semi-conservative replication

Evidence for the semi-conservative method of replication was provided by experiments carried out by the American biochemists Meselsohn and Stahl, as follows:

- Many generations of the bacterium, *Escherichia coli*, were grown in cultures containing ^{15}N (heavy nitrogen), to ensure that their DNA contained ^{15}N.
- They were then transferred to a medium containing only ^{14}N (light nitrogen). New DNA material would therefore be made up of ^{14}N (light nitrogen).
- The next two generations of *Escherichia coli* each had their DNA separated by centrifugation and the presence of light and heavy nitrogen in the DNA was detected by its absorption of ultra-violet light.
- The results showed that the first generation of bacteria had DNA containing equal proportions of ^{14}N and ^{15}N.
- In the second generation, half the bacteria had DNA containing only ^{14}N, and the remainder had DNA with equal proportions of ^{14}N and ^{15}N.

This process is illustrated in figure 3.9.

SUMMARY TEST 3.3

DNA replication involves the separation of the two **(1)** chains that make up the molecule. To allow the helix to untwist, one chain is broken by the enzyme **(2)**. The helix then splits at points called **(3)** and new DNA is made simultaneously at these points. The sections are then joined by the enzyme **(4)**. Any incorrectly paired bases are removed using the enzyme **(5)**. As the new DNA contains half of the original DNA, the process is known as **(6)** replication, as proved by experiments by two scientists called **(7)** and **(8)** using the bacterium **(9)** and the radioactive isotope **(10)**.

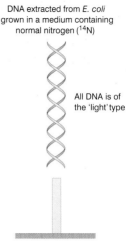

DNA made with ^{14}N (normal nitrogen)
DNA made with ^{15}N (heavy nitrogen)

DNA extracted from *E. coli* grown in a medium containing normal nitrogen (^{14}N)

All DNA is of the 'light' type

Light

DNA extracted from *E. coli* grown in a medium containing heavy nitrogen (^{15}N) and then transferred to a medium containing normal nitrogen (^{14}N)

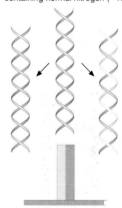

Intermediate

DNA extracted from *E. coli* grown in a medium containing heavy nitrogen (^{15}N)

Heavy

Relative weight of DNA as determined by centrifugation

Fig 3.9 Interpretation of experiments on semi-conservative replication of DNA

<table>
<tr><td>

3.4

</td><td>

Genetic code

</td></tr>
</table>

AQA.A	EDEXCEL
AQA.A (Human)	EDEXCEL (Human)
AQA.B	OCR

3.4.1 Evidence that DNA is the hereditary material

DNA is now widely recognised as the hereditary material of cells. Evidence for this comes from a number of sources:

- **Chromosome analysis** – chromosomes are made up of DNA and protein only, and as chromosomes can be seen to play a part in cell division, it seems likely that one or other of these substances is the hereditary material.
- **Constancy of DNA in the cell** – the amount of DNA remains constant for all cells except gametes, which have half the usual amount. These facts fit in with the expected changes in the quantity of hereditary material during cell division.
- **Metabolic stability of DNA** – DNA is extremely stable, a characteristic that is essential to any material which is passed from generation to generation over millions of years.
- **Mutagenic effects and DNA** – agents such as X-rays and certain chemicals which are known to cause inherited mutations can also be shown to alter the structure of DNA.
- **Bacterial transformation experiments** – if the bacteria that cause pneumonia are killed and injected into mice, they cause no ill effects. In the same way, if a harmless related species is injected, the mice are unaffected. If both species are injected together, however, the mice die of pneumonia. The explanation is this. The pneumonia-causing bacteria have the genetic information to make the toxin that causes pneumonia, but because they are dead they cannot make it. The live but harmless species has the manufacturing ability, but no instructions on how to make the toxin. The information must therefore have been passed to the living harmless species, enabling them to make the toxin and transforming them into a pneumonia-causing variety. When different extracts from the dead harmful bacteria were purified and tested for their ability to transform the living types, only DNA was found to be effective. Also, the use of enzymes which break down DNA prevented transformation.
- **Use of viral DNA** – viruses are known to inject material into host cells, which causes the cells to produced new viruses. This material clearly possesses the genetic information for making viruses. By radioactively labelling the DNA of viral cells, it can be shown that, when the viral cells infect bacteria, both the host bacterial cells and the new viruses become radioactively labelled. This means that DNA is the material which possesses the genetic information.

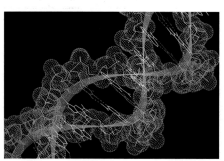

Computer representation of part of a DNA molecule

3.4.2 The triplet code

With the structure of DNA worked out, its role in heredity established and its replication explained, the question remained – how exactly does it mastermind the construction of new cells and organisms? The answer lay in the sequence of bases on the DNA molecule. Although there are only four bases – adenine, guanine, cytosine and thymine – the great length of the DNA molecule means that there is an almost unlimited variety of combinations of these bases. DNA is thus a set of instructions written in a language made up of just four letters (A, G, C and T). How, then, are these instructions translated into cell structure?

Cells are made up of a variety of chemicals, but most of these are the same regardless of the organism or the cell in question. It is only in a cell's proteins that real differences arise. Each cell has its own, often unique, mixture of proteins which defines its individual structure and function. Most chemicals in cells, including proteins, are produced by the action of enzymes and these, too, are proteins. If it is DNA that determines which proteins, especially enzymes, are produced, this would explain how DNA determines the structure of cells and, therefore, of organisms.

Proteins are extremely large molecules of almost unlimited variety, and yet they are made up of just 20 different amino acids. If the bases on the DNA were codes for amino acids, the sequence of these bases in the DNA would determine their precise order in a protein, and therefore the properties of the protein. With only four different bases, though, how can 20 amino acids be coded for? Clearly, the code for each must include more than just one base – in fact three is the minimum required. It is therefore a **triplet code** of three bases which determines each amino acid.

3.4.3 Unravelling the triplet code

Having established that each amino acid was coded by three bases on the DNA molecule, the question was which triplet of bases coded for which amino acid? The problem was solved as follows:

- It is known that, in protein synthesis, amino acids are lined up on a template of mRNA which is produced by DNA (see unit 3.6).
- Messenger RNA was synthesised which had a repeating sequence of one triplet code, e.g. AGC AGC AGC etc.
- Cell-free extracts of bacterial cell components were set up which had all the necessary biochemical requirements for protein synthesis.
- Twenty such tubes were prepared, each one having a different radioactively labelled amino acid.
- Messenger RNA with one particular repeating sequence was added to each of the 20 test tubes.
- Whichever test tube produced a polypeptide must contain the amino acid which is coded for by the triplets of bases on the mRNA in that tube. In the case of the AGC AGC AGC example, only the test tube with the amino acid serine produced a polypeptide. The mRNA triplet of AGC therefore codes for the amino acid serine.
- By repeating the process with all 64 possible combinations of three bases on mRNA, the complete dictionary of codes was established (table 3.3).

3.4.4 Features of the triplet code

Further experiments, including **frame-shift** ones carried out by Watson and Crick, have revealed the following features of the triplet code. In each case, the codon referred to is the triplet of bases found on mRNA.

- A few amino acids have only a single triplet code, e.g. tryptophan is coded only by UGG.
- Most amino acids have up to six codons, e.g. leucine is coded for by UUA, UUG, CUU, CUC, CUA and CUG.
- The code is a **degenerate code**, because most amino acids have more than one triplet code.
- Three codons, UAA, UAG and UGA, do not code for any amino acid. These are called **stop** or **nonsense codes**, which mark the end of a polypeptide chain (section 3.6.4).
- The code is **non-overlapping**, i.e. each base in the sequence is read only once: six bases numbered 123456 are read as triplets 123 and 456, rather than triplets 123, 234, 345, 456. Non-overlapping codes need more bases, but are less likely to be affected by error.
- The code is **universal**, i.e. it is the same in all organisms.

Table 3.3 *The genetic code. The base sequences shown are those on mRNA*

First position	Second position				Third position
	U	C	A	G	
U	Phe	Ser	Tyr	Cys	U
	Phe	Ser	Tyr	Cys	C
	Leu	Ser	Stop	Stop	A
	Leu	Ser	Stop	Trp	G
C	Leu	Pro	His	Arg	U
	Leu	Pro	His	Arg	C
	Leu	Pro	Gln	Arg	A
	Leu	Pro	Gln	Arg	G
A	Ile	Thr	Asn	Ser	U
	Ile	Thr	Asn	Ser	C
	Ile	Thr	Lys	Arg	A
	Met	Thr	Lys	Arg	G
G	Val	Ala	Asp	Gly	U
	Val	Ala	Asp	Gly	C
	Val	Ala	Glu	Gly	A
	Val	Ala	Glu	Gly	G

Protein synthesis – transcription

Proteins, especially enzymes, are essential to all aspects of life. Every organism needs to make its own, sometimes unique, proteins. The biochemical machinery in the cytoplasm of each cell has the capacity to make any and every protein from just 20 amino acids. Exactly which proteins it manufactures depends upon the instructions that are provided at any given time, by the DNA in the cell's nucleus. The process can be thought of as a bakery where the basic equipment and ovens can manufacture any variety of bread or cake from relatively few basic ingredients. Which particular ones are made depends on the recipe the baker uses on any particular day. By choosing different recipes at different times, rather than making everything all the time, the baker can meet seasonal demands, adapt to changing customer needs, and avoid waste. The publication of many copies of the recipe book can be likened to DNA replication; taking a photocopy to use in the bakery is therefore **transcription**. Making the cakes, using the photocopied recipe, is **translation**. If the book is not removed from the library, many copies of the recipe can be made, and the same cakes produced in many places at the same time or over many years.

3.5.1 Production of messenger RNA (Transcription)

Transcription (Fig 3.10) is the process of making **messenger RNA** (section 3.1.5) from part of the DNA molecule. This mRNA then carries the information out of the nucleus to the ribosomes in the cytoplasm, which are the site of protein synthesis. The process is as follows:

- The enzyme **helicase** acts on a specific region of the DNA molecule called a **cistron**.

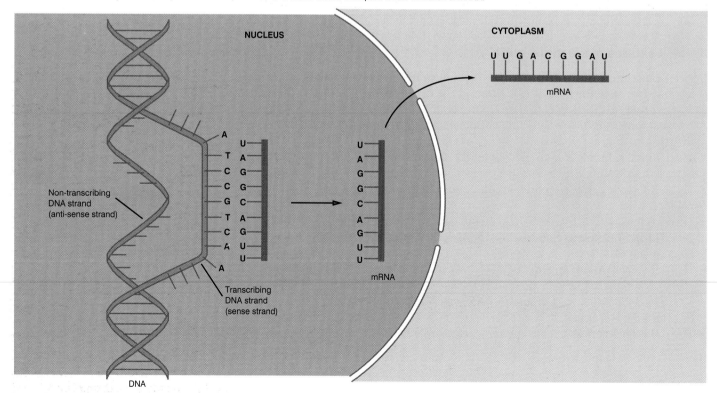

Fig 3.10 *Summary of transcription*

- The hydrogen bonds between the bases are broken, causing the two strands to separate and expose the bases in that region.
- The enzyme **RNA polymerase** moves along one of the two DNA strands known as the **coding (transcribing** or **sense) strand**, causing the bases on this strand to join with the individual complementary nucleotides from the pool which is present in the nucleus.
- In this way, an exposed guanine base on the DNA is linked to the cytosine base of a free nucleotide. Similarly, cytosine links to guanine and thymine joins to adenine. The exception is adenine, which links to uracil rather than thymine.
- As the RNA polymerase adds the nucleotides one at a time, to build a strand of mRNA, so the DNA strands rejoin behind it. As a result, only around 12 base pairs on the DNA are exposed at any one time.
- When the RNA polymerase reaches a particular sequence of bases on the DNA which it recognises as a 'stop' code, it detaches, and the production of mRNA is then complete.

3.5.2 Modification of RNA (Processing)

Before leaving the nucleus, the mRNA produced during transcription is **modified** as follows:
- A guanine nucleotide is added to one end of the mRNA. This 'cap' is used to set off the process of translation when the mRNA reaches a ribosome.
- Around 100 adenine nucleotides are added to the other end of the mRNA. It is thought that this 'tail' may prevent the breakdown of the mRNA in the cytoplasm, because mRNA without a 'tail' is rapidly destroyed.
- Portions of mRNA called **introns**, which have no functional value, are found in **eukaryotic cells**. These are removed from the mRNA (Fig 3.11).

3.5.3 Translocation of mRNA

The mRNA molecules are too large to diffuse out of the nucleus and so, having been processed, they leave via a nuclear pore. Protected from enzymatic action by its adenine 'tail', the mRNA is attracted to the ribsomes, to which it attaches itself.

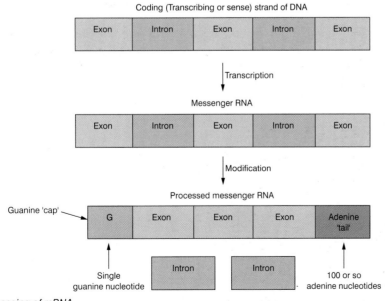

Fig 3.11 Processing of mRNA

SUMMARY TEST 3.5

DNA controls protein synthesis by the formation of a template known as (**1**), which is formed when a specific region of the DNA called a (**2**) is opened up by the enzyme (**3**). Along the strand called the (**4**) strand moves the enzyme (**5**), causing the bases on the strand to link with their corresponding nucleotides, which in the case of adenine is the nucleotide containing the base (**6**). The template is then modified by the addition of a (**7**) nucleotide at one end and about 100 (**8**) nucleotides at the other. Non-functional portions of the template, called (**9**), are removed before it leaves the (**10**) and enters the cytoplasm.

3.6

AQA.A	EDEXCEL
AQA.A (Human)	EDEXCEL (Human)
AQA.B	OCR

Protein synthesis – translation

Translation is the process whereby the messenger RNA from the nucleus of a cell forms a polypeptide, in accordance with the sequence of organic bases along its length (see also section 12.6.1 on cystic fibrosis). The process begins with the activation of the amino acids which will make up the polypeptide.

3.6.1 Amino acid activation

The amino acids present in cells must first be **activated** before they can be assembled into a polypeptide. This occurs in two stages:

* The amino acid first forms an intermediate with **ATP**, which provides the energy for the next stage.
* The intermediate then combines with transfer RNA to form an amino acid–tRNA complex called **amino-acyl tRNA** (Fig 3.12). The reaction is controlled by the enzyme, amino-acyl tRNA synthetase.

Although the base structure of tRNA is always the same (unit 3.1), a sequence of three bases on the anticodon loop varies. There are at least 60 variants, which correspond to a codon of three bases on the messenger RNA. At the other end of the tRNA molecule, there is always the sequence of bases adenine-cytosine-cytosine, and it is to this end that the amino acid attaches. Each amino acid therefore has its own tRNA molecule, with its own unique anticodon of bases.

3.6.2 Starting polypeptide construction

* A ribosome becomes attached to one end of the mRNA molecule.
* The starting point on the mRNA is normally the triplet of bases **(codon)**, AUG.
* The amino-acyl tRNA molecule with the anticodon sequence of UAC moves to the ribosome and pairs up with the AUG sequence on the mRNA (Fig 3.13).
* As the tRNA which pairs with the AUG sequence on the mRNA always carries the amino acid methionine, polypeptides initially have methionine as the first amino acid.
* However, if methionine does not make up part of the finished polypeptide, it is removed at the end of the synthesis.

WHERE ARE PROTEINS ASSEMBLED?

Proteins for use within the cell, e.g. haemoglobin, are simply released from the ribosomes into the cytoplasm.

Proteins that are to be exported from the cell, e.g. digestive enzymes or mucus, are assembled on the rough endoplasmic reticulum and transported to the Golgi apparatus (section 4.6.3), where they may have other molecules attached, e.g. carbohydrates. They are then secreted by exocytosis (section 4.11.4)

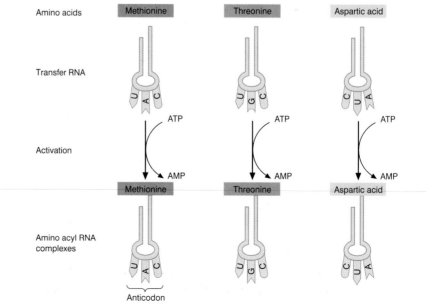

Fig 3.12 Amino acid activation

3.6.3 Making the polypeptide

The following explanation of how a polypeptide is made is illustrated in figure 3.13.
- The ribosome moves along the mRNA, bringing together two tRNA molecules at any one time, each pairing up with the corresponding two codons on the mRNA.
- By means of an enzyme (peptidyl transferase), the two amino acids on the tRNA are joined by a peptide bond.
- The ribosome moves on to the third codon in the sequence on the mRNA, thereby linking the amino acids on the second and third tRNA molecules.
- As this happens, the first tRNA is released from its amino acid (methionine) and is free to collect another methionine molecule from the amino acid pool in the cell.
- The process continues in this way, with up to 15 amino acids being linked each second, until a complete polypeptide chain is built up.
- Up to 50 ribosomes can pass immediately behind the first, so that many identical polypeptides can be assembled simultaneously (Fig 3.14). A group of ribosomes acting in this way is known as a **polysome**.

3.6.4 Finishing the polypeptide

The process described in section 3.6.3 continues until the ribosome reaches a **stop (nonsense) codon**. These are UGA, UAG and UAA, and do not attract a tRNA. At this point, therefore, the ribosome, mRNA and the last tRNA molecule all separate and the polypeptide chain is complete. It now needs to be assembled into the final protein.

3.6.5 Assembling the protein

What happens to the polypeptide next depends upon the protein being made, but usually involves the following:
- The polypeptide is made into its secondary structure by being either coiled into an α-helix or folded into a β-pleated sheet (unit 1.6).
- The secondary structure is folded into its tertiary structure (unit 1.6).
- Different polypeptide chains are linked to form the quaternary structure, along with associated (non-protein) prosthetic groups (unit 1.6).

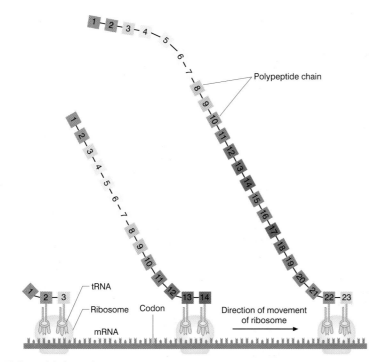

Fig 3.14 Polypeptide formation

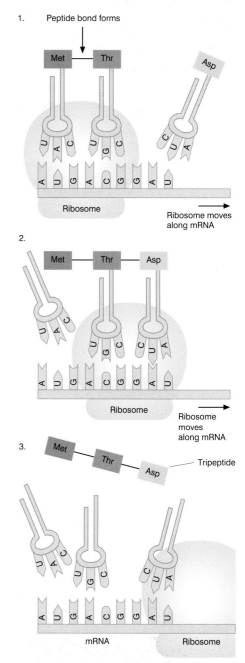

Fig 3.13 Translation

SUMMARY TEST 3.6

The formation of a polypeptide from the sequence of bases on messenger RNA is known as (**1**). It begins with the addition of (**2**) to amino acids, in a process known as activation. These amino acids then combine with transfer RNA at a sequence of three bases known as the (**3**), to form a complex called (**4**). To form the polypeptide, a (**5**) becomes attached to the mRNA molecule.

1 DNA is made up of two polynucleotide strands, the sense strand and the anti-sense strand. Messenger RNA is transcribed from the DNA sense strand, which contains the genetic code.

a The graph shows the number of bases found in the sense strand and the anti-sense strand of a short piece of DNA, and the mRNA transcribed from it.

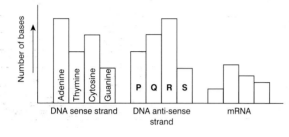

(i) Identify the base represented by each of the following letters.

P

Q

R

S *(2 marks)*

(ii) Explain why the total number of bases in the DNA sense strand and the total number of bases in the DNA anti-sense strand are the same. *(1 mark)*

(iii) Explain why the total number of bases in the DNA sense strand and the total number of bases in the mRNA are different. *(1 mark)*

b The mRNA has a sequence of 1824 bases. How many amino acids will join to form the polypeptide chain? *(1 mark)*

c Although DNA is double-stranded, only the sense strand determines the specific amino acid sequence of a polypeptide. Suggest a role of the anti-sense strand.

(1 mark)

(Total 6 marks)

AQA Jan 2001, HB (A) BYA3, No.3

2 a **Table 1** shows the percentage of different bases in DNA from different organisms.

Source of DNA	Adenine %	Guanine %	Thymine %	Cytosine %
Human	30	20	30	20
Rat	28	22	28	22
Yeast	31	19	31	19
Turtle	28	22	28	22
E.coli	24			
Salmon	29	21	29	21
Sea urchin	33	17	33	17

(i) What information about the ratios of the different bases in DNA can you work out from the table?

(2 marks)

(ii) Give the results that you would expect for DNA from the *E.coli* bacterium. Explain how you arrived at your answer. *(3 marks)*

(iii) Turtles have the same percentages of the four different bases as rats. Explain why they can still be very different animals. *(1 mark)*

b **Table 2** shows the percentage of different bases in the DNA from a virus.

Adenine %	Guanine %	Thymine %	Cytosine %
25	24	33	18

(i) Describe how the ratios of the different bases in this virus differ from those in **Table 1**. *(1 mark)*

(ii) The structure of DNA in this virus is not the same as DNA in other organisms. Suggest what this difference in DNA structure might be. *(1 mark)*

c Describe how proteins are synthesised using the DNA code. *(7 mark)*

(Total 15 marks)

AQA Jan 2001, B (A) BYA2, No.9

3 The diagram below shows part of a molecule of deoxyribonucleic acid (DNA).

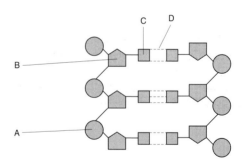

a Name A, B, C and D. *(4 marks)*

b Analysis of a molecule of DNA showed that cytosine accounted for 42% of the content of the nitrogenous bases. Calculate the percentage of bases in the molecule which would be thymine. Show your working.

(3 marks)

c During the process of **transcription**, one of the DNA strands is used as a template for the formation of a complementary strand of messenger RNA (mRNA). The diagram below shows the sequence of bases in part of a strand of DNA.

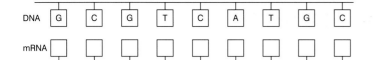

- (i) Write the letters of the complementary bases in the boxes of the mRNA strand. *(2 marks)*
- (ii) How many amino acids are coded for by this part of the strand of mRNA? *(1 mark)*

(Total 10 marks)

Edexcel 6101/01 Jan 2001, B B(H) AS, No.8

4 a Name the organelle where proteins are synthesised from amino acids. *(1 mark)*

b The diagram shows a transfer RNA molecule (tRNA).

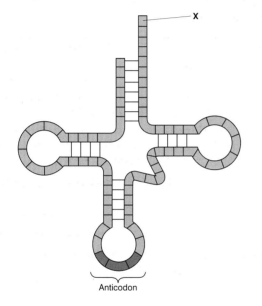

Anticodon

- (i) During protein synthesis, which molecule is attached to the tRNA molecule at **X**? *(1 mark)*
- (ii) What is an anticodon? *(1 mark)*
- (iii) Give **one** way in which the structure of a tRNA molecule is different from an mRNA molecule. *(1 mark)*

(Total 4 marks)

AQA June 2001, B (B) BYB2, No.5

5 a Draw a labelled diagram to show the structure of an RNA nucleotide. *(2 marks)*

b The diagram shows a molecule of an enzyme called ribonuclease. Each amino acid in the protein is indicated by a 3-letter symbol e.g. Arg = arginine.

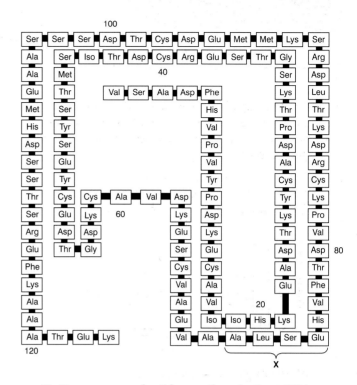

- (i) How many nucleotides are there in the mRNA molecule that codes for this enzyme? *(1 mark)*
- (ii) The table gives the mRNA code for the four amino acids in the part of the enzyme labelled **X**. Give the DNA code for the part of the enzyme labelled **X**. *(1 mark)*

Amino acid	Symbol	mRNA code
Alanine	Ala	GCU
Glutamine	Glu	GAG
Leucine	Leu	UUA
Serine	Ser	AGU

c (i) Where does translation occur in a cell? *(1 mark)*
- (ii) Describe what happens during translation. *(3 marks)*

d Explain how the structure of DNA is related to its function. *(6 marks)*

(Total 14 marks)

AQA Jan 2001, B (B) BYB2, No.7

4.1

Microscopy

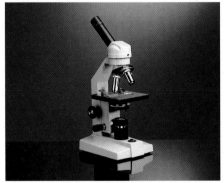

Light microscope

Microscopes are instruments which produce a magnified image of an object. Their use has opened up a world of detail which was hidden from our ancestors. There are two basic types of microscope:

- **The light (optical) microscope**, which uses a beam of light focused by glass lenses.
- **The electron microscope**, which uses a beam of electrons focused by electromagnets.

4.1.1 The light microscope

In its simplest form, a light microscope consists of a single lens which operates as a magnifying glass. More effective is the **compound light microscope**, which has three systems of lenses:

- **The condenser lenses** are located beneath the microscope stage and can be adjusted in height to ensure that light is focused on the specimen being examined. This allows the resolving power of the microscope to be used to its full effect.
- **The objective lenses** produce an initial magnified image of the specimen.
- **The eyepiece lenses** further magnify the image produced by the objective lenses.

4.1.2 Magnification

The magnification of an object is how many times bigger the image is when compared with the orginal object:

$$\text{magnification} = \frac{\text{size of image}}{\text{size of object}}$$

In practice, it is more likely that you will be asked to calculate the size of an object when you know the size of the image and the magnification. In this case:

$$\text{size of object} = \frac{\text{size of image}}{\text{magnification}}$$

The important thing to remember in calculating the magnification is to ensure that the units of length are the same for both the object and the image.

4.1.3 Resolution

The **resolution** or **resolving power** of a microscope is the minimum distance apart that two objects can be for them to appear as separate items. Whatever the type of microscope, the resolving power depends on the wavelength or form of radiation used. In a light microscope it is about 2µm – any two objects which are 2µm or more apart will be seen separately, but any objects closer than 2µm will appear as a single item.

Increasing the magnification increases the size of an object, but does not increase its resolution, i.e. the object looks bigger, but the detail is not changed.

4.1.4 The electron microscope

The relatively poor resolving power of the light microscope limits its use. The electron microscope, rather than using a beam of light, utilizes a beam of electrons. Electron beams have a much smaller wavelength and so can resolve objects as close

Table 4.1 *Comparison of advantages and disadvantages of the light and electron microscopes*

Light microscope	Electron microscope
Advantages	**Disadvantages**
Cheap to purchase and operate	Expensive to purchase and operate
Small and portable – can be used almost anywhere	Very large and must be operated in special rooms
Unaffected by magnetic fields	Affected by magnetic fields
Preparation of material is relatively quick and simple, requiring only a little expertise	Preparation of material is lengthy and requires considerable expertise and sometimes complex equipment
Material rarely distorted by preparation	Preparation of material may distort it
Natural colour of the material can be observed	All images are in black and white
Disadvantages	**Advantages**
Magnifies objects up to 1000× only	Magnifies objects more than 500 000×
The depth of field is restricted	It is possible to investigate a greater depth of field

as 2nm apart – a thousand times closer than a light microscope can resolve. However, since the molecules that make up air absorb electrons, a near vacuum has to be created inside the instrument. There are two types of electron microscope:

- the transmission electron microscope (TEM)
- the scanning electron microscope (SEM).

4.1.5 The transmission electron microscope (TEM)

The **transmission electron microscope** has an **electron gun** which produces a beam of electrons. The beam is then 'focused' onto the specimen by means of the electromagnets which make up the condenser. The greater the electrical current applied to these electromagnets, the more the beam is deflected. After passing through the condenser, the image is enlarged by passing through objective and projector lenses. The human eye cannot detect electrons, and so the image is made visible by directing the electron beam onto a fluorescent screen. In a TEM, the beam passes through a thin section of the specimen. Parts of this specimen absorb electrons, and therefore appear dark. Other parts of the specimen allow the electrons to pass through and hit the screen, which fluoresces, and so these parts appear bright. The image produced on the screen can be photographed to give a **photoelectron micrograph.**

4.1.6 The scanning electron microscope (SEM)

One disadvantage of the TEM is that specimens must be extremely thin to allow electrons to penetrate. The result is therefore a flat, two-dimensional image. Although basically similar to a TEM, the SEM overcomes this problem by directing a beam of electrons onto the surface of the specimen from above, rather than penetrating it from below. The specimen must first be dried and coated with a metal (to produce secondary electrons). The beam is then passed back and forth across a portion of the specimen in a regular pattern. The electrons are scattered by the specimen and the pattern of this scattering depends on the contours of the specimen surface. By careful analysis of the pattern of scattered electrons and secondary electrons that is produced, a three-dimensional image can be formed.

Electron microscope

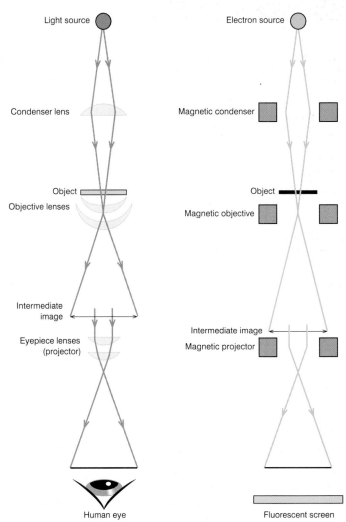
Fig 4.1 *Comparison of radiation pathways in light and electron microscopes*

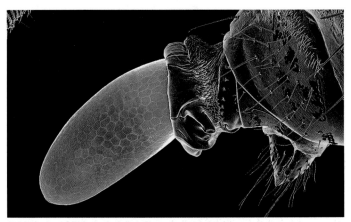
False-colour SEM of a fruit fly laying an egg (×80)

4.2

AQA.A   EDEXCEL
AQA.A (Human)   EDEXCEL (Human)
AQA.B   OCR

Cell structure

The cell is the basic unit of life and current cell theory states that

- all living organisms are made up of one **(unicellular)** or more cells **(multicellular)**
- metabolic processes take place within cells
- new cells are derived from existing ones
- cells possess the genetic material of an organism which is passed from parent to daughter cells
- a cell is the smallest unit of an organism capable of surviving independently.

4.2.1 The animal cell

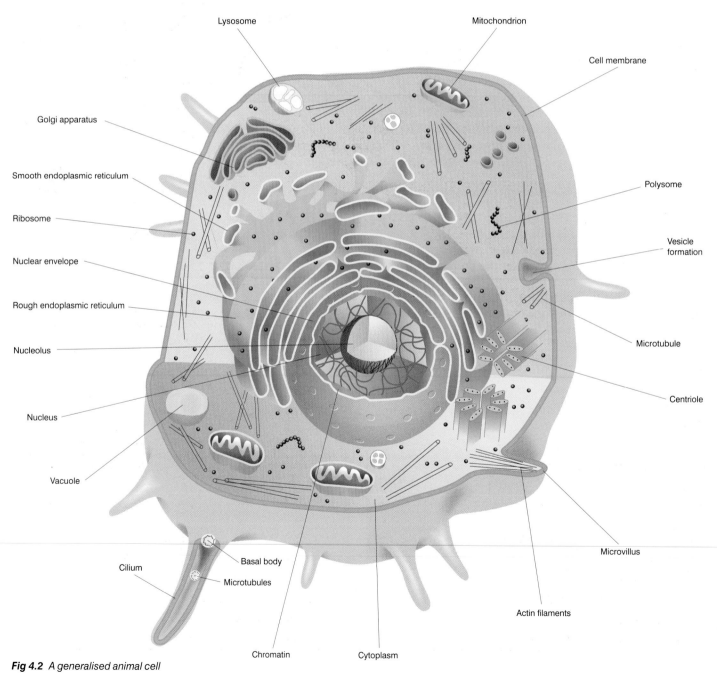

Fig 4.2 *A generalised animal cell*

4.2.2 The plant cell

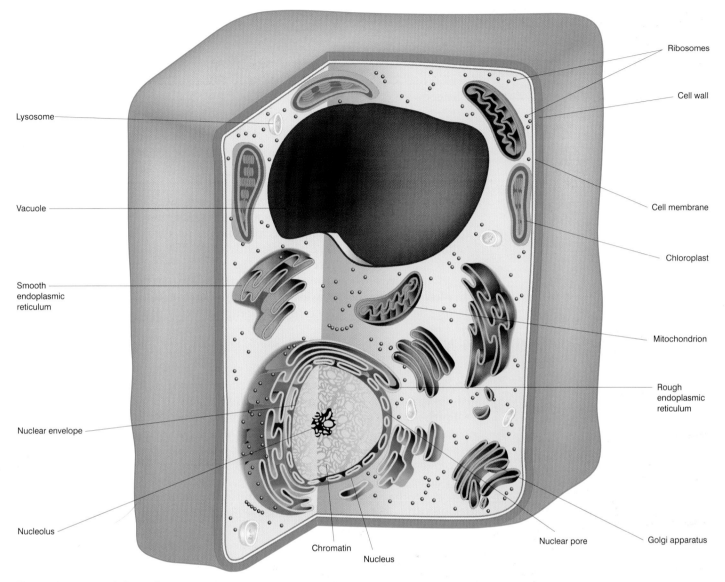

Fig 4.3 *A generalised plant cell*

Table 4.2 *Differences between plant and animal cells*

Plant cells	Animal cells
Tough, slightly elastic cellulose cell wall present (in addition to the cell membrane)	Cell wall absent – only a membrane surrounds the cell
Pits and plasmodesmata present in the cell wall	No cell wall and therefore no pits or plasmodesmata
Chloroplasts present in large numbers in most cells	Chloroplasts absent
Mature cells normally have a large single, central vacuole filled with cell sap	Vacuoles, e.g. contractile vacuoles, if present, are small and scattered throughout the cell
Cytoplasm normally confined to a thin layer at the edge of the cell	Cytoplasm present throughout the cell
Nucleus at the edge of the cell	Nucleus anywhere in the cell, but often central
Centrioles absent in higher plants	Centrioles present
Cilia and flagella absent in higher plants	Cilia or flagella often present
Starch grains used for storage	Glycogen granules used for storage
Only some cells are capable of division	Almost all cells are capable of division

51

Cells, tissues and organs

AQA.A	EDEXCEL
AQA.A (Human)	EDEXCEL (Human)
AQA.B	OCR

In multicellular organisms, cells are specialised to perform specific functions. Similar cells are then grouped together into tissues and organs for increased efficiency.

4.3.1 Cell differentiation

Unicellular organisms such as **protozoa** and some algae perform all the essential life functions inside the boundaries of a single cell. Although they perform all functions adequately, they cannot be totally efficient at all of them, because each function will require a different type of cellular structure. One activity may be best carried out by a long thin cell, while another might suit a spherical shape. No one cell can provide optimum conditions for all functions. For this reason, the cells of multicellular organisms are each differently adapted to perform a particular role. This is known as **cell differentiation**. The cells of a multicellular organism have therefore evolved to become more and more suited to one specialised function. As they have done so, they have lost the ability to carry out certain functions and so become dependent on other cells to perform these activities for them.

4.3.2 Tissues

For working efficiency, cells are normally grouped together. Such a collection of similar cells which perform a specific function is known as a **tissue**. Examples of tissues include:

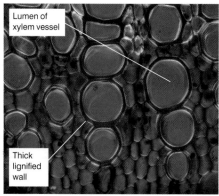

Lumen of xylem vessel

Thick lignified wall

Xylem (TS) (×400 approx)

- **Epithelial tissues** (section 6.1.3), which are found in animals and consist of sheets of cells on a basement membrane held together by intercellular substance. They line the surfaces of organs and often have a protective or secretory function. There are many types, including **squamous epithelium** made up of thin, flat cells which line organs where diffusion takes place (e.g. alveoli of the lungs), and **ciliated columnar epithelium**, which lines ducts such as the oviduct and trachea, where their cilia are used to move mucus over the epithelial surface.
- **Xylem** (section 7.2.1) occurs in plants and is made up of a number of cell types. It is used to transport water and mineral ions throughout the plant and also gives mechanical support.
- **Phloem** (section 7.2.3) also occurs in plants and has a number of different cell types. It is used to transport organic materials around the plant.

4.3.3 Organs

Just as cells are grouped into tissues, so tissues are grouped together into organs. An **organ** is a combination of tissues which are coordinated to perform a variety of functions, although they often have one predominant major physiological function. Organs are themselves grouped into **organ systems**.

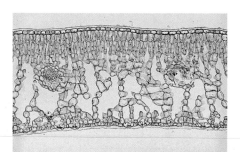

A leaf is an example of an organ

In animals, for example, the stomach is an organ which carries out the digestion of certain types of food (section 6.9.2). It is made up of tissues such as:

- **muscle** to churn and mix the stomach contents
- **epithelium** to protect the stomach wall and produce secretions
- **connective tissue** to hold together the other tissues.

The stomach, along with the other organs like the duodenum, ileum, pancreas and liver, forms the organ system known as the digestive system.

In plants, a leaf (section 7.4.1) is an organ made up of the following tissues:
- **palisade mesophyll** which carries out photosynthesis
- **spongy mesophyll** adapted for gaseous diffusion
- **epidermis** to protect the leaf and allow gaseous diffusion
- **phloem** to transport organic materials away from the leaf
- **xylem** to transport water and **ions** into the leaf.

It is not always easy to determine which structures are organs. Blood capillaries, for example, are *not* organs whereas arteries and veins both are. All three structures have the same major physiological function – namely the transport of blood. However, capillaries are made up of just one tissue – squamous epithelium – whereas arteries and veins are made up of many tissues, e.g. epithelial, muscle and connective tissues.

4.3.4 Cell fractionation and ultracentrifugation

In order to study the structure and function of the various organelles which make up cells, it is necessary to obtain large numbers of isolated organelles. There are two stages in achieving this:
- **Cell fractionation** involves cells being placed in a cold, **isotonic**, buffered solution, for the following reasons:
  - cold – to reduce enzyme activity which might break down the organelles
  - isotonic – to prevent organelles bursting due to the osmotic influx of water
  - buffered – to maintain a constant pH.
  They are then broken up using a pestle and mortar or an electrical homogeniser (blender) to break the cell membrane and/or wall and release the organelles. The resultant fluid, known as homogenate, is then filtered to remove any complete cells and large pieces of debris.
- **Ultracentrifugation** is the process by which fragments in the filtered homogenate are separated in a machine called a centrifuge. This spins tubes of homogenate at very high speed, to create a centrifugal force. At slower speeds, only the very heaviest organelles are forced out of suspension, into a thin deposit at he bottom of the tube. This deposit includes the nuclei, which can be isolated by removing the **supernatant liquid**. This supernatant liquid can then be spun in the centrifuge at a faster speed, separating out the next most heavy component. By continuing in this way, smaller and smaller fragments can be separated out (table 4.3).

A centrifuge

Table 4.3 Separation of organelles by ultracentrifugation

Organelles to be separated out	Speed of centrifugation expressed as gravitation force / g	Duration of centrifugation / mins
Nuclei	1000	10
Mitochondria	3500	10
Lysosomes	16500	20
Ribosomes	100000	60

SUMMARY TEST 4.3

The cells of multicellular organisms become adapted to a particular function in a process called cell (**1**). A group of cells performing a specific (**2**) is called a tissue. An example of a tissue is epithelium, which takes a number of forms. (**3**) epithelium consists of thin flat cells found in the (**4**) of the lungs and is adapted for (**5**). In plants, the tissue that transports water and minerals is called (**6**) and the one transporting organic materials is (**7**). Organs are groups of tissues combined to perform one major function. An example is the stomach, which has (**8**) tissue to mix the food, (**9**) tissue to hold tissues together and (**10**) to produce secretions. An example in plants is the leaf, which performs the function of (**11**). It possesses, amongst other tissues, (**12**) mesophyll for carrying out diffusion and palisade mesophyll to carry out (**13**). To separate out organelles, cells are disrupted in a blender to form a liquid known as (**14**).

Prokaryotic cells (bacteria)

Although cells come in a bewildering variety of size, shape and function, they nevertheless fall into two basic groups:

- **Prokaryotic cells** ('pro' = before, 'karyote' = nucleus) have no nucleus or nuclear membrane.
- **Eukaryotic cells** ('eu' = true, 'karyote' = nucleus) have a nucleus bounded by a nuclear membrane.

Other differences between prokaryotic and eukaryotic cells are listed in table 4.4.

Table 4.4 *Comparison of prokaryotic and eukaryotic cells*

Prokaryotic cells	Eukaryotic cells
No true nucleus, only diffuse area(s) of nucleoplasm with no nuclear envelope	Distinct nucleus, with a nuclear envelope
No nucleolus	Nucleolus is present
Circular strands of DNA but no chromosomes	Chromosomes present in which DNA is located
No membrane-bounded organelles	Membrane-bounded organelles such as mitochondria are present
No chloroplasts, only photosynthetic lamellae in some bacteria	Chloroplasts present in plants and algae
Ribosomes are smaller (70S type)	Ribosomes are larger (80S type)
Flagella (if present) lack internal 9+2 microtubule arrangement	Flagella, where present, have a 9+2 internal microtubule arrangement
No endoplasmic reticulum or associated Golgi apparatus and lysosomes	Endoplasmic reticulum present along with Golgi apparatus and lysosomes
Cell wall made of peptidoglycan	Where present, cell wall is made mostly of cellulose or chitin

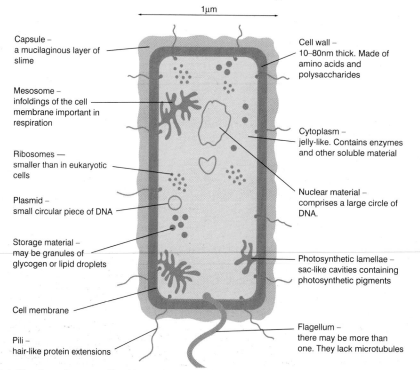

Fig 4.4 *Structure of a generalised bacterial cell*

4.4.1 Structure of a bacterial cell

Bacteria occur in every habitat in the world; they are versatile, adaptable and successful. Much of their success is a result of their small size, normally in the range 0.1–10μm in length. Their cellular structure is relatively simple (Fig 4.4). All bacteria possess a cell wall which is made up of **peptidoglycan** (murein) – a polysaccharide cross-linked by peptide molecules. Around this wall, many bacteria further protect themselves by secreting a **capsule** of mucilaginous slime. Hair-like structures made of protein and called **pili** extend through the cell wall in some species. These enable the bacteria to stick to one another or to other surfaces. **Flagella** occur in certain types of bacteria. These lack microtubules and so do not beat, as they do in eukaryotic cells. Their rigid corkscrew shape and rotating base, however, cause bacteria to spin through fluids.

Within the cytoplasm of bacterial cells are scattered **ribosomes** (70S type). These are smaller than those of eukaryotic cells (80S type), but nevertheless serve the same function in protein synthesis. **Glycogen granules** and **oil droplets** are used for storage. Infoldings of the cell, known as **mesosomes**, often occur. These provide a large surface area for the attachment of respiratory enzymes. In photosynthetic bacteria there are lamellae called **thylakoids**, which contain the enzymes and bacterial chlorophyll essential to photosynthesis. The genetic material in bacteria is in the form of a circular strand of DNA. Separate from this, and not necessary for growth and metabolism, are smaller circular pieces of DNA called **plasmids**. These can reproduce themselves independently and may give the bacterium resistance to harmful chemicals such as antibiotics. Plasmids are used extensively as vectors (carriers of genetic information) in **genetic engineering** (section 12.3.3).

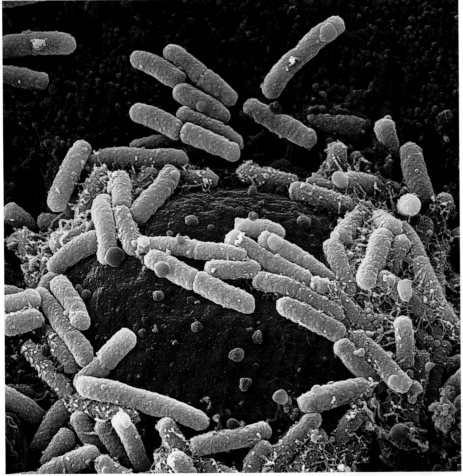

SEM of a colony of the rod-shaped bacterium – Escherichia coli

Table 4.5 Roles of structures found in a bacterial cell

Cell structure	Role
Cell wall	Physical barrier which protects against mechanical damage and excludes certain substances
Capsule	Protects bacterium from other cells, e.g. white blood cells, and also helps groups of bacteria to stick together for further protection
Cell surface (plasma) membrane	Acts as a differentially permeable layer which controls the entry and exit of chemicals
Mesosome	Provides a large surface area for the attachment of respiratory enzymes
Flagellum	Aids movement of bacterium because its rigid, corkscrew shape and rotating base help the cell spin through fluids
Pili	Help cells stick to one another or to other surfaces
Circular DNA	Possesses the genetic information for the replication of bacterial cells
Plasmids	Possess genes which aid the survival of bacteria in adverse conditions, e.g. produce enzymes which break down antibodies
Ribosomes (70S type)	Site of protein synthesis
Glycogen granules	Store carbohydrates for breakdown during respiration to provide energy
Lipid droplets	Store lipids as a more concentrated, longer-term, store for conversion to carbohydrate and use in respiration
Photosynthetic lamellae (thylakoids)	Contain enzymes and bacterial chlorophyll and therefore carry out photosynthesis

SUMMARY TEST 4.4

Prokaryotic cells lack a distinct (**1**) and include the group of organisms called (**2**). Their DNA is (**3**) in shape and known as a (**4**) and their ribosomes are smaller than in (**5**) cells and known as the (**6**) type. Where photosynthesis takes place it does so in lamellae called (**7**) rather than in a true chloroplast.

Nucleus, chloroplast and mitochondrion

AQA.A	EDEXCEL
AQA.A (Human)	EDEXCEL (Human)
AQA.B	OCR

4.5.1 The nucleus

The nucleus, the most prominent feature of a **eukaryotic** cell when viewed under the microscope, may vary in shape, size and position, from cell to cell. In contrast, its function remains essentially the same – to retain the organism's hereditary material and to control the cell's activities. Usually spherical and between 10 and 20µm in diameter, the nucleus has a number of parts:

- **The nuclear envelope** is a double membrane which surrounds the nucleus. Its outer membrane is continuous with the endoplasmic reticulum of the cell and often has ribosomes on its surface. It serves to control the entry and exit of materials in and out of the nucleus and to contain the reactions taking place within it.
- **Nuclear pores** allow the passage of large molecules such as messenger RNA (unit 3.5) out of the nucleus. There are typically around 3000 in each nucleus, each being 40–100nm in diameter.
- **Nucleoplasm** is the granular, jelly-like material which makes up the bulk of the nucleus.
- **Chromatin** is found within the nucleoplasm and is composed of DNA and associated proteins. This is the diffuse form that chromosomes take up when the cell is not dividing. When the cell divides, the chromatin condenses into chromosomes (unit 5.1). Chromatin comprises **heterochromatin** – densely packed DNA which stains darkly, and **euchromatin**, which stains less heavily.
- **The nucleolus** is a small spherical body (occasionally there is more than one) within the nucleoplasm. It manufactures ribosomal RNA (section 3.1.3) and assembles the ribosomes.

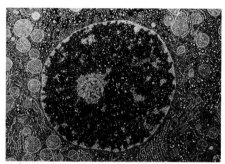

Liver cell nucleus

The functions of the nucleus are to:

- act as the control centre of the cell through the production of mRNA and protein synthesis
- retain the genetic material of the cell in the form of DNA/chromosomes
- manufacture rRNA and ribosomes
- start the process of cell division.

4.5.2 Chloroplasts

Chloroplasts are found in eukaryotic cells which photosynthesise. They are flat discs usually 2–10µm in diameter and 1µm thick (Fig 4.5), and are made up of a number of parts:

- **The chloroplast envelope** is a double membrane, the inner one of which is folded into a series of lamellae. It controls the entry and exit of substances in and out of the chloroplast.
- **The stroma** is a colourless, gelatinous matrix which contains the enzymes necessary for the light-independent stage of photosynthesis. Small amounts of DNA and oil droplets are also found in the stroma.
- **The grana** are structures that look like a stack of coins. There are typically 50 grana in a chloroplast, and each is made up of up to 100 stacked, flattened sacs called **thylakoids** or **lamellae** (Fig 4.5). It is to the thylakoids that the chlorophyll molecules are attached. The grana therefore carry out the light-dependent stage of photosynthesis.
- **Starch grains** act as temporary stores of the carbohydrate which is produced during photosynthesis.

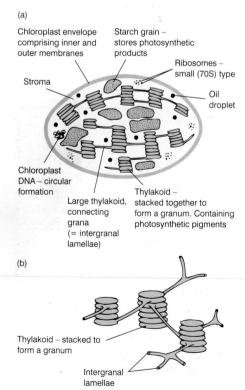

(a)

Chloroplast envelope comprising inner and outer membranes

Starch grain – stores photosynthetic products

Ribosomes – small (70S) type

Stroma

Oil droplet

Chloroplast DNA – circular formation

Large thylakoid, connecting grana (= intergranal lamellae)

Thylakoid – stacked together to form a granum. Containing photosynthetic pigments

(b)

Thylakoid – stacked to form a granum

Intergranal lamellae

Fig 4.5 *Structure of chloroplasts*

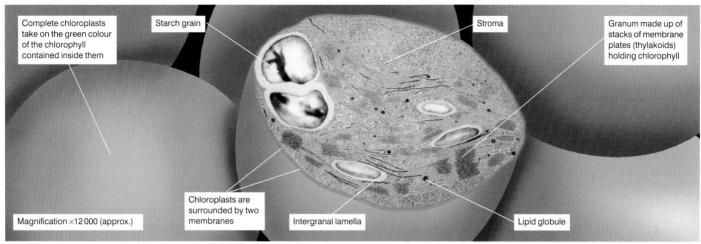

Fig 4.6 *Chloroplast structure*

4.5.3 The mitochondrion

Present in all but a few eukaryotic cells, mitochondria (Fig 4.7) are rod-shaped and 1–10μm in diameter. They are made up of a number of parts:

- **A double membrane** surrounds the organelle, the outer one controlling the entry and exit of material. The inner membrane is folded to form extensions known as cristae.
- **Cristae** are shelf-like extensions of the inner membrane, some of which extend across the whole width of the mitochondrion. These provide a large surface area for the attachment of structures called **stalked (elementary) particles**, which are about 4nm high (Fig 4.8). The stalked particles contain enzymes involved in the synthesis of ATP.
- **The matrix** makes up the remainder of the mitochondrion. It is a semi-rigid material containing protein, lipids and traces of DNA allowing them to control the production of their own proteins. The enzymes involved in **Krebs cycle** are found in the matrix, as are mitochondrial ribosomes.

Functions of mitochondria

Mitochondria act as the sites for the Krebs cycle and **electron transport** stages of respiration. They are therefore responsible for the production of energy-rich ATP molecules from carbohydrates. Because of this, the number of mitochondria, their size and the number of cristae all increase in cells which have a high level of metabolic activity and therefore need a good supply of ATP. Such cells include those of the muscles and the liver.

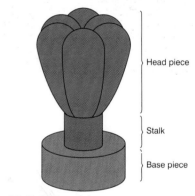

Fig 4.8 *Stalked particle*

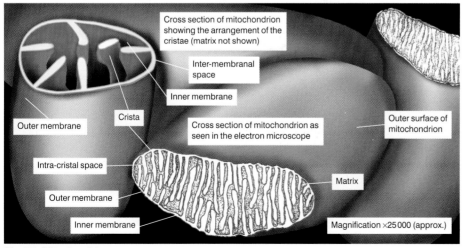

Fig 4.7 *Mitochondria*

SUMMARY TEST 4.5

The nucleus is surrounded by a double membrane called the **(1)**, which has pores in it with a diameter of **(2)**. Within the nucleus is a small spherical body called the **(3)** which manufactures **(4)** RNA. Chloroplasts have within them structures that look like a stack of coins. These are called **(5)** and contain **(6)** molecules. Mitochondria are the sites for the two stages of respiration known as **(7)** and **(8)**. The enzymes for these processes are found either in the **(9)** or attached to **(10)** particles that are fixed to folds of the inner membrane, called **(11)**.

4.6 Endoplasmic reticulum and ribosomes, Golgi apparatus and lysosomes

AQA.A	EDEXCEL
AQA.A (Human)	EDEXCEL (Human)
AQA.B	OCR

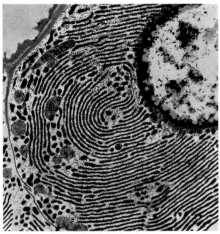

Endoplasmic reticulum

4.6.1 Endoplasmic reticulum

The endoplasmic reticulum (ER) is an elaborate, three-dimensional system of sheet-like membranes spreading through the cytoplasm of cells. It is continuous with the nuclear membrane. The membranes enclose flattened sacs called cisternae (Fig 4.9). There are two types of ER:

- **Rough endoplasmic reticulum (RER)** has ribosomes present on the outer surfaces of the membranes.
- **Smooth endoplasmic reticulum (SER)** lacks ribosomes on its surface and is often more tubular in appearance.

Functions of endoplasmic reticulum

- Provides a large surface area for the synthesis of proteins (RER).
- Provides a pathway for the transport of materials, especially proteins, throughout the cell (RER).
- Synthesises, stores and transports lipids (SER).
- Synthesises, stores and transports carbohydrates (SER).
- Contains lytic enzymes (SER of liver cells).

It follows that cells which need to manufacture and store large quantities of carbohydrates, proteins and lipids have very extensive ER. Such cells include liver and secretory cells.

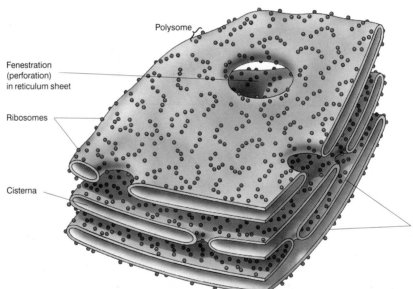

Fig 4.9 Structure of rough endoplasmic reticulum

4.6.2 Ribosomes

Ribosomes are small cytoplasmic granules found in all cells. They may occur in the cytoplasm or be associated with the RER. There are two types, depending on the cells they are found in:

- **80S type**, found in **eukaryotic cells**, is around 25nm in diameter
- **70S type**, found in **prokaryotic cells** (unit 4.4), is slightly smaller.

Each ribosome has two sub-units – one large and one small (Fig 4.10) – each of which contains ribosomal RNA and protein. Despite their small size, they occur in such vast numbers that they can account for up to 25% of the dry mass of a cell. Ribosomes are important in protein synthesis (units 3.5 and 3.6).

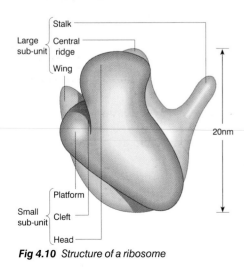

Fig 4.10 Structure of a ribosome

4.6.3 Golgi apparatus

The **Golgi apparatus (Golgi body)** occurs in almost all eukaryotic cells and is similar to SER in structure (section 4.6.1), except that it is more compact. It consists of a stack of membranes which make up flattened sacs or **cisternae** and associated hollow **vesicles** (Fig 4.11). The proteins and lipids produced by the ER are passed through the Golgi apparatus in strict sequence. The Golgi modifies these proteins, often adding non-protein components such as carbohydrate to them. It also 'labels' them, allowing them to be accurately sorted and sent to their correct destinations. Once sorted, the modified proteins and lipids are transported in vesicles which are regularly pinched off from the ends of the Golgi cisternae (Fig 4.11). These vesicles move to the cell surface, where they fuse with the membrane and release their contents to the outside.

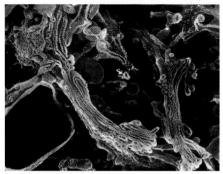

Golgi apparatus of an olfactory bulb cell

Functions of the Golgi apparatus

In general, the Golgi apparatus acts as the cell's post office, receiving, sorting and delivering proteins and lipids. More specifically, it
- adds carbohydrates to proteins to form glycoproteins such as mucin
- produces secretory enzymes such as those secreted by the pancreas
- secretes carbohydrates such as those used in making cell walls in plants
- transports, modifies and stores lipids
- forms lysosomes (section 4.6.4).

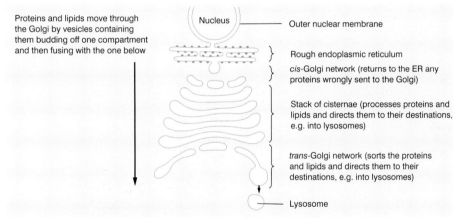

Fig 4.11 The Golgi apparatus and its relationship to the nucleus, ER and lysosomes

4.6.4 Lysosomes

Lysosomes are formed when the vesicles produced by the Golgi apparatus include within them enzymes such as proteases and lipases. Up to 50 such enzymes may be contained in a single lysosome. Up to 1.0 μm in diameter, lysosomes isolate these potentially harmful enzymes from the rest of the cell, before releasing them, either to the outside or into a **phagocytic** vesicle within the cell (Fig 4.12).

Functions of lysosomes

Lysosomes are used to destroy foreign material inside or outside the cell. More particularly, they
- break down material ingested by phagocytic cells such as white blood cells or *Amoeba*
- release enzymes to the outside of the cell **(exocytosis)** in order to destroy material around the cell
- digest worn out organelles **(autophagy)** so that the useful chemicals of which they are made can be re-used
- completely break down cells after they have died **(autolysis)**.

Given the roles that lysosomes perform, it is not surprising that they are especially abundant in secretory and phagocytic cells.

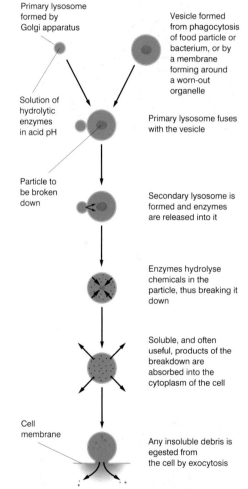

Fig 4.12 The functioning of a lysosome

4.7

Other cellular structures

AQA.A	EDEXCEL
AQA.A (Human)	EDEXCEL (Human)
AQA.B	OCR

In addition to the major organelles already dealt with, there are other, often smaller but no less important, structures found in cells.

4.7.1 Centrioles

Found in almost all animal cells and certain algae and fungi, centrioles
- arise in a distinct area of the cytoplasm called the centrosome
- occur in pairs
- are hollow cylinders, 0.5μm long and with a diameter around 0.2μm
- have an internal structure comprising nine sets of three microtubules (Fig 4.13)
- have a role in cell division (units 5.2 and 5.3), in which they are involved in **spindle** formation
- may be involved in the formation of the microtubules which make up the **cytoskeleton** of the cell (section 4.7.2).

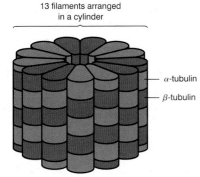

Fig 4.13 Centrioles

4.7.2 Microtubules

Occurring widely throughout eukaryotic cells, microtubules are
- slender, straight, unbranched tubes of protein
- 25nm in diameter, but of varying length
- made up of α and β forms of the protein **tubulin**, which is arranged helically to form the wall (Fig 4.14).

Along with microfilaments and other fibres, the microtubules make up the intricate **cytoskeleton** of cells, which acts as a physical framework for the cytoplasm. The roles of the microtubules are therefore to
- provide an internal skeleton and so determine cell shape
- form channels along which vesicles and other materials are transported within cells
- make up the spindle fibres during cell division and so control the movement of chromosomes
- form part of cilia and flagella, thereby contributing to movement of materials by cells
- provide a template on which cellulose fibres are lined up during the laying down of the cellulose cell wall in plant cells.

13 filaments arranged in a cylinder

α-tubulin
β-tubulin

Fig 4.14 Arrangement of proteins in a microtubule

4.7.3 Cilia

Cilia (Fig 4.15), and the longer but similar **flagella**, occur only in certain specialised cells. Cilia are
- fine threads extending from the cell surface
- around 0.2μm in diameter with a length typically between 5 and 10μm
- found in large numbers in those cells possessing them (e.g. approximately 200 on a ciliated epithelial cell in the trachea of humans)
- made up of a 9+2 arrangement of microtubules.

The functions of cilia are
- to move an entire organism, e.g. cilia on the surface of the **protozoan**, *Paramecium*
- to move material within an organism, e.g. cilia lining the respiratory tract of mammals move mucus towards the throat, and cilia lining the oviduct move the ovum towards the uterus.

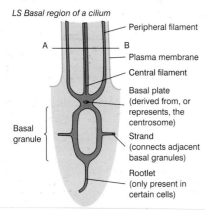

LS Basal region of a cilium

Peripheral filament
Plasma membrane
Central filament
Basal plate (derived from, or represents, the centrosome)
Strand (connects adjacent basal granules)
Rootlet (only present in certain cells)

Basal granule

TS Cilium (section A/B)

One of nine paired peripheral filaments
Extension of paired peripheral filaments – increasingly well developed towards base of the cilium
Plasma membrane
One of two central filaments

Fig 4.15 Structure of a cilium

4.7.4 Microvilli

Found in certain specialised animal cells such as those lining the intestines and kidney tubules, microvilli are
- finger-like projections of the cell surface membrane (plasma membrane)
- around 0.6μm in length
- able to contract due to the presence of **actin** and **myosin** filaments
- used to increase the surface area of cells to allow them to absorb material rapidly
- collectively termed a **brush border** because, under the microscope, they look like the bristles on a brush.

4.7.5 Vesicles

Vesicles are structures formed by the Golgi apparatus (section 4.6.3). They are liquid-filled sacs surrounded by a membrane, which functions to isolate the contents of the vesicle from the rest of the cell. Vesicles are used to transport 'cocktails' of chemicals around the cell or to the cell surface, where they are exported to the outside.

4.7.6 Cellulose cell wall

Characteristic of all plants cells, the cellulose cell wall consists of microfibrils of the polysaccharide, cellulose, embedded in a matrix. Cell walls have the following features:
- They consist of a number of polysaccharides such as cellulose and **glycoproteins**.
- The matrix is composed of substances which include lignin and pectin.
- There is a thin layer of pectates, called the **middle lamella**, which marks the boundary between adjacent cell walls.
- There is a **primary wall** of randomly arranged cellulose microfibrils.
- In older cells, a **secondary cell** wall is also laid down, in which the cellulose is organised in sheets of parallel microfibrils (Fig 4.16).

The functions of the cellulose cell wall are to
- provide mechanical strength to prevent the cell bursting under the pressure created by the **osmotic** influx of water
- give strength to the plant as a whole
- allow water to pass along it and so contribute to the movement of water through the plant (**apoplast pathway**) (section 7.5.3)
- retain water within xylem vessels, in which the walls are impregnated with waterproof materials such as lignin.

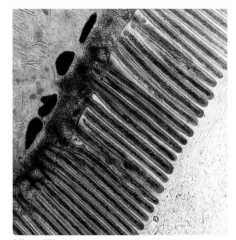

Microvilli

SUMMARY TEST 4.7

Microtubules are made up of the protein **(1)** and form the structural framework of the cell known as the **(2)**. They form a 9+2 arrangement within the organelle called the **(3)**, which is important in cell division, and within **(4)**, which occur on specialised epithelial cells lining the **(5)** and **(6)** in mammals. Finger-like projections of cells that increase their surface area are called **(7)** and are known collectively as a **(8)**. The Golgi apparatus produces **(9)**, which transport chemicals around the cell. The cellulose cell wall in plants has a matrix that may be impregnated with pectin and **(10)**. Pectin is also found in the boundary between adjacent cell walls, known as the **(11)**.

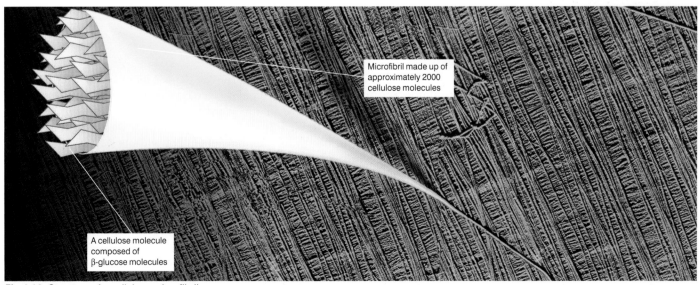

Microfibril made up of approximately 2000 cellulose molecules

A cellulose molecule composed of β-glucose molecules

Fig 4.16 *Structure of a cellulose microfibril*

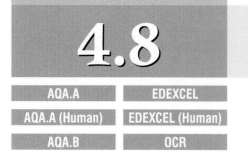

AQA.A	EDEXCEL
AQA.A (Human)	EDEXCEL (Human)
AQA.B	OCR

Cell surface membranes (plasma membrane)

The cell surface membrane is the boundary between the cell cytoplasm and the environment. It controls the movement of substances into and out of the cell, permanently excluding some and permanently containing others; yet others may cross the membrane on one occasion but be prevented from doing so at another time. The cell surface membrane is therefore said to be **partially permeable**. It is made up of proteins (45%) and phospholipids (45%), the remainder (10%) being cholesterol, glycolipids and glycoproteins. The arrangement of each of these molecules is described below and illustrated in figure 4.17; table 4.6 summarises their functions.

4.8.1 Phospholipids

Phospholipids have a molecular structure which is made up of two parts (section 1.4.3):
- a **hydrophilic 'head'** which is attracted to water but not to fat
- a **hydrophobic 'tail'** which is repelled by water but mixes readily with fat.

The phospholipids therefore form a double row or **bilayer**, arranged as follows:
- one layer of phospholipids has its hydrophilic heads pointing inwards (attracted by the water in the cell cytoplasm)
- the other layer of phospholipids has its hydrophilic heads pointing outwards (attracted by the water which surrounds all cells)
- the hydrophobic tails of both phospholipid layers point into the centre of the membrane – protected, as it were, from the water on both sides. Lipid soluble material moves through the membrane via the phospholipid portion.

4.8.2 Proteins

The proteins of the cell surface are arranged more randomly than the regular pattern of phospholipids. They are embedded in the phospholipid bilayer in two main ways:
- **Extrinsic (peripheral) proteins** occur either on the surface of the bilayer or only partly embedded in it, but never extending completely across it. They act either to give mechanical support to the membrane or, in conjunction with glycolipids, as cell receptors.
- **Intrinsic (integral) proteins** completely span the phospholipid bilayer from one side to the other. Some act as carriers to transport water-soluble material across the membrane, while some are enzymes.

4.8.3 Cholesterol

Cholesterol molecules occur throughout the cell surface membrane, where they are almost as numerous as phospholipid molecules. Cholesterol molecules are very hydrophobic and therefore have an important role in preventing leakage of water and dissolved ions out of the cell. They also pull together the fatty acid 'tails' of the phospholipid molecules, limiting their movement but without making the membrane as a whole too rigid.

4.8.4 Glycolipids

Glycolipids occur where a carbohydrate chain is associated with phospholipids in the cell surface membrane. The carbohydrate portion extends from the phospholipid bilayer into the watery environment outside the cell, where it acts as a recognition site for specific chemicals, e.g. the human ABO blood system operates as a result of glycolipids on the cell surface membrane.

***Table 4.6** Summary of functions of the components of the cell surface membrane*

Proteins

- provide structural support
- act as carriers transporting water-soluble substances
- function as enzymes
- form ion channels for sodium, potassium, etc
- act as energy transducers
- form recognition sites by identifying cells
- help cells adhere together
- act as receptors, e.g. for hormones

Phospholipids

- allow lipid-soluble substances to enter and leave the cell
- prevent water-soluble substances entering and leaving the cell
- give the membrane fluidity

Cholesterol

- reduces lateral movement of phospholipids
- makes membrane less fluid at high temperatures
- prevents leakage of water and dissolved ions from the cell

Glycolipids

- act as recognition sites, e.g. ABO blood system
- help maintain stability of the membrane
- help cells attach to one another and so form tissues

Glycoproteins

- act as recognition sites for hormones and neurotransmitters
- help cells attach to one another and so form tissues
- act as antigens allowing cells to recognise one another, e.g. lymphocytes can recognise an organism's own cells.

4.8.5 Glycoproteins

Carbohydrate chains are attached to many extrinsic proteins on the outer surface of the cell membrane. These glycoproteins also act as recognition sites, more particularly for hormones and neurotransmitters. The term **glycocalyx** is applied collectively to the glycoproteins and the glycolipids.

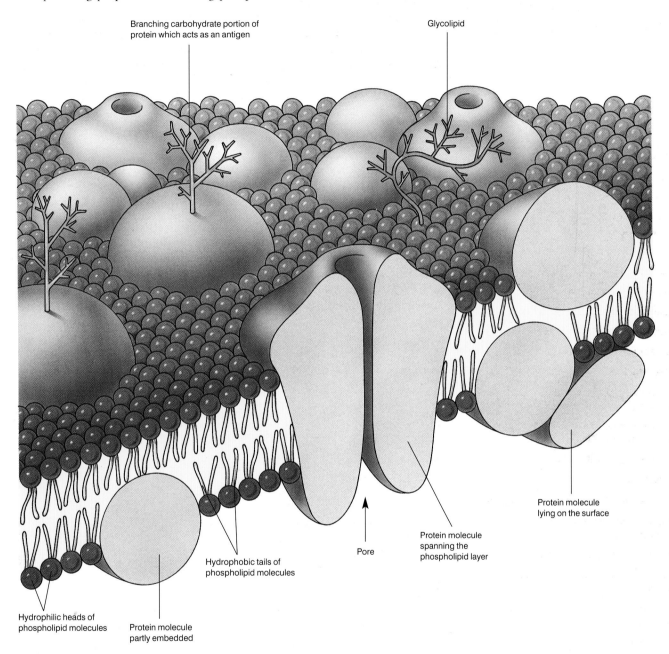

Branching carbohydrate portion of protein which acts as an antigen

Glycolipid

Protein molecule lying on the surface

Pore

Protein molecule spanning the phospholipid layer

Hydrophobic tails of phospholipid molecules

Hydrophilic heads of phospholipid molecules

Protein molecule partly embedded

Fig 4.17 *Structure of the cell surface (plasma) membrane*

4.8.6 Fluid-mosaic model of the cell surface membrane

The term **fluid-mosaic model** was first coined by Singer and Nicholson in 1972 and is applied to the cell surface membrane for the following reasons:

- **Fluid** – because the individual molecules of the cell surface membrane can move relative to one another. This gives the membrane a flexible structure which is constantly changing in shape.
- **Mosaic** – because the proteins embedded in the phospholipid bilayer vary in shape, size and pattern in the way that the stones or tiles of a mosaic do.

4.9 Diffusion

The movement of material into and out of cells occurs in a number of ways (Table 4.7), some of which require energy **(active transport)** and some of which do not **(passive transport)**. Diffusion is an example of passive transport.

4.9.1 Explanation of diffusion

As all movement requires energy, it is possibly confusing to describe diffusion as passive transport. What is meant by passive, in this sense, is that the energy comes from the natural, inbuilt motion of particles, rather than from some outside source. To help understand diffusion and other passive forms of transport it is necessary to understand that

- all particles are constantly in motion due to the kinetic energy that they possess
- this motion is random, with no set pattern to the way the particles move around
- particles are constantly bouncing off one another as well as other objects, e.g. the sides of a vessel in which they are contained.

Given those facts, figure 4.18 shows in a series of diagrams how particles concentrated together in part of a closed vessel will, of their own accord, distribute themselves evenly throughout the vessel, due to diffusion. Diffusion is therefore defined as **'the net movement of molecules or ions from a region where they are more highly concentrated to one where their concentration is lower'**.

1. If 10 particles occupying the left-hand side of a closed vessel are in random motion, they will collide with each other and the sides of the vessel. Some particles from the left-hand side move to the right, but initially there are no available particles to move in the opposite direction, so the movement is in one direction only. There is a large concentration gradient and diffusion is rapid.

3. Some time later, the particles will be evenly distributed throughout the vessel and the concentrations will be equal on each side. The system is in equilibrium. However, the particles are not static but remain in random motion. With equal concentrations on each side, the probability of a particle moving from left to right is equal to the probability of one moving in the opposite direction. There is no concentration gradient and no net diffusion.

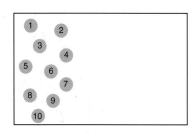

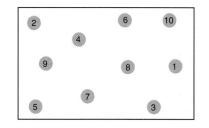

2. After a short time the particles (still in random motion) have spread themselves more evenly. Particles can now move from right to left as well as from left to right. However, with a higher concentration of particles (7) on the left than on the right (3), there is a greater probability of a particle moving to the right than in the reverse direction. There is a smaller concentration gradient and diffusion is slower.

4. At a later stage, the particles remain evenly distributed and will continue to do so. Although the number of particles on each side remains the same, individual particles are continuously changing position. This situation is called **dynamic equilibrium.**

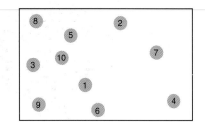

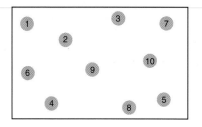

Fig 4.18 Diffusion

64

4.9.2 Rate of diffusion

A number of factors affect the rate at which molecules or **ions** diffuse. These include:
- **The concentration gradient** – the greater the difference in concentration between two regions of the molecules or ions, the faster the rate of diffusion.
- **The area over which diffusion takes place** – the larger the area, the faster the rate of diffusion.
- **The distance over which diffusion occurs** – the shorter the distance, the faster the rate of diffusion.

The relationship between these three factors is expressed in **Fick's Law**, which states:

Diffusion is proportional to
$$\frac{\text{surface area} \times \text{difference in concentration}}{\text{length of diffusion path}}$$

Although Fick's Law gives a good guide to the rate of diffusion, it is not wholly applicable to cells, because diffusion is also affected by
- **the nature of the cell membrane** – its composition and number of pores
- **the size and nature of the diffusing molecule** – for example:
  - small molecules diffuse faster than large ones
  - fat-soluble molecules such as glycerol diffuse faster than water-soluble ones
  - polar molecules diffuse faster than non-polar ones.

4.9.3 Facilitated diffusion

Facilitated diffusion is a passive process relying only on the kinetic energy of the diffusing molecules. Like diffusion, it occurs along a concentration gradient, but it differs in that it occurs at specific points on the membrane where there are special protein molecules. These proteins form water-filled channels **(protein channels)** across the membrane and therefore allow water-soluble ions and molecules such as glucose and amino acids to pass through. Such molecules would usually diffuse only very slowly through the phospholipid bilayer of the membrane. The channels are selective, each opening only in the presence of a specific molecule. When the particular molecule is not present, the channel remains closed. In this way, some control is kept over the entry and exit of substances. An alternative form of facilitated diffusion involves **carrier proteins** which also span the membrane. When a particular molecule specific to the protein is present it binds with the protein, causing it to change shape in such a way that the molecule is released to the inside of the membrane (Fig 4.19). Again, there is no use of external energy, and the molecules move from a region where they are highly concentrated to one of lower concentration, using only the kinetic energy of the molecules themselves.

SUMMARY TEST 4.9

Diffusion is the net movement of molecules or ions from where they are in a **(1)** concentration to a region where their concentration is **(2)**. The energy for this movement comes from the **(3)** energy of the molecules themselves and the process is therefore said to be a **(4)** one. If the area over which diffusion takes place is made smaller, its rate becomes **(5)**. If the concentration gradient is reduced, the rate becomes **(6)** and if the distance over which diffusion takes place is made shorter, its rate becomes **(7)**. Facilitated diffusion, which is faster than diffusion, may involve molecules known as **(8)** that span the cell membrane.

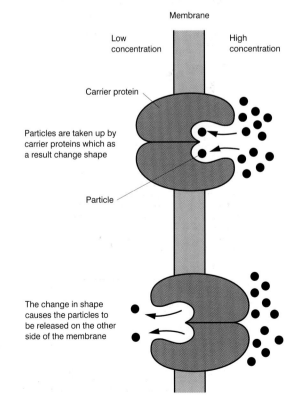

Membrane

Low concentration

High concentration

Carrier protein

Particles are taken up by carrier proteins which as a result change shape

Particle

The change in shape causes the particles to be released on the other side of the membrane

Fig 4.19 *Facilitated diffusion*

AQA.A	EDEXCEL
AQA.A (Human)	EDEXCEL (Human)
AQA.B	OCR

Osmosis is a special form of diffusion involving only water molecules. It is defined as **the passage of water from a region where it is more highly concentrated (higher water potential) to a region where its concentration is lower (lower water potential), through a partially permeable membrane.** Cell surface membranes and those surrounding organelles are **partially permeable**, i.e. they are permeable to water and certain solutes, but not to many other molecules.

4.10.1 Explanation of osmosis

Consider the hypothetical situation in part (a) of figure 4.20:
- Both the solute (sucrose) and the solvent (water) molecules are in random motion.
- The partially permeable membrane, however, only allows water molecules across it, and not the sucrose molecules.
- The water molecules diffuse across until their concentration is equal on both sides.
- At this point, a dynamic equilibrium should be established and there should be no net movement of water.

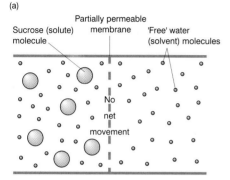

(a)

Number of free water molecules is the same on both sides of the partially permeable membrane and so there is no net movement of water.

Now consider the actual situation as shown in part (b) of figure 4.20:
- Water molecules tend to cluster around the sucrose molecules.
- These water molecules are not free to cross the membrane, because they adhere to the sucrose molecules.
- There are therefore more 'free' water molecules on the right-hand side of the membrane than on the left.
- More water molecules move from right to left than in the reverse.
- There is a net flow of water from right to left, due to osmosis.

4.10.2 Water potential

Water potential is represented by the Greek letter psi (Ψ), and is measured in units of pressure, usually kiloPascals (kPa). It is the pressure created by 'free' water molecules. Under standard conditions of temperature and pressure (25°C and 100kPa), pure water is said to have a water potential of zero. It follows that
- the addition of solute to pure water will lower its water potential because some water molecules will be attracted to the solute molecules, and will no longer be 'free'
- the water potential of a solution (water + solute) must always be less than zero, i.e. a negative value
- the more solute that is added (i.e., the more concentrated a solution), the more negative (lower) its water potential
- water will move by osmosis from a region of higher (less negative) water potential (e.g. –10kPa) to one of lower (more negative) water potential (e.g. –20kPa).

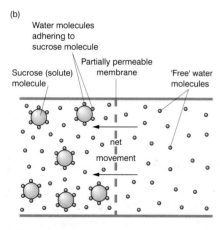

(b)

Some water molecules are attracted to the sucrose molecules and so are not free to diffuse. There are more 'free' water molecules on the right of the membrane and so there is a net flow of water molecules from right to left.

Fig 4.20 *Osmosis*

4.10.3 Understanding water potential

Water potential can be confusing, not least because its highest value, that of pure water, is zero, and so all other values are negative. The more negative the value, the lower is the water potential. It may be easier to understand if you think of water potential as an overdraft at the bank. The bigger the overdraft, the more negative is the amount of money you have. The smaller the overdraft, the less negative the money you have. An account with a big overdraft (more negative value) has a greater need of money than an account with a small overdraft (less negative value).

In the same way, a solution with a high water potential (less negative value) has less need of water than a solution with a lower water potential (more negative value). Water therefore moves from solutions with a high water potential (e.g. those containing less solute) to ones with a low water potential (e.g. those containing more solute)– but only when separated by a partially permeable membrane.

4.10.4 Hypotonic, isotonic and hypertonic solutions

Osmosis occurs not only when a solution is separated from pure water by a partially permeable membrane, but also when two solutions of different solute concentrations are similarly separated. The following terms are applied to this situation:
- **hypotonic solution** – the more dilute of the two solutions
- **hypertonic solution** – the more concentrated of the two solutions
- **isotonic solutions** – both solutions have the same concentrations.

There is always net movement of water from the hypotonic solution to the hypertonic solution, because there are more free water molecules in the hypotonic solution (Fig 4.21). When the solutions are isotonic, the amount of water diffusing one way is exactly offset by that moving the other, so there is no net movement of water.

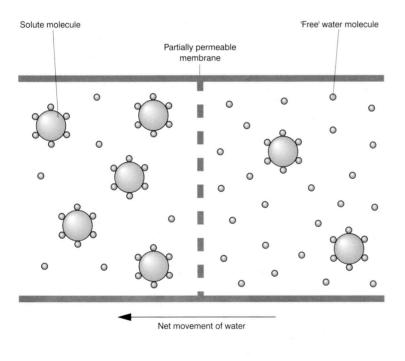

More solute molecules
Hypertonic solution
Low concentration of 'free' water molecules
Low (more negative) water potential

Fewer solute molecules
Hypotonic solution
High concentration of 'free' water molecules
High (less negative) water potential

Fig 4.21 *Movement of water from a hypotonic to a hypertonic solution*

SUMMARY TEST 4.10

Osmosis is the passage of **(1)** from a region where it is more highly concentrated to a region where its concentration is lower, through a **(2)** membrane. Water potential is the **(3)** created by 'free' water molecules and its value for pure water is **(4)**. The addition of a solute to water makes its water potential **(5)**. If solution A has a water potential of –10kPa and solution B has a water potential of –20kPa, water will move into solution **(6)**. Solution B is said to be **(7)** to solution A because it is less dilute. Solution A is therefore **(8)** to solution B. If both solutions had the same solute concentration, they would be said to be **(9)**.

Active transport

Active transport differs from passive forms of transport such as diffusion and osmosis, in the following ways:
- metabolic energy in the form of **ATP** is needed
- materials are moved **against** a concentration gradient
- carrier protein molecules which act as 'pumps' are involved
- the process is very selective, with specific substances being transported.

Materials may be transported either as molecules or in bulk as larger particles, in a process called **cytosis**. There are two forms of cytosis (Fig 4.22):
- **endocytosis** – the bulk movement of material **into** the cell
- **exocytosis** – the bulk movement of the material **out of** the cell.

Table 4.7 *Comparison of different forms of transport in cells*

Process	Occurs against a conc. gradient	Needs energy (ATP)	may use carrier molecules
Diffusion	No	No	No
Faciliated diffusion	No	No	Yes
Osmosis	No	No	No
Active transport	Yes	Yes	Yes

4.11.1 Mechanism of active transport

There are a number of possible explanations for how active transport occurs. One theory involves the proteins which span the phopholipid bilayer of a membrane. These proteins accept the molecules or ions to be transported on one side of the membrane and then change shape, using energy from ATP to carry the molecules or ions to the other side (Fig 4.23). Sometimes more than one substance may be moved in the same direction at the same time. Occasionally, one substance is moved into a cell at the same time as a different one is removed from it. One example of this is the **sodium–potassium pump**. Here, sodium ions are actively removed from the cell while potassium ions are taken in from the surroundings. This process is essential to the creation of a nerve impulse.

4.11.2 Requirements for active transport

Certain conditions are necessary if a cell is to carry out active transport effectively. These include:
- the presence of numerous mitochondria
- a ready supply of ATP
- a high respiratory rate.

Clearly, any factor which affects respiratory rate will affect active transport, and therefore higher temperatures or an increased supply of oxygen will increase the rate of active transport. Lower temperatures, less oxygen or the presence of respiratory inhibitors such as cyanide will slow the rate of active transport.

4.11.3 Endocytosis

Endocytosis is the bulk movement of material into a cell by active means. It takes two forms:
- **Phagocytosis** involves the invagination of the cell to form a cup-shaped depression in which large particles or even whole organisms are contained. The depression is then pinched off to the inside of the cell, forming a **vacuole**. The process occurs in organisms such as *Amoeba*, in which it is used as a method of feeding. A few specialised cells in higher organisms also carry out phagocytosis. These cells are called **phagocytes**, and include certain types of white blood cells which ingest harmful bacteria in this way to fight infection or destroy worn-out cells such as red blood cells. In both cases, lysosomes fuse with the vacuole formed during phagocytosis, releasing their enzymes into the vacuole. The particles inside are therefore broken down and any useful soluble products are absorbed.

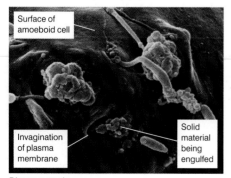

Phagocytosis

- **Pinocytosis** is very similar to phagocytosis, except that the vacuoles formed are smaller and are called **vesicles**. Pinocytosis is used more for the uptake of liquids or large molecules rather than solid particles, and is therefore also called 'cell drinking'.

4.11.4 Exocytosis

Exocytosis is the bulk movement of material out of the cell. It is the reverse of phagocytosis and pincytosis: vacuoles and/or vesicles within the cell fuse with the cell surface membrane and their contents are expelled into the medium outside. Undigested material from the food vacuoles in *Ameoba* are removed in this way. In higher organisms, exocytosis is used to release hormones, e.g. insulin, from the cells that manufacture them.

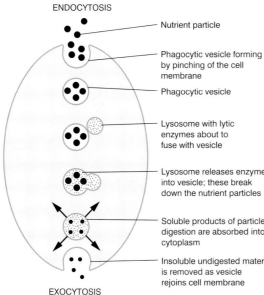
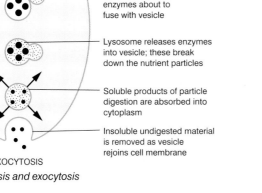

Fig 4.22 *Endocytosis and exocytosis*

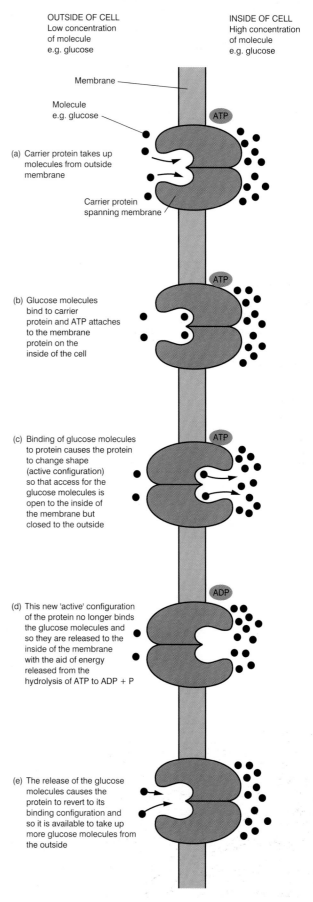

Fig 4.23 *Active transport*

SUMMARY TEST 4.11

Active transport occurs **(1)** a concentration gradient. It requires energy in the form of **(2)** and cells exhibiting active transport therefore have numerous **(3)** and a high **(4)** rate. The bulk movement across the cell surface membrane of molecules and particles is called **(5)**. Where the movement is into the cell it is called **(6)**, and where the movement is out of the cell it is called **(7)**. The invagination of a cell surface membrane to ingest particles and form a vacuole is known as **(8)**, whereas the invagination of the membrane to ingest liquids and form a vesicle is known as **(9)**.

1 The diagram shows part of a cell that secretes enzymes.

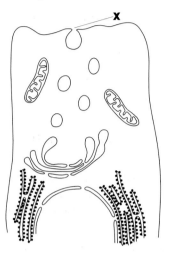

a Give **one** piece of evidence, visible in the diagram, which shows that this cell is a eukaryotic cell. *(1 mark)*

b Some cells similar to that shown in the diagram were grown in a culture. Radioactive amino acids were added to the solution in which they were being grown. The radioactivity acts as a label on the amino acid so that it can be detected wherever it is. This radioactive label allows amino acids to be followed through the cell. At various times, samples of the cells were taken and the amount of radioactivity in different organelles was measured. The results are shown in the table.

Time after radioactive amino acids were added to the solution / minutes	Amount of radioactivity present / arbitrary units		
	Golgi apparatus	Rough endoplasmic reticulum	Vesicles
1	21	120	6
20	42	68	6
40	86	39	8
60	76	28	15
90	50	27	28
120	38	26	56

  (i) What happens to the amino acids in the rough endoplasmic reticulum? *(2 marks)*
  (ii) Use the information in the table to draw arrows on the diagram showing the path of radioactivity through and out of the cell at **X**. *(3 marks)*
  (iii) Name the process which is occurring at point **X** on the diagram. *(1 mark)*

(Total 7 marks)

AQA Jan 2001, B/HB (A) BYA1, No.5

2 Ultracentrifugation was used to separate the components of cells from lettuce leaves. The flow chart summarises the steps in the process.

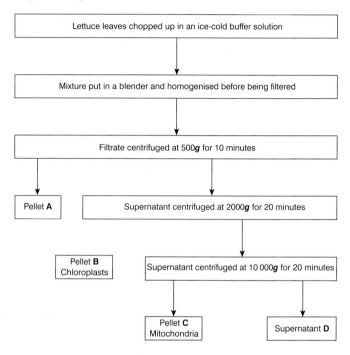

a Explain why the mixture was filtered before it was centrifuged. *(1 mark)*

b (i) Name the organelle present in the largest numbers in pellet **A**. *(1 mark)*

  (ii) Name an organelle likely to be present in supernatant **D**. *(1 mark)*

c The mitochondria from pellet **C** were observed with an electron microscope. They had all burst and appeared as shown in the diagram.

What does this suggest about the water potential of the solution in which the chopped leaves were put? Explain your answer. *(2 marks)*

(Total 5 marks)

AQA Jan 2001, B/HB (A) BYA1, No.7

3 The table below refers to four membrane transport processes: diffusion, facilitated diffusion, osmosis and active transport. If the statement is correct, place a tick (✔) in the appropriate box and if the statement is incorrect, place a cross (✘) in the appropriate box.

Process	Takes place against a concentration gradient	Requires energy in the form of ATP
Diffusion		
Facilitated diffusion		
Osmosis		
Active transport		

(Total 4 marks)

Edexcel 6101/01 June 2001, B B(H) AS/A, No.1

4 The table below refers to three organelles commonly found in eukaryotic cells. Complete the table by writing the name of the organelle, its description or **one** function, as appropriate, in each of the five boxes provided.

Name of organelle	Description	ONE function
Golgi apparatus		
	Cylindrical organelles made up of microtubules	Involved in spindle organisation during cell division in animal cells
	Rod-shaped structures with a double membrane, the inner one folded to form cristae	

(Total 5 marks)

Edexcel 6101/01 June 2001, B B(H) AS/A, No.3

5 The diagram shows an organelle from a palisade mesophyll cell, as seen with an electron microscope.

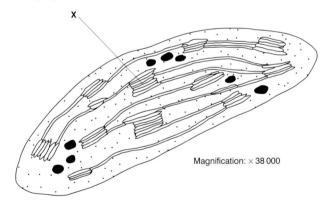

Magnification: × 38 000

a Name:
   (i) the organelle;
   (ii) the part labelled **X**. *(2 marks)*

b Calculate the maximum length of the organelle in micrometres. Show your working. *(2 marks)*

c Give **two** ways in which the structure of this organelle is adapted for its function. *(2 marks)*

(Total 6 marks)

AQA June 2001, B (B) BYB1, No.4

6 The diagram shows a mitochondrion.

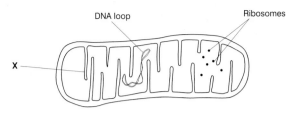

a (i) Name the part labelled **X**. *(1 mark)*
   (ii) A human liver cell contains several hundred mitochondria. A cell from a plant root has only a small number. Suggest an explanation for this difference. *(2 marks)*
   (iii) Mitochondria contain some DNA and ribosomes. Suggest the function of these. *(2 marks)*

b Mitochondria may be separated from homogenised cells by differential centrifugation. During this process the cells must be kept in an isotonic solution. Explain why. *(2 marks)*

c Ribosomes in bacterial cells differ from those in the cytoplasm of eukaryotic cells. When centrifuged at high speed, the eukaryotic cell ribosomes sediment more rapidly than bacterial ribosomes. Explain what this tells you about the difference between bacterial and eukaryotic ribosomes. *(1 mark)*

(Total 8 marks)

AQA Jan 2001, B (B) BYB1, No.5

7 The figure below is an electron micrograph of part of a leaf mesophyll cell.

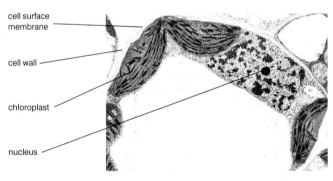

a With reference to this figure, state **two** reasons why the cell is considered to be a eukaryotic cell. *(2 marks)*

b State the functions of the following structures in this cell
   (i) chloroplast; *(1 mark)*
   (ii) cell surface membrane. *(1 mark)*

c Calculate the magnification of the figure. (Show your working.) *(2 marks)*

d State **one** way in which a typical animal cell differs from the cell shown in the figure. *(1 mark)*

(Total 7 marks)

OCR 2801 June 2001, B (BF), No.1

Nuclear division and reproduction

5.1 Chromosomes and the cell cycle

AQA.A	EDEXCEL
AQA.A (Human)	EDEXCEL (Human)
AQA.B	OCR

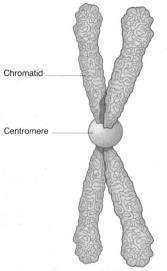

Fig 5.1 *Structure of a chromosome*

Chromatid

Centromere

The cells which make up organisms are always derived from existing cells, by the process of division, which occurs in two main stages:

- **Nuclear division**, is the process by which the nucleus divides. There are two types of nuclear division:
  - **Mitosis**, results in 2 daughter nuclei having the same number of chromosomes as the parent nucleus. The nuclei formed are normally genetically identical to the parent one.
  - **Meiosis**, results in 4 daughter nuclei having half the number of chromosomes as the parent nucleus. The nuclei formed have a genetic composition different from the parent one.
- **Cell division (Cytokinesis)** follows nuclear division, and is the process by which the whole cell divides.

5.1.1 The cell cycle

Cells do not divide continuously, but undergo a regular cycle of division separated by periods of cell growth. This is known as the **cell cycle** and has three stages:

- **Interphase** – occupies most of the cell cycle, and is sometimes known as the resting phase, because no division takes place. In one sense, this could hardly be further from the truth, as interphase is a period of intense chemical activity, divided into three parts:
  - **First growth (G₁) phase**, when the proteins from which cell organelles are synthesised are produced.
  - **Synthesis (S) phase**, when DNA is replicated.
  - **Second growth (G₂) phase**, when organelles grow and divide and energy stores are increased.
- **Nuclear division**, when the nucleus divides into 2 (mitosis) or 4 (meiosis).
- **Cell division (Cytokinesis)**, when the cell divides into 2 (mitosis) or 4 (meiosis).

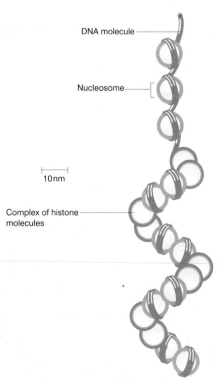

DNA molecule

Nucleosome

10 nm

Complex of histone molecules

Fig 5.2 *Detailed structure of a chromosome*

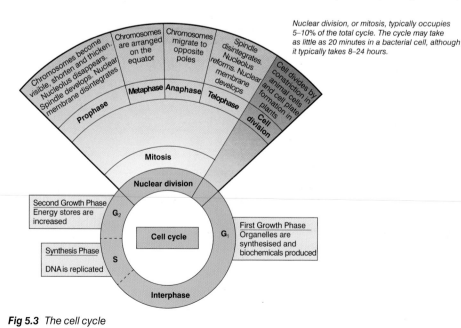

Nuclear division, or mitosis, typically occupies 5–10% of the total cycle. The cycle may take as little as 20 minutes in a bacterial cell, although it typically takes 8–24 hours.

Fig 5.3 *The cell cycle*

5.1.2 Chromosome structure

Chromosomes have a characteristic shape, occur in pairs, and carry the hereditary material of the cell. Chromosomes are only visible as discrete structures when a cell is dividing. The rest of the time, they consist of widely spread areas of darkly staining material called **chromatin** and heterochromatin (section 4.5.1). When they are visible, chromosomes appear as long, thin threads around 50 µm long. They are made up of two strands called **chromatids**, joined at a point called the **centromere** (Fig 5.1).

Chromosomes are made up of three basic materials:
- **proteins** (70%), mostly in the form of **histones**, scaffold proteins and **polymerases**
- **deoxyribonucleic acid** – DNA (15%)
- **ribonucleic acid** – RNA (10%).

To fit in, the considerable length of DNA found in each cell (around 2 metres in humans) is highly coiled and folded. This DNA is held in position by proteins called **histones**, which together form a complex known as **chromatin**. The chromatin has a beaded appearance due to the presence of **nucleosomes** (Fig 5.2). These make up a portion of DNA which is 146 base pairs in length and wrapped around eight histone molecules.

For convenience, and to make study of them easier, photographs of chromosomes are cut out and pasted into a logical format where they are arranged in their pairs and given numbers to identify them. These arrangements are called **karyotypes**.

5.1.3 Chromosome number

Although the number of chromosomes is always the same for normal individuals of a species, it varies from one species to another (table 5.1). The number of chromosomes is no indication of the level of organisation, complexity or evolutionary status of a species. It has no significance.

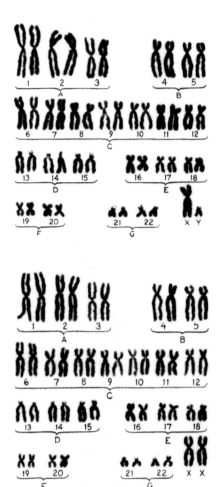

Human karyotypes (male above and female below)

Table 5.1 Chromosome numbers of various organisms

Species	Chromosome number
Crocus	6
Locust	24
Tomato	24
Cat	38
Human	46
Potato	48
Dog	78
Some protozoa	300

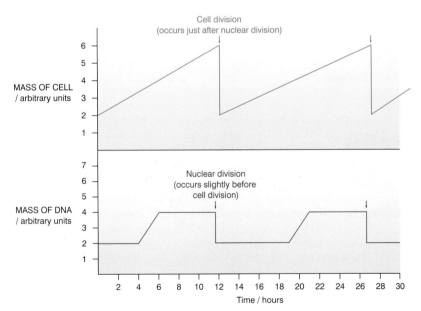

Fig 5.4 Variations in cell and DNA mass during cell cycle

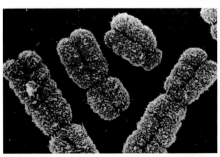

False-colour scanning electron micrograph of a group of human chromosomes

Mitosis

By mitosis, the nucleus of a cell divides in such a way that the two resulting nuclei both have the same number and type of chromosomes as the parent nucleus. Except in the rare event of a mutation, the genetic make-up of the two daughter nuclei is also identical to that of the parent nucleus. Mitosis is always preceded in the cell cycle by a period during which the cell is not dividing. This period is called **interphase** (section 5.1.1). Although mitosis is a continuous process, it can be divided into four stages for convenience:

- **Prophase** – chromosomes become visible and the nuclear envelope disappears.
- **Metaphase** – chromosomes arrange themselves at the centre (equator) of the cell.
- **Anaphase** – chromatids migrate to opposite poles.
- **Telophase** – the nuclear envelope reforms.

The complex process is illustrated in the photographs and in figure 5.5.

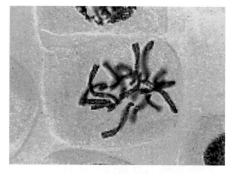

5.2.1 Prophase

In prophase, the chromosomes first become visible, to begin with as long thin threads, which later shorten and thicken. Animal cells contain two cylindrical organelles called **centrioles** (section 4.7.1), each of which moves to opposite ends (called **poles**) of the cell. From each of the centrioles, **microtubules** develop (section 4.7.2) which span the cell from pole to pole. Collectively, these microtubules are called the **spindle apparatus**. As plant cells lack centrioles but do develop a spindle apparatus, centrioles are clearly not essential to microtubule formation. The nucleolus disappears and the nuclear envelope breaks down, leaving the chromosomes free in the cytoplasm of the cell. These chromosomes are drawn towards the equator of the cell.

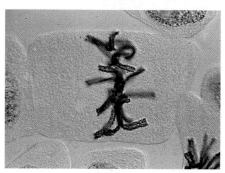

5.2.2 Metaphase

By metaphase, the DNA has replicated and the chromosomes are made up of two **chromatids** joined by the **centromere** (section 5.1.2). It is to this centromere that some microtubules from the poles are attached, and the chromosomes are pulled along the spindle apparatus and arrange themselves across the equator of the cell.

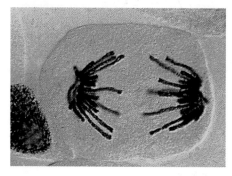

5.2.3 Anaphase

In anaphase, the centromeres divide into two and the microtubules joined to each contract, causing the individual chromatids that make up the chromosome to separate and move to opposite poles of the cell. Now called **daughter chromosomes**, the chromatids move rapidly to their respective poles. The energy for the process is provided by mitochondria, which gather around the spindle fibres. If cells are treated with chemicals that destroy the spindle, the chromosomes remain at the equator, unable to reach the poles.

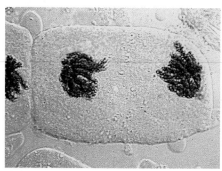

5.2.4 Telophase

In this stage, the daughter chromatids reach their respective poles and become longer and thinner, finally disappearing altogether, leaving only widely spread chromatin. The spindle fibres disintegrate, and the nuclear envelope and nucleolus re-form.

The main stages of mitosis: prophase, metaphase, anaphase and telophase

5.2.5 The significance of mitosis

Mitosis is significant because it produces exact copies of existing cells. This is essential for three reasons:

- **Growth** – tissues develop as a result of mitosis of existing cells. The new cells have metabolic processes and systems identical to those of the parent cells.
- **Repair** – damaged cells must be replaced by exact copies of those which are affected.
- **Asexual reproduction** – organisms can make best use of a suitable habitat by producing new individuals identical to those which are already thriving there. Mitosis achieves this, and is the basis of natural and artificial **cloning** (see section 12.7.4), e.g. **vegetative propagation** and the splitting of embryos.

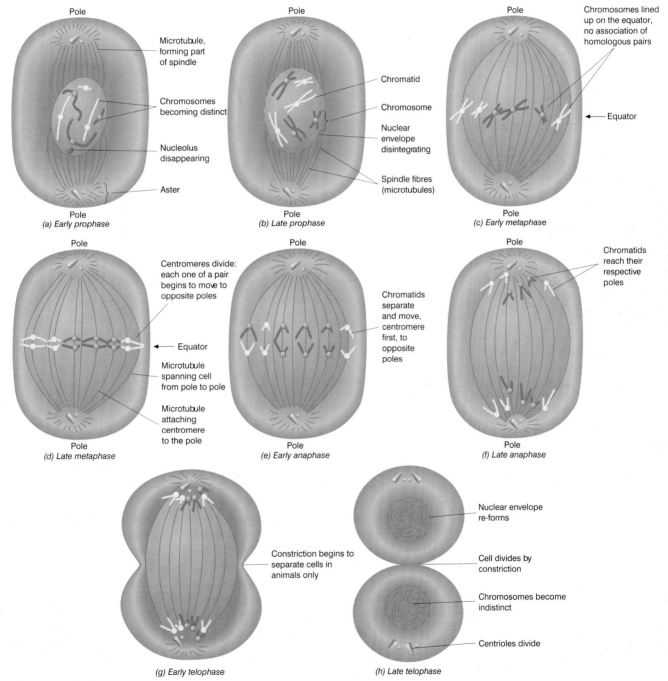

Fig 5.5 *Stages of mitosis*

5.3 Meiosis

Meiosis consists of one division of the chromosomes, followed by two divisions of the nucleus and the cell. This results in the number of chromosomes being halved, from the **diploid (2n)** number found in the parent cell to the **haploid (n)** number found in the four daughter nuclei that are produced in meiosis. Meiosis takes place in the formation of sperm and ova in animals, and spores in most plants. It involves two nuclear divisions:

- **first meiotic division** has a highly modified prophase stage, but is otherwise similar to mitosis
- **second meiotic division** is a typical mitotic division.

Although a continuous process, meiosis is divided into stages for convenience.

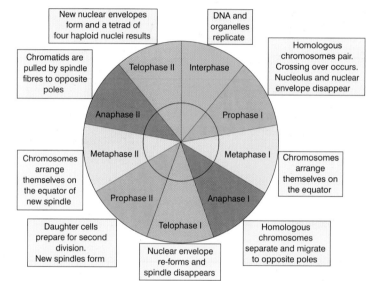

Fig 5.6 *Meiosis*

5.3.1 First meiotic division

As in mitosis, the division is made up of prophase I, metaphase I, anaphase I and telophase I. (The I is used to distinguish this stage from the second meiotic division for which the suffix II is used.)

- **Prophase I** – the chromosomes shorten and fatten and come together in their homologous pairs (section 5.4.2) to form what is called a **bivalent**. The process of pairing is known as **synapsis**. The chromatids of the homologous chromosomes wrap around one another and attach at points called **chiasmata** (singular = **chiasma**). The chromosomes may break at these points, and swap similar sections of chromatids with one another, in a process called **crossing over**. Finally, the nucleolus disappears, the nuclear envelope disintegrates, and the spindle apparatus forms.
- **Metaphase I** – the centromeres become attached to the spindle and the bivalents arrange themselves on the equator of the cell, with each of a pair of homologous chromosomes facing opposite poles. The order of the chromosomes on the equator is entirely random.
- **Anaphase I** – one of each pair of homologous chromosomes is pulled by the spindle fibres to opposite poles.
- **Telophase I** – in most animal cells, a nuclear envelope forms around the chromosomes at each pole, but in most plant cells there is no telophase I and the cell goes directly into metaphase II.

Table 5.2 *Differences between mitosis and meiosis*

Mitosis	Meiosis
A single division of the chromosome and the nucleus	A single division of the chromosome, but a double division of the nucleus
The number of chromosomes remains the same	The number of chromosomes is halved
Homologous chromosomes do not associate	Homologous chromosomes associate to form bivalents in prophase I
Chiasmata are never formed	Chiasmata may be formed
Crossing over never occurs	Crossing over may occur
Daughter cells are identical to parent cells (in the absence of mutations)	Daughter cells are genetically different from parental ones
Two daughter cells are formed	Four daughter cells are formed, although in females usually only one is functional

5.3.2 Second meiotic division

The same four phases are repeated, except that this time it is chromatids rather than chromosomes that are separated out, and the process occurs simultaneously in the two cells produced during the first meiotic division.

- **Prophase II** – where the nuclear membrane has re-formed, it breaks down again, the nucleolus disappears, the chromosomes shorten and thicken, and the spindle re-forms.
- **Metaphase II** – the chromosomes arrange themselves on the equator of each cell.
- **Anaphase II** – the centromeres divide and the chromatids are pulled to opposite poles by the spindle fibres.
- **Telophase II** – the chromatids reach their respective poles and they become less distinct. The nuclear envelope and nucleolus re-form. The spindle disappears, leaving four cells known as a **tetrad**. The process of meiosis is summarised in figures 5.6 and 5.7.

SUMMARY TEST 5.3

Meiosis results in a halving of the number of chromosomes from the **(1)** number to the **(2)** number. During the first prophase stage, the chromosomes come together in their **(3)** pairs to form a **(4)** in a process known as **(5)**. The chromatids of each pair attach at points called **(6)** and may exchange equivalent portions in a process called **(7)**.

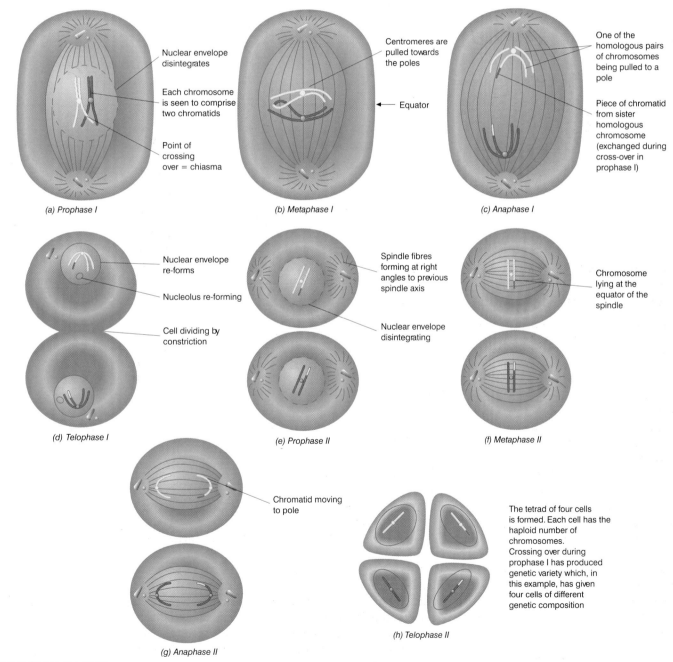

Nuclear envelope disintegrates

Each chromosome is seen to comprise two chromatids

Point of crossing over = chiasma

(a) Prophase I

Centromeres are pulled towards the poles

Equator

(b) Metaphase I

One of the homologous pairs of chromosomes being pulled to a pole

Piece of chromatid from sister homologous chromosome (exchanged during cross-over in prophase I)

(c) Anaphase I

Nuclear envelope re-forms

Nucleolus re-forming

Cell dividing by constriction

(d) Telophase I

Spindle fibres forming at right angles to previous spindle axis

Nuclear envelope disintegrating

(e) Prophase II

Chromosome lying at the equator of the spindle

(f) Metaphase II

Chromatid moving to pole

(g) Anaphase II

The tetrad of four cells is formed. Each cell has the haploid number of chromosomes. Crossing over during prophase I has produced genetic variety which, in this example, has given four cells of different genetic composition

(h) Telophase II

Fig 5.7 *Stages of meiosis*

In the process of sexual reproduction, offspring are formed as a result of the fusion or **syngamy** of gametes. If each gamete contained the full number of chromosomes (e.g. 46 in humans), when two gametes fused the resulting organism would have twice as many chromosomes as each of the parents. It follows therefore, that at some stage in the life cycle the number of chromosomes must be halved. This halving of the chromosome number is brought about by meiosis, which is a feature of the life cycle of all sexually reproducing organisms.

5.4.1 Haploid and diploid cells

Each species has a fixed number of chromosomes in each of its cells (section 5.1.3). This is usually made up of two of each type of chromosome, i.e. there are two chromosomes that determine the same characteristics. Any cell that contains this double set of chromosomes is called **diploid**, and this is indicated by the symbol **2n**. When a diploid cell divides by mitosis, it produces another diploid cell:

$$\text{diploid (2n) cell} \xrightarrow{\text{mitosis}} \text{diploid (2n) cell}$$

However, when a diploid cell divides by meiosis, it produces a cell with half the number of chromosomes, and this is known as a haploid cell, which is indicated by the symbol **n**:

$$\text{diploid (2n) cell} \xrightarrow{\text{meiosis}} \text{haploid (n) cell}$$

In most organisms, it is the gametes (sperm and ova) which are the haploid cells.

5.4.2 Homologous chromosomes

All diploid cells of organisms have two sets of chromosomes, one set provided by each parent. There are therefore always two sets of genetic information for each characteristic of an individual. Any two chromosomes which determine the same characteristics are termed a **homologous pair**. 'Determining the same characteristics' is not the same as being identical. For instance, a homologous pair of chromosomes may each possess information on eye colour and hair texture, but one chromosome may carry the code for blue eyes and curly hair, while the other carries the code for brown eyes and straight hair.

5.4.3 Life cycle

All life cycles are, in effect, an alternation of a haploid stage and a diploid stage. Although there are other life cycles, the main types are those of animals and plants. The animal life cycle (Fig 5.8) has the following features:
- gametes are produced by meiosis
- gametes fuse almost immediately they are formed, and so the interval between meiosis and fertilisation is short
- the diploid phase is long and dominant
- the haploid phase is short.

The plant life cycle (Fig 5.9) has the following features:
- gametes are produced by mitosis
- spores (**pollen grains** and **embryo sac**) are produced by meiosis
- the interval between meiosis and fertilisation is longer than in animals
- the haploid stage can sometimes be long (some pollen grains remain viable for many years).

ADVANTAGES AND DISADVANTAGES OF ASEXUAL AND SEXUAL REPRODUCTION

ASEXUAL ADVANTAGES
Only one parent needed
Rapid process
Many offspring produced

ASEXUAL DISADVANTAGES
Little or no variety amongst offspring
Little evolutionary potential

SEXUAL ADVANTAGES
Much genetic variety amongst offspring
Greater evolutionary potential

SEXUAL DISADVANTAGES
Two parents required
Slower process
Fewer offspring

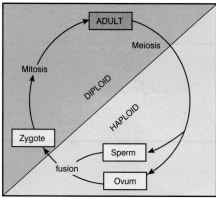

Fig 5.8 Life cycle of most animals

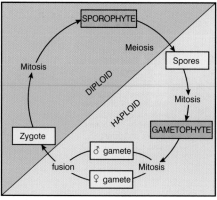

Fig 5.9 Life cycle of most plants

5.4.4 The importance of meiosis

Meiosis is important because it

- halves the number of chromosomes in the gametes, so that when two haploid gametes fuse, the diploid number is restored
- leads to increased genetic variation.

Variation is achieved by meiosis because

- it produces haploid gametes which, when they fuse at fertilisation, combine the different genetic material of the two parents
- during metaphase I, homologous chromosomes arrange themselves randomly and so they are re-sorted in the daughter cells. This is similar to shuffling a pack of cards and dealing them into two hands. The combinations will be different each time (Fig 5.11)
- crossing over during prophase I allows equivalent sections of chromatids to be exchanged, thus separating linked genes and creating new genetic combinations.

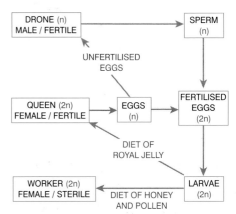

Unfertilised haploid (n) eggs develop into haploid males (drones), while fertilised eggs develop into diploid females. Female larvae fed on royal jelly develop into fertile queen bees, and those fed on pollen and honey develop into sterile worker bees.

Fig 5.10 Life cycle of the honey bee

In arrangement 1, the two pairs of homologous chromosomes orientate themselves on the equator in such a way that the chromosome carrying the allele for brown eyes and the one carrying the allele for blood group A migrate to the same pole. The alleles for blue eyes and blood group B migrate to the opposite pole. Cell 1 therefore carries the alleles for brown eyes and blood group A, while cell 2 carries the alleles for blue eyes and blood group B.

In arrangement 2, the left-hand homologous pair of chromosomes is shown orientated the opposite way around. As this orientation is random, this arrangement is equally as likely as the first. The result of this different arrangement is that cell 3 carries the alleles for blue eyes and blood group A, whereas cell 4 carries the alleles for brown eyes and blood group B.

All four resultant cells are different from one another. With more homologous pairs, the number of possible combinations becomes enormous. A human, with 23 such pairs, has the potential for $2^{23} = 8\ 388\ 608$ combinations.

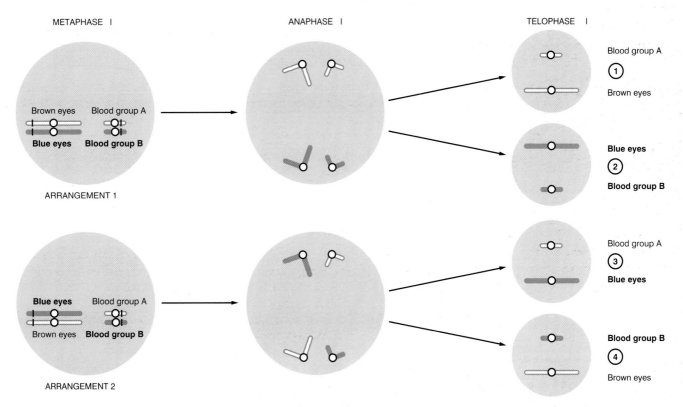

Fig 5.11 Variety brought about by meiosis

Flower structure

EDEXCEL

Sexual reproduction in flowering plants is carried out by structures which are located in flowers. Although very varied in arrangement, flowers typically have four functional sets of structures – the **calyx**, **corolla**, **androecium** and **gynaoecium**. All the four sets of floral parts are attached to the swollen end of the flower stalk, known as the **receptacle**. The number, shape and size of each set of parts vary from species to species, with particularly noticeable differences between flowers using insects for pollination and those relying on the wind. Figure 5.12 illustrates the structure of a generalised insect-pollinated flower.

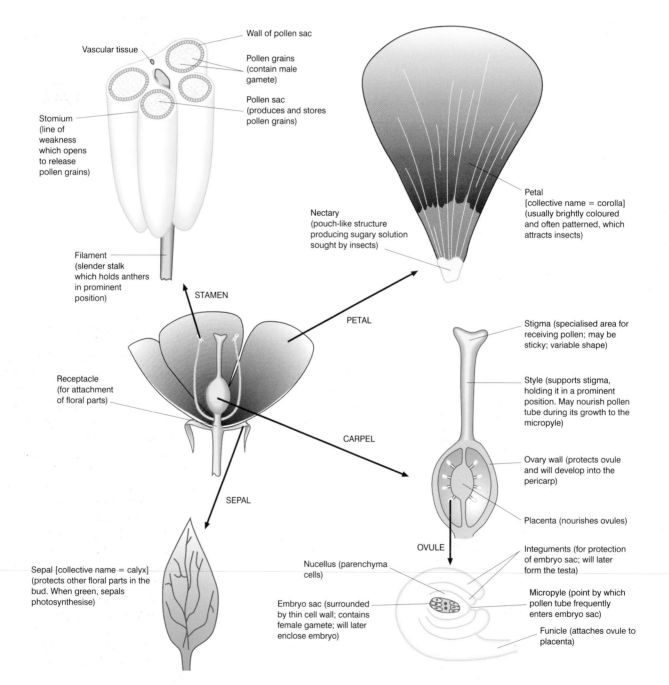

Vascular tissue

Wall of pollen sac

Pollen grains (contain male gamete)

Pollen sac (produces and stores pollen grains)

Stomium (line of weakness which opens to release pollen grains)

Filament (slender stalk which holds anthers in prominent position)

STAMEN

Nectary (pouch-like structure producing sugary solution sought by insects)

Petal [collective name = corolla] (usually brightly coloured and often patterned, which attracts insects)

PETAL

Stigma (specialised area for receiving pollen; may be sticky; variable shape)

Style (supports stigma, holding it in a prominent position. May nourish pollen tube during its growth to the micropyle)

Receptacle (for attachment of floral parts)

CARPEL

Ovary wall (protects ovule and will develop into the pericarp)

Placenta (nourishes ovules)

SEPAL

OVULE

Integuments (for protection of embryo sac; will later form the testa)

Micropyle (point by which pollen tube frequently enters embryo sac)

Funicle (attaches ovule to placenta)

Nucellus (parenchyma cells)

Embryo sac (surrounded by thin cell wall; contains female gamete; will later enclose embryo)

Sepal [collective name = calyx] (protects other floral parts in the bud. When green, sepals photosynthesise)

Fig 5.12 *Structure of generalised insect-pollinated flower*

5.5.1 The calyx

Calyx is the collective term for all the modified leaves or **sepals** which form the outer part of the flower. The sepals protect the other parts of the flower when they are developing within the bud. They are usually green, and so produce carbohydrate by photosynthesis, but in a few plants, such as lilies and tulips, they are brightly coloured, which helps to attract insects for **pollination**.

5.5.2 The corolla

Corolla is the collective name for all the petals of a flower. Petals are brightly coloured in flowers which attract insects for pollination but are reduced or absent in wind-pollinated ones. The collective name for the calyx and the corolla is the **perianth**, a term especially used where the petals and sepals look alike, e.g. tulips.

5.5.3 The androecium

Androecium is the collective name for all the male reproductive parts of the flower and includes the following:
- The **anther** is made up of **pollen sacs** (usually four), within which are produced the **pollen grains** that contain the male gamete.
- The **filament** is a long slender stalk which holds the anthers aloft.
- The **stamen** is the collective name for the anther and filament.

5.5.4 The gynoecium

Gynoecium is the collective name for all the female reproductive parts of the flower and includes the following:
- **The ovary** is a hollow structure containing one or more ovules, each of which encloses the female gamete or **egg cell**.
- **The style** is a long slender stalk, along which the pollen tube grows towards the ovule.
- **The stigma** is the expanded, and often sticky, tip of the stigma, to which pollen adheres.
- **The carpel** is the collective name for the ovary, style and stigma.

Anther, showing dehiscence (SEM) (×60)

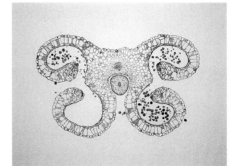

Dehiscent anther (TS)

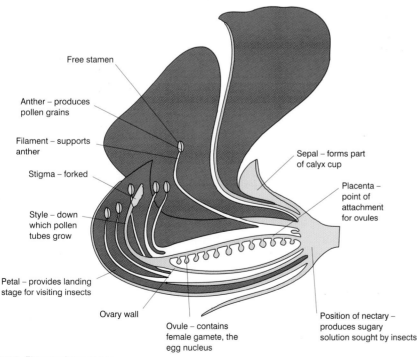

Free stamen

Anther – produces pollen grains

Filament – supports anther

Stigma – forked

Style – down which pollen tubes grow

Petal – provides landing stage for visiting insects

Ovary wall

Ovule – contains female gamete, the egg nucleus

Sepal – forms part of calyx cup

Placenta – point of attachment for ovules

Position of nectary – produces sugary solution sought by insects

Fig 5.13 *Flower of sweet pea*

SUMMARY TEST 5.5

All parts of the flower are attached to the swollen tip of the flower stalk known as the **(1)**. The outer parts that protect the bud are collectively called the **(2)**, which comprises individual parts called **(3)**. The collective name for the brightly coloured flower parts that attract insects is the **(4)**. The collection of male reproductive structures, or **(5)**, comprises the **(6)**, which contains the pollen sacs within which pollen **(7)** are produced. These contain the **(8)**. The female parts or **(9)** comprise the ovary, which contains one or more **(10)**, within which is the egg cell. The slender long extension from the ovary is called the **(11)**, which has a sticky tip called the **(12)**, on which pollen lands.

Pollination

Pollination is the transfer of pollen from the anther to the stigma. There are two types:

- **Self pollination** occurs when the pollen is transferred from an anther to a stigma on the same flower, or to a different flower but on the same plant. In either case, there is little genetic variety introduced, because the male and female gametes arise from the same genotype.
- **Cross pollination** occurs when pollen is transferred from the anther of one plant to the stigma of a flower on a different plant of the same species. As the gametes arise from parents with different genotypes, there is greater variety amongst the offspring than with self pollination.

5.6.1 Ensuring cross pollination

As there are considerable evolutionary advantages in having genetic variability amongst offspring, plants have evolved mechanisms which prevent self pollination and increase the chance of cross pollination. These include the following:

- **Protandry** occurs when all the anthers (male organs) on a plant ripen and release their pollen before the stigma and ovules (female organs) have matured. There is therefore very little chance of any pollen reaching the stigma of the same plant.
- **Protogyny** is the condition in which the stigmas ripen before the anthers on the same plant have matured. Again, self pollination is avoided because, by the time the pollen is released, the ovules on the same plant will already have been fertilised by pollen from other plants.
- **Dioecious** species are ones in which all the flowers of an individual plant are either male or female. It is therefore impossible for self pollination to take place. Completely dioecious plants are rare, but willows are examples.
- **Monoecious** species have separate male and female flowers on the same plant. By making self pollination less likely, this increases the prospects of cross pollination, e.g. in maize.
- **Incompatibility** occurs when the pollen tube will develop only if the stigma on which it lands has a genetic make-up different from its own, e.g. in pear.
- **Structural modifications** to the flowers of certain plants make self pollination difficult, and so increase the likelihood of cross pollination.

As pollen grains have no means of moving themselves, they rely on either insects or the wind to transport them. Flowers are specially adapted to suit one or other method of transfer. In general, most **dicotyledonous plants** such as trees, shrubs and broad-leaved herbaceous plants are insect pollinated, while most **monocotyledonous plants** such as grasses and other narrow-leaved varieties are wind pollinated.

5.6.2 Adaptations to insect pollination

Insects and flowering plants have influenced each other's evolution so much that they are dependent on each other for survival. Plants provide nectar, which is an essential food for many insect species, while insects transfer pollen, without which a new generation of plants could not arise. Insect-pollinated plants have evolved brightly coloured petals and powerful scents, which make their flowers conspicuous and recognisable to insects at a distance. Nectar, and sometimes extra pollen, are provided as food, which is both a reward and, consequently, an inducement for insects to visit other flowers of the same species – ideally, transferring pollen in the process. The structure of an insect-pollinated flower is illustrated in figure 5.12 in unit 5.5, and their adaptations are listed in table 5.3.

Wild sage about to be pollinated by a bee

Wind dispersal of pollen from male catkins of alder tree

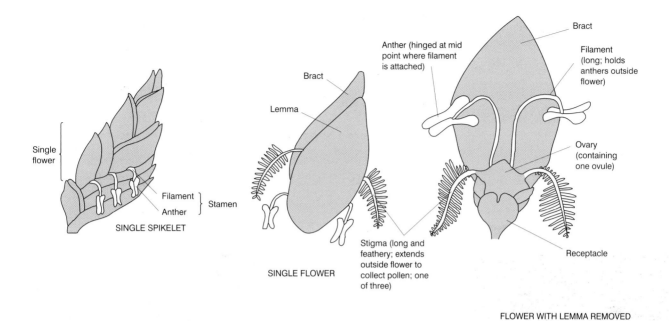

Fig 5.14 *Wind-pollinated flower*

5.6.3 **Adaptations to wind pollination**

Scent, nectar and brightly coloured petals have no influence on the wind, and flowers pollinated by wind have therefore remained dull, unattractive and without scent or nectar. Their petals are often reduced in size or absent altogether, as their presence would only shelter the remaining flower parts from the wind. The anthers and stigmas protrude from the flower, where they catch the wind, and the stigmas are often large and feathery, giving an increased surface area and increasing their chances of filtering the pollen carried by the wind. Much of the pollen will be lost because the plant cannot control where it goes, so copious amounts are produced. Figure 5.14 illustrates the structure of a typical wind-pollinated flower, and table 5.3 lists its various adaptations.

Table 5.3 *Comparison of wind- and insect-pollinated flowers*

Wind-pollinated flowers	Insect-pollinated flowers
Plants often occur in dense groups covering large areas	Plants often solitary or in small groups
Flowers are often unisexual, with an excess of male flowers	Mostly bisexual (hermaphrodite) flowers
Petals are dull and much reduced in size	Petals are large and brightly coloured, which makes them conspicuous to insects
No scent or nectar is produced	Flowers produce scent and/or nectar, which attracts insects
Stigmas often protrude outside the flower on long styles	Stigmas lie deep within the corolla
Stigmas are often feathery, giving them a large surface area to filter pollen from the air	Stigmas are relatively small, as the pollen is deposited accurately by the pollinating insects
Anthers dangle outside the flower on long filaments, so the pollen is easily released into the air	Anthers lie inside the corolla, where the pollinating insect brushes against them when collecting the nectar
Enormous amounts of pollen are produced, which offsets the high degree of wastage during dispersal	Less pollen is produced, as pollen transfer is more precise and there is less wastage
Pollen is smooth, light and small, and sometimes has 'wing-like' extensions, which aid wind transport	Pollen is larger and often bears projections, which mean it sticks to the insect

SUMMARY TEST 5.6

The transfer of pollen from the anther to the stigma on the same plant is called **(1)** pollination. Transfer between different plants is **(2)** pollination. The advantage of genetic variety produced when transfer is between different plants has led plants to develop means of ensuring this happens. In some plants the **(3)** release pollen before the female organs have matured. This is known as **(4)**. Where the female parts mature first it is called **(5)**. In **(6)** species there are separate male and female flowers on the same plant. Where all the flowers on a plant are either male or female the term used is **(7)**. Insect-pollinated plants produce a sugary **(8)** or excess **(9)**, which induces insects to seek out similar flowers.

Fertilisation in flowering plants

With the transfer of pollen from the anther to the stigma complete, the next stage in the production of offspring is **fertilisation**.

5.7.1 Events leading to fertilisation

Having reached the stigma, the pollen grain (and the male gamete it contains) is still some distance from the female gamete, the egg cell, which is protected within the ovule. Therefore, before fertilisation can take place, the following processes occur:

- On landing on the stigma, the pollen grains (there are usually many) absorb water.
- A solution of sucrose is secreted by the stigma, which stimulates the pollen grain to germinate.
- A pollen tube emerges from each pollen grain.
- Being negatively **aerotropic** and positively **hydrotropic**, the pollen tube pushes down into the loosely packed cells of the style, its growth controlled by the tube nucleus.
- Attracted by chemicals produced by the micropyle of the ovule, the pollen tube grows down the style.
- **Pectases** produced by the pollen tube help separate the cells of the style, and so assist the passage of the pollen tube.
- The pollen tubes enter the ovules of the ovary through the micropyle and release the two male nuclei each contains.

Groundsel pollen (SEM) (× 370)

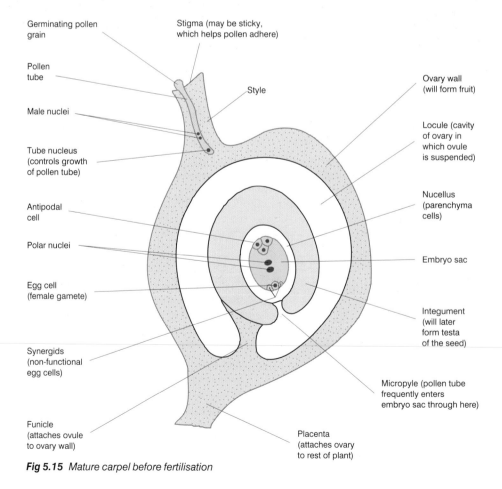

Fig 5.15 *Mature carpel before fertilisation*

5.7.2 Fertilisation

The process of fertilisation in plants is a double fertilisation, which involves:
- one male nucleus fusing with the egg nucleus to form a diploid (2n) zygote, which will develop into the embryo
- the other male nucleus fusing with the two polar nuclei to give a triploid (3n) primary endosperm nucleus which develops into endosperm – a nutrient supply for the developing embryo.

Once fertilised, each of the ovules forms a seed and these are protected within the ovary, which forms the fruit.

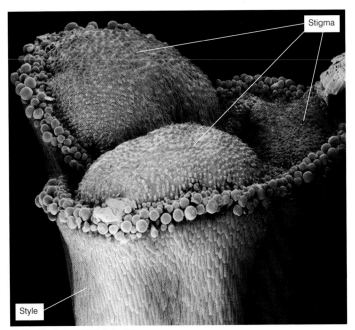

Daffodil stigma (SEM) (× 23)

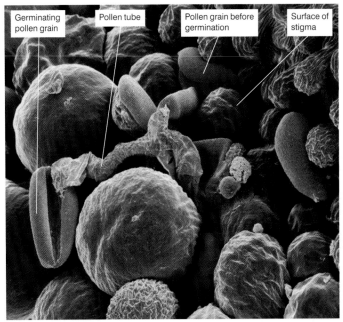

Daffodil pollen (SEM) (× 270)

SUMMARY TEST 5.7

In angiosperms the female gamete, called the (1), is protected within the (2). In order to reach it, the male gametes have to travel along the pollen tube. A few minutes after landing on the (3), the pollen grain germinates, assisted to do so by the secretion of (4) from the surface on which it landed. The pollen tube emerges from the pollen grain and pushes between the loosely packed cells of the (5), assisted by the secretion of (6) by the tip of the pollen tube. The growth of the tube is controlled by (7), although its direction of growth is the result of the tube being positively (8) and negatively (9). The pollen tube therefore grows to the (10), where it enters the ovule of the (11) and releases its two male nuclei. One male nucleus fuses with the (12) to give a diploid (13) which will develop into the (14). The other male nucleus fuses with the other two (15) to give a triploid (16) nucleus that forms food for the developing embryo. Each fertilised ovule will form a (17), while the (18) forms the fruit.

5.8

Gametes and gametogenesis

EDEXCEL

EDEXCEL (Human)

AQA.B

THE FEMALE GAMETE

The **ovum** is often referred to as the female gamete. Strictly speaking, this is incorrect, because the second meiotic division which produces the ovum only takes place after the head of the sperm enters the secondary oocyte. The ovum is therefore formed after, rather than before, fertilisation. The female gamete is, more accurately, the secondary oocyte

Sexual reproduction leads to genetic diversity amongst the offspring it produces. This diversity allows organisms to evolve in response to the changing world around them. Essential to the sexual process are gametes, which are produced by a mechanism called **gametogenesis**. In mammals, gametogenesis takes two forms, **oogenesis** and **spermatogenesis**.

5.8.1 Gametes

Gametes are **haploid** reproductive cells which fuse with another gamete of the opposite sex to form a **diploid** zygote, which marks the beginning of a new individual. In flowering plants the two types of gamete are:
- **the egg cell** – the female gamete
- **the male nucleus** – the male gamete.

In mammals, the two types are:
- **the secondary oocyte** (Fig 5.16) – the female gamete (see note in margin)
- **the spermatozoa (sperm)** (Fig 5.17) – the male gamete.

In mammals, the two types of gamete are markedly different. The sperm are small, motile and produced in vast numbers, whereas the secondary oocytes are larger, non-motile, food-storing cells produced in much small quantities.

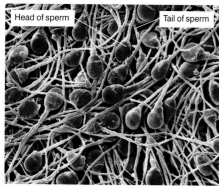

Mass of human sperm (SEM) (×108)

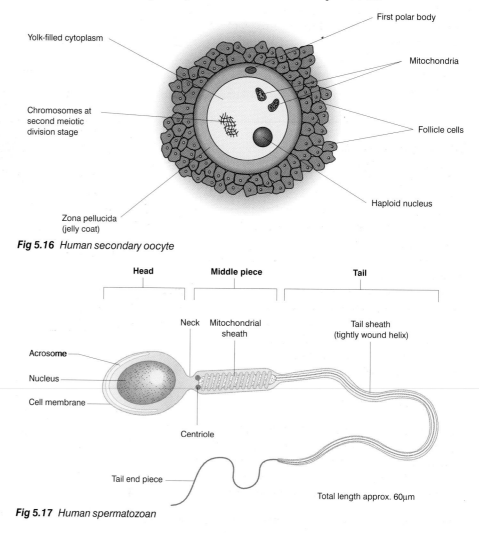

Fig 5.16 *Human secondary oocyte*

Fig 5.17 *Human spermatozoan*

5.8.2 Oogenesis

Oogenesis is the production of oocytes in the ovaries of females. The process, which is summarised in figure 5.18, has the following stages in humans:

- Before birth, the **germinal epithelium**, which lines the ovaries, divides by **mitosis** to form **oogonia**.
- The oogonia grow and enlarge to form **primary oocytes**.
- By birth, the ovaries contain several hundred thousand primary oocytes which have reached the prophase stage of the first meiotic division (unit 5.3).
- The primary oocytes develop follicle cells around them, to form **primary follicles**.
- At puberty, follicle stimulating hormone (unit 5.11) causes successive primary oocytes to continue their development, so that ovaries have oocytes at various stages of maturity.
- Every month or so, one (occasionally more) primary oocyte completes its first meiotic division and develops into a **secondary oocyte** plus a smaller, non-functional structure called the **first polar body**.
- At the same time, the primary follicle grows into a larger structure called the **Graafian follicle**.
- The Graafian follicle ruptures the ovary wall and releases its secondary oocyte into the oviduct, in a process called **ovulation**.
- This secondary oocyte does not complete its second meiotic division, and so develop into an **ovum**, until a sperm fertilises it.
- The remains of the Graafian follicle form the **corpus luteum** (yellow body) which produces, temporarily, the hormone progesterone (unit 5.11).

5.8.3 Spermatogenesis

Spermatogenesis is the production of spermatozoa in the testes of males. The process, which is summarised in figure 5.19, includes the following stages in humans:

- At puberty, under the influence of the hormone testosterone, the **germinal epithelium**, which lines the **seminiferous tubules** of the testes, divides by mitosis to form spermatogonia.
- The spermatogonia grow and enlarge to form **primary spermatocytes**.
- Primary spermatocytes undergo the first stage of **meiosis**, to form **secondary spermatocytes**.
- Secondary spermatocytes undergo the second stage of meiosis, to form **spermatids**.
- Spermatids enter the lumen of the seminiferous tubules, where they mature into spermatozoa.

SUMMARY TEST 5.8

Gametes are reproductive cells which contain the **(1)** number of chromosomes. In a male mammal they are called **(2)**, while in flowering plants the male gamete is the **(3)**. The female gamete in mammals is produced by a process called **(4)** and is known as the **(5)**.

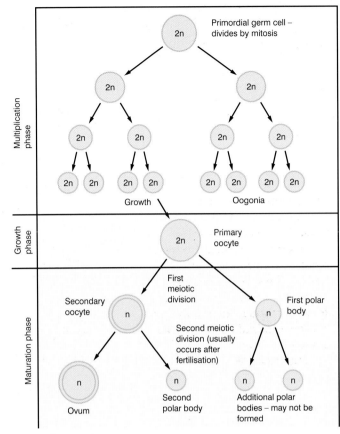

Fig 5.18 *Oogenesis*

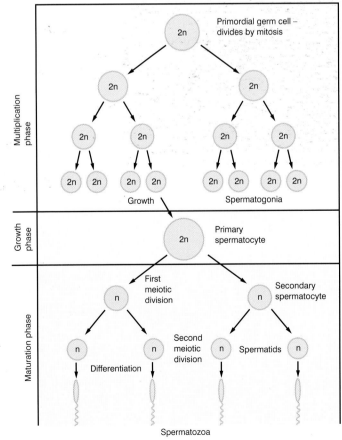

Fig 5.19 *Spermatogenesis*

Male human reproductive system

EDEXCEL

EDEXCEL (Human)

The structure of the urinogenital system (reproductive and urinary system) of a human male is illustrated in figure 5.20.

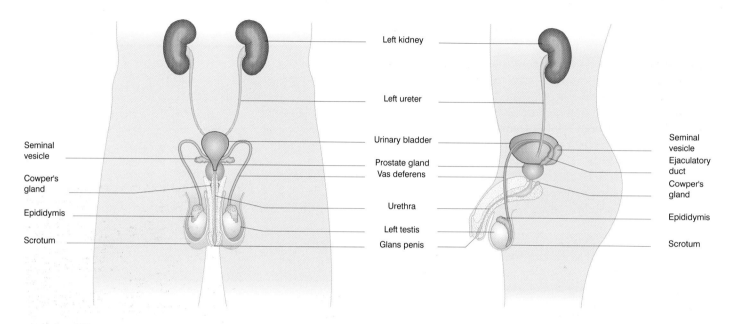

Fig 5.20 *Male human urinogenital system*

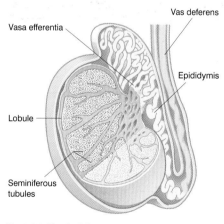

Fig 5.21 *Testis (LS)*

5.9.1 The testes

The **testes** are two egg-shaped structures which develop in the abdominal cavity, descending into the scrotum before birth. They produce the male gametes (spermatozoa), and are located outside the trunk of the body, because the optimum temperature for sperm development is 35°C – some 2°C below normal body temperature. Each testis is suspended by a spermatic cord made up of the sperm duct, an artery, vein and nerve, all bound together by connective tissue. Around each testis is a fibrous coat, and internally it is separated by septa into a series of lobules (Fig 5.21). Within each lobule lie the **seminiferous tubules** (Fig 5.23, opposite) – tiny tubes lined with germinal ephithelium that gives rise to the sperm, which are nourished by the Sertoli cells. Between the tubules are **interstitial cells**, which secrete the hormone, testosterone.

5.9.2 The penis

The **penis** is an external cylinder of spongy tissue which transfers semen into the vagina of the female. The penis has three separate columns of erectile tissue in its shaft (Fig 5.22). Under suitable arousal, this tissue becomes engorged with blood, causing it to become erect. The end of the penis is expanded to form the **glans penis**, a sensitive region covered by retractable skin called the **foreskin**. This foreskin is sometimes removed surgically, for medical or religious reasons, in an operation called **circumcision**. The **urethra**, which carries (at different times) urine and semen, emerges at the tip of the glans penis.

5.9.3 Other male reproductive structures

- **Epididymis** – a single, tightly coiled tube around 6 metres in length and continuous with the seminiferous tubules. The epididymis stores the sperm, which mature and develop the ability to swim as they pass through it.
- **Vas deferens** – a thick, muscular tube which carries the now fully motile sperm to the urethra, via the ejaculatory duct.
- **Seminal vesicles** – paired, pouch-like glands which contribute up to 60% of the volume of the semen. They secrete a thick, yellowish fluid, rich in fructose, which provides nourishment for the sperm.
- **Prostate gland** – a doughnut-shaped structure lying beneath the urinary bladder which secretes a thin, whitish fluid making up around 30% of the semen. This fluid increases sperm mobility and, being alkaline, neutralises any acidity due to urine.
- **Cowper's glands** – a pair of pea-shaped glands located below the prostate gland, they secrete a mucus-like fluid which makes up around 5% of semen and lubricates the urethra.
- **Scrotum** – an external sac of skin-covered muscle into which the testes descend before birth. Muscles in its wall allow the testes to be pulled towards the body to raise the temperature of the testes, or lowered to cool them. In this way their temperature is kept at around 35°C.

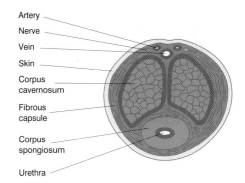

Artery
Nerve
Vein
Skin
Corpus cavernosum
Fibrous capsule
Corpus spongiosum
Urethra

Fig 5.22 *Penis (TS)*

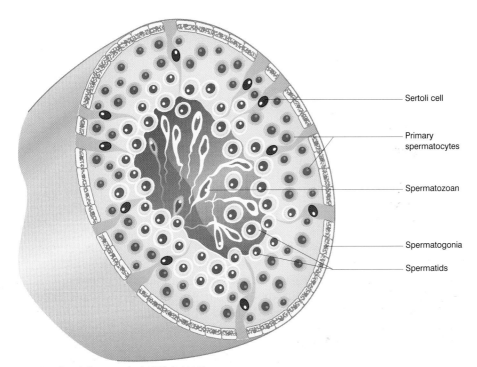

Sertoli cell

Primary spermatocytes

Spermatozoan

Spermatogonia

Spermatids

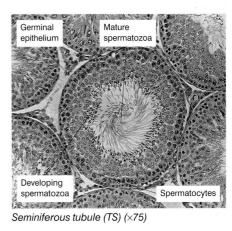

Germinal epithelium
Mature spermatozoa
Developing spermatozoa
Spermatocytes

Seminiferous tubule (TS) (×75)

Fig 5.23 *Seminiferous tubule (TS) (×1000)*

SUMMARY TEST 5.9

Sperm are produced by the walls of over 1km of tiny tubes called (**1**). These tubes make up the bulk of the two egg-shaped structures called the (**2**) that are suspended within a sac called the (**3**) by the (**4**) that comprises the sperm duct, blood vessels and (**5**), bound together by (**6**). The tiny tubes combine to form a 6m-long tube called the (**7**) that stores sperm. The sperm are initially nourished by (**8**) but later, as they pass along the (**9**), they are nourished by the thick, yellow secretion of the (**10**) that is rich in (**11**). The sperm are maintained at a favourable pH by secretions from the (**12**). The sperm plus the secretions added to them are known as (**13**), which is ejaculated via the tube called the (**14**) at the centre of the organ whose sensitive tip is the (**15**).

Female human reproductive system

EDEXCEL

EDEXCEL (Human)

The structure of the urinogenital (reproductive and urinary) system of a human female is illustrated in figure 5.24.

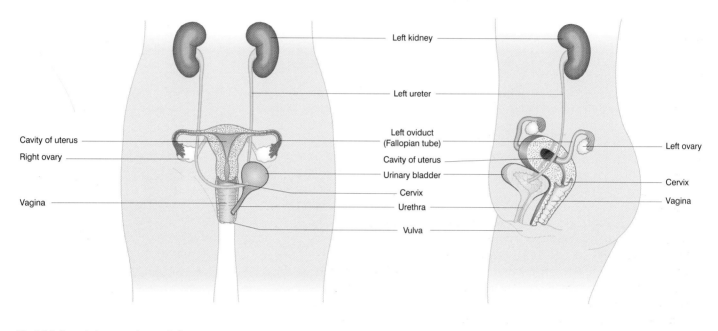

Fig 5.24 *Female human urinogenital system*

5.10.1 The ovaries

There are two ovaries (Fig 5.25), each about the size and shape of a large almond. They are situated in the abdominal cavity, held in position by the ovarian ligaments. The external coat is the **germinal epithelium**, which divides to form **oocytes** when the female is still a foetus. By birth, each female has several hundred thousand oocytes which have reached the prophase stage of the first meiotic division. The ovaries function to produce the female gametes by the process of oogenesis (section 5.8.2), and to produce the hormones **oestrogen** and **progesterone** which control the reproductive processes, including the **menstrual cycle** (unit 5.11).

5.10.2 The vulva and vagina

The **vulva** is the collective name for the external genitalia of females. It consists of two outer lips of skin called the **labia majora**, which protect the urinary and reproductive openings, and between which are two more delicate and sensitive folds of tissue called the **labia minora**. In front of the vaginal opening is a small bud-like organ, the **clitoris**. The equivalent of the penis in the male, the clitoris is the most sensitive part of the female sex organs, and swells during arousal. Between the vaginal opening and the clitoris is the opening of the urethra, which carries urine from the bladder. The **vagina** is a tube of **smooth muscle**, some 10–15 cm long but capable of considerable distension, especially during childbirth.

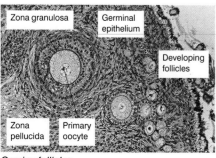

Ovarian follicles

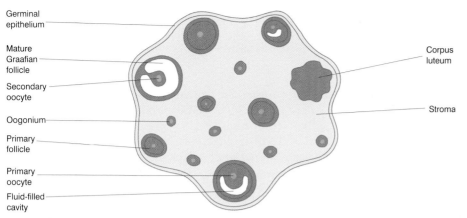

Germinal epithelium

Mature Graafian follicle

Secondary oocyte

Oogonium

Primary follicle

Primary oocyte

Fluid-filled cavity

Corpus luteum

Stroma

Fig 5.25 *Section through ovary*

It is lined internally with stratified epithelium and a mucus-secreting layer. The vagina both accommodates the penis during sexual intercourse and allows the passage of the baby during childbirth.

5.10.3 Other female reproductive structures

- **Cervix** – a ring of muscle at the top of the vagina. Normally having a small tubular opening into the uterus, this expands greatly during childbirth, to allow the passage of the baby.
- **Uterus** – a pear-shaped organ with walls made up of smooth muscle, it is lined by a mucus membrane called the endometrium. The uterus, also known as the **womb**, is where the foetus develops during pregnancy.
- **Fallopian tubes** – also known as the **oviducts**, are two small tubes around 10cm long. They have feathery, funnel-like openings that partly envelope the ovaries and from where they collect the secondary oocytes released at ovulation. The walls of the Fallopian tubes are made up of smooth muscle lined by a mucus-secreting layer of ciliated epithelium. The cilia aid the movement of the secondary oocyte from the ovary to the uterus.

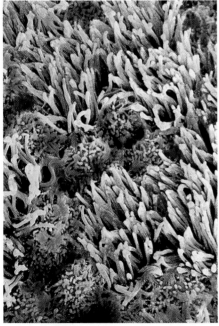

False-colour SEM of epithelium of Fallopian tube

SUMMARY TEST 5.10

The ovaries produce female gametes called **(1)** that are made by the ovaries' external coat, the **(2)**. Each month a gamete passes into one of the two 10cm-long tubes called the **(3)** that are made up of walls of **(4)** lined with **(5)**, which move the gametes towards the uterus. The inner lining of the uterus is a soft mucous membrane called the **(6)**. The uterus opens into the **(7)** through a ring of muscle called the **(8)**. The external genitalia of a female are known as the **(9)** and have two folds of skin, the outer of which is called the **(10)**, beneath which is a bud-like organ, the **(11)**, that is highly sensitive and swells during sexual arousal.

The menstrual cycle

In humans, and certain other primates, the reproductive system of females undergoes a regular cycle of changes known as the **menstrual cycle**. Beginning at puberty, it continues until the **menopause** at 45–50 years of age, and is controlled through the interaction of a number of hormones.

5.11.1   Changes during the menstrual cycle

Usually around 28 days, the length of each menstrual cycle varies from individual to individual. It is divided into four stages (Fig 5.27):

- **The menstrual phase** (days 1–5) occurs when the **endometrium** (uterus lining) is shed along with some blood **(menstruation)**.
- **The follicular phase** (days 6–13) is when a **Graafian follicle** develops within the ovary and matures, ready for its secondary oocyte to be released. At the same time, the endometrium, which was lost in the menstrual phase, is repaired and thickened.
- **The ovulatory phase** (day 14) is the release of the secondary oocyte from the Graafian follicle in the ovary **(ovulation)**.
- **The luteal phase** (days 15–28) is so called because the now empty Graafian follicle develops into a **corpus luteum** (yellow body). If fertilisation does not take place and/or the **blastocyst** (section 5.12.4) does not implant in the endometrium, then the corpus luteum degenerates and the endometrium breaks down, marking the start of the next menstrual cycle. The unfertilised secondary oocyte passes out of the body via the vagina.

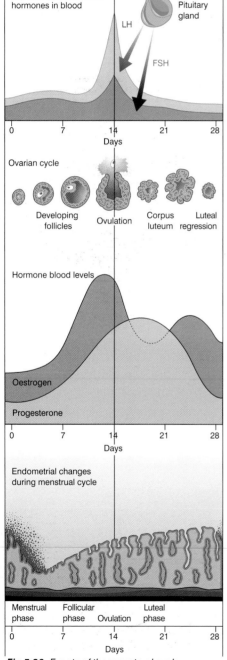

Fig 5.26 *Events of the menstrual cycle*

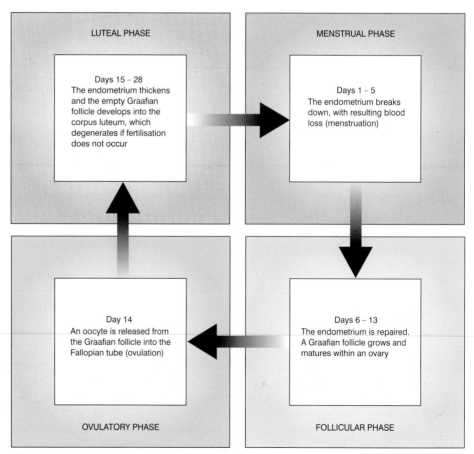

Fig 5.27 *Summary of changes during the menstrual cycle*

5.11.2 Hormonal control of the menstrual cycle

There are five hormones which control the menstrual cycle, each interacting with the other in a way that ensures a regular cycle of events. The cycle begins with production, by the hypothalamus, of a hormone which is called **gonadotrophin releasing hormone** (GnRH) because it stimulates the anterior lobe of the pituitary gland to produce its hormonal secretions. The two hormones produced by the anterior lobe of the pituitary gland (which lies at the base of the brain) are known as **gonadotrophic stimulating hormones** and function as follows:

- **Follicle stimulating hormone (FSH)** causes the Graafian follicles in the ovary to develop, and stimulates the ovaries to produce oestrogen.
- **Luteinising hormone (LH)** causes ovulation to occur, and stimulates the ovary to produce progesterone from the corpus luteum.

The remaining two hormones are produced by the ovaries. These are known as **ovarian hormones** and function as follows:

- **Oestrogen** causes the rebuilding of the endometrium of the uterus after menstruation, and stimulates the pituitary gland to produce LH.
- **Progesterone** maintains the endometrium of the uterus in readiness to receive the blastocyst (young embryo), and inhibits the production of FSH from the pituitary gland.

In a simplified form the sequence of operation is:

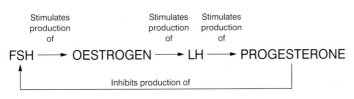

- Progesterone, at the end of the sequence, inhibits the FSH at the beginning.
- In the absence of FSH, production of oestrogen, LH and progesterone also ceases.
- In the absence of progesterone, the inhibition of FSH ceases.
- FSH production resumes and the cycle repeats itself.

This alternate switching on and off of these hormones is responsible for the regular sequence of events in the menstrual cycle (Fig 5.28).

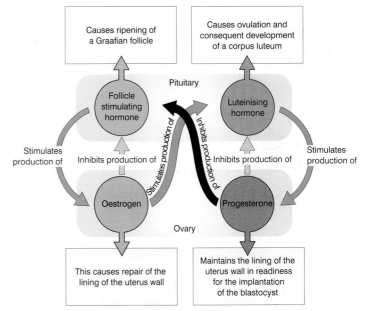

Human secondary oocyte in Fallopian tube (SEM) (×300)

Fig 5.28 *Hormone interaction in the menstrual cycle*

SUMMARY TEST 5.11

The menstrual cycle in humans lasts around **(1)** days. It begins about the age of 12 years and continues until the **(2)** at around the age of 45–50 years. The cycle begins with the initial discharge of blood called menstruation that lasts around **(3)** days. This is followed by the **(4)** phase, when a Graafian follicle develops in an ovary as a result of the influence of the hormone called **(5)**. This hormone is produced by the **(6)** gland when it is stimulated by **(7)** hormone from the hypothalamus. The ovulatory phase usually occurs on day **(8)** of the cycle and involves the release of a **(9)** from the Graafian follicle, an event stimulated by the hormone known as **(10)**. The final phase is called the **(11)** phase during which the Graafian follicle develops into a **(12)** which degenerates if **(13)** does not occur, leading to the breakdown of the **(14)** – the lining of the uterus – and the start of the next cycle.

5.12 Fertilisation and early development

EDEXCEL

EDEXCEL (Human)

Before further development can take place, the female gamete, or secondary oocyte, must be fertilised by the male gamete or spermatozoan. In humans this takes place in the female reproductive tract and it is therefore **internal fertilisation**.

5.12.1 Mating

Mating (or **copulation**) is the process whereby male gametes are transferred into the female of the species. In humans it is more often called **sexual intercourse**. It is preceded by a period of courtship – a process designed to ensure compatibility of the two partners and so make it more likely that they will remain together to bring up the offspring, thus increasing its chances of survival. Immediately before successful mating, a number of changes to the reproductive organs of both sexes are initiated by a variety of erotic stimuli –

In males:
- the blood supply to the penis increases, filling its spongy tissue with blood
- this causes the penis to become hard and erect.

In females:
- the blood supply to the genitalia increases, causing the clitoris and labia to swell
- the vagina secretes mucus, assisting penetration by the penis.

After the penis has been inserted into the vagina, repeated thrustings stimulate sensory cells on the glans penis. This in turn causes reflex **peristaltic** contractions of the muscular walls of the epididymis and vas deferens, which push stored sperm towards the urethra. Secretions from the seminal vesicles, prostate gland and Cowper's glands are added before the mixture, now known as **semen**, is forcefully ejaculated near the top of the vagina. This ejaculation is accompanied by a sensation of extreme pleasure, known as an **orgasm**. The female orgasm is similarly intense, and results from the contraction of the vagina and uterus.

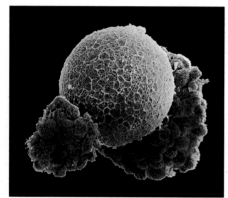

Human oocyte with cumulus cells (SEM) (×500 approx.)

5.12.2 Semen

Apart from water, which makes up the majority of the 5cm³ of semen produced in a single ejaculation, the semen also contains:
- **sperm** – around 500 million in a single ejaculation
- **mucus** – which makes the semen more viscous
- **alkaline chemicals** – to neutralise the acid conditions encountered in the vagina, which would otherwise kill the sperm
- **sugars** – such as fructose – which provide energy for the sperm and so make them more mobile
- **prostaglandins** – hormones which cause the uterus and Fallopian tubes to contract in a way that helps sperm reach the secondary oocyte.

5.12.3 Fertilisation

Spermatozoa in the semen are either ejaculated directly through the cervix into the uterus, or swim there, propelled by the lashing of their long tails. They then swim through the uterus and into the Fallopian tubes, assisted by contraction of the uterus walls. During this journey, the spermatozoa undergo changes called **capacitation**, which are started by secretions from the genital tract of the female. Without capacitation, the sperm are incapable of fertilising the oocyte. Meanwhile, a secondary oocyte released from the ovary may be travelling down the Fallopian tube, helped by the movement of cilia on the tube walls. This

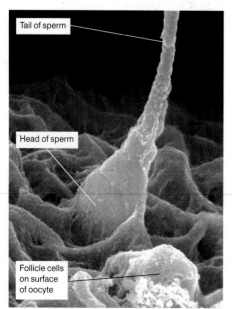

Tail of sperm

Head of sperm

Follicle cells on surface of oocyte

Human sperm penetrating oocyte (SEM) (×4000)

secondary oocyte is surrounded by up to 2000 **cumulus cells** (see photo), which both provide nutrients for the oocyte and make it easier for the cilia to move it towards the uterus. The secondary oocyte will survive only 24 hours unless fertilised, so this process must take place near the top of the Fallopian tube if it is to be successful. The meeting of sperm and oocyte appears to be pure chance. Of the 100 or so sperm that survive this far, only one will fertilise the oocyte. The act of fertilisation involves the following sequence of events:

- The sperm releases, from the **acrosome** at its tip, a trypsin-like enzyme called **acrosin**.
- Acrosin softens the plasma membrane covering the oocyte.
- The acrosome inverts to form a needle-like filament, which pierces the now softened plasma membrane.
- On being pierced, the plasma membrane thickens to form the **fertilisation membrane**, which prevents entry of more sperm.
- The sperm tail is discarded, while the head and middle piece (section 5.8.1) enter the oocyte.
- The second meiotic division of the oocyte (which has been temporarily suspended) occurs immediately, and the oocyte becomes an **ovum**.
- The nuclei of the spermatozoan and ovum fuse to form a diploid zygote.

5.12.4 Implantation and early development

Immediately fertilisation has taken place, the ovum repeatedly divides mitotically to give a hollow ball of cells called the **blastocyst**. The blastocyst is moved down the Fallopian tubes by the cilia lining them, reaching the uterus some 3 days later. Within the next 3–4 days it will start to implant into the lining of the uterus – a process which takes a further 7 days to complete. The outer layers of the blastocyst form two embryonic membranes:

- **The chorion** forms villi which grow into the uterus lining, from which they absorb nutrients. The villi later make up part of the placenta.
- **The amnion** forms a membrane around the embryo and encloses a watery fluid, which protects the embryo from physical damage. The events of early development are illustrated in figure 5.29.

SUMMARY TEST 5.12

The movement of an erect penis within the vagina stimulates sensory cells on the **(1)** that in turn lead to **(2)** contractions of the muscular walls of the epididymis and **(3)**. Around 5 cm³ of semen is thereby forcefully ejected from the **(4)** at the tip of the penis in a process called **(5)**. The sperm in the semen swim up through the **(6)** into the uterus and then the **(7)**, both of which contract as a response to **(8)** in the semen.

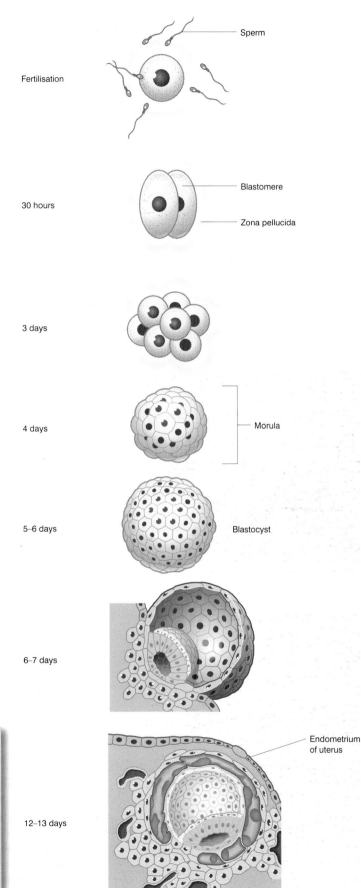

Fig 5.29 Embryo development

Foetal development and birth

Once implanted in the uterus lining, the blastocyst develops into an embryo which obtains its nutrients from a special structure called the **placenta**.

5.13.1 The placenta

The placenta is a disc-like structure which develops from the chorionic villi of the outer layer of the blastocyst. It allows the blood of the mother and foetus to come close together without actually mixing. This is essential, because the foetus will have tissues developed, in part, from the genetic material of the father. If the two were to come in contact, the immune system of the mother would reject the foetus. The placenta is attached to the uterus lining and is linked to the foetus by the umbilical cord. The functions of the placenta are as follows:

- **Exchange of materials** between the mother and foetus, including:
  - **dissolved gases** – oxygen from the mother to the foetus, and carbon dioxide in the reverse direction. To aid exchange, the haemoglobin of the foetus has a greater affinity for oxygen than that of the mother (section 8.2.3)
  - **nutrients** – glucose, amino acids, minerals and vitamins from the mother to the foetus
  - **waste products** – urea from the foetus to the mother.
- **Protection** – the placenta protects the foetus by:
  - **filtering out pathogens** – many harmful disease-causing agents and their toxins are prevented from crossing the placenta from the mother to the foetus. However, some viruses, such as rubella (German measles) and HIV, are capable of entering the foetus in this way
  - **excluding hormones** – the maternal hormones are prevented from crossing the placenta. If they were allowed to cross, they could upset the growth, metabolism and development of the foetus
  - **excluding harmful chemicals** – many toxic chemicals in the mother's blood are prevented from reaching the foetus. Some notable exceptions, however, include nicotine, alcohol and some narcotic drugs such as heroin
  - **providing passive immunity** – maternal **antibodies** can pass across the placenta, giving a temporary immunity which persists for the first few months of the baby's life
  - **permitting different blood pressures** to exist between the foetus and the mother. If the blood supplies were directly linked, the higher maternal blood pressure would damage the delicate vessels and organs of the foetus.
- **Hormone production** – the placenta produces a number of hormones.
  - **Human chorionic gonadotrophin (HCG)** is produced during the first 12 weeks of pregnancy. It prevents the **corpus luteum** in the ovary from degenerating, and allows it to continue to produce progesterone, which maintains the pregnancy (see below).
  - **Progesterone** is produced directly by the placenta from about the 6th week of pregnancy, gradually taking over this role from the corpus luteum. Progesterone prevents ovulation and menstruation, and thereby ensures there are no more foetuses to complicate the pregnancy, and that the established foetus is not aborted.
  - **Oestrogen** production is also taken over from the corpus luteum from around the 6th week of pregnancy.
  - **Lactogen** causes the milk-producing glands in the breasts (fig 5.30) to develop and enlarge.

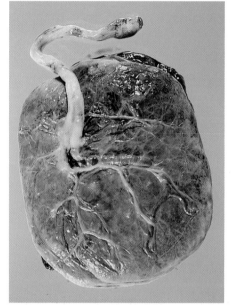

Human placenta and umbilical cord

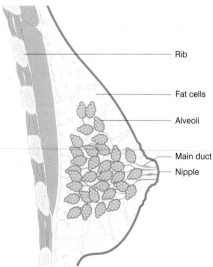

Rib

Fat cells

Alveoli

Main duct
Nipple

Fig 5.30 *Human breast*

5.13.2 Birth

Birth (also known as **parturition**) and its timing are controlled by a series of hormones. While the process is not yet fully understood, the likely sequence of events is listed here:

- At around the 38th week of pregnancy, the foetal pituitary gland produces **adrenocorticotrophic hormone (ACTH)**. The trigger for its production may be a stress response to the increasing pressure of the uterus wall on the foetus.
- The ACTH stimulates the foetal adrenal glands to produce steroid hormones, which are transported by foetal blood to the placenta.
- The placenta in turn secretes **prostaglandins** and reduces its production of progesterone.
- The prostaglandins cause the uterus wall to begin contracting, pushing the foetus against the cervix.
- Receptors in the cervix send nervous messages to the maternal brain which stimulate the posterior lobe of the pituitary gland to produce the hormone, **oxytocin**. The reduced level of progesterone in the blood also encourages oxytocin production.
- Oxytocin causes further contractions of the uterus wall, creating more pressure on the foetus, which in turn is forced further against the cervix.
- Both actions stimulate more ACTH production and so the whole sequence of events is reinforced. This is an example of positive feedback, whereby a response, in this case the contraction of the uterus wall, is progressively increased.

The actual birth takes place in three stages:
- The amniotic membranes rupture ('breaking of the waters') and the cervix dilates, causing the mucus plug it encircles to break free.
- The uterine contractions force the baby out through the cervix and vagina **(the delivery)**.
- The placenta ('afterbirth') is expelled from the uterus.

5.13.3 Lactation

Lactation is the production of milk from the mammary glands (breasts). It is influenced throughout by hormones.
- **Progesterone, oestrogen** and **lactogen** stimulate the development of the lactiferous glands in the breasts during pregnancy.
- **Prolactin** is produced by the anterior lobe of the pituitary gland immediately after birth and causes the lactiferous glands to begin milk production.
- **Oxytocin** is produced by the posterior lobe of the pituitary gland in response to suckling by the infant. Carried by the blood, oxytocin stimulates the muscles around the lactiferous glands to contract, thus forcibly ejecting the milk they contain out through the nipple. This is known as the **milk ejection reflex** or **let-down reflex**.

The first-formed milk is known as **colostrum**. It is mildly laxative, and helps the baby to remove some of the wastes, e.g. bile juice, that have collected in the intestines during

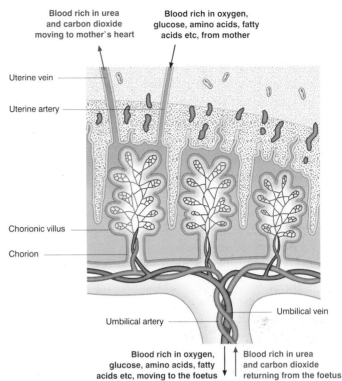

Fig 5.31 *The human placenta*

pregnancy. Milk provides both nutrition and antibodies to the newly born. The antibodies provide passive immunity until such time as the baby is able to develop its own.

In addition to providing nutrition, the baby's suckling suppresses ovulation, and so reduces fertility in the period after birth. This helps to ensure that mammals do not have too many offspring close together and is thus a form of birth control in humans. Indeed, it has been estimated that breast-feeding prevents more pregnancies, worldwide, than all artificial contraceptives put together.

> **SUMMARY TEST 5.13**
>
> The placenta develops from **(1)** on the outer layer of the **(2)**. One of its functions is to produce hormones such as human chorionic gonadotrophin that is made during the first **(3)** weeks of pregnancy. This hormone prevents the **(4)** from degenerating and so allows it to continue producing progesterone, which prevents both **(5)** and **(6)** from occurring. Progesterone, oestrogen and **(7)** from the placenta cause the breasts to develop, while the production of milk from the **(8)** glands is stimulated by **(9)** from the **(10)** lobe of the pituitary gland. These glands forcefully eject milk, initially called **(11)**, through the nipple in response to the hormone **(12)**.

The menstrual cycle (unit 5.11) in humans is an example of an **oestrous cycle**. As the breeding of domesticated animals, for meat, milk or other products, is affected by their oestrous cycles, control of these can bring increased yields. At the same time, knowledge of the menstrual cycle has allowed humans to control fertility artificially.

5.14.1 Oestrous – its detection and control in cattle

The oestrous cycle of a cow (Fig 5.32) usually lasts for 21 days. It is controlled by hormones in basically the same way as the menstrual cycle. One difference is that a cow is receptive to mating for only a very short period in the cycle – some 12–18 hours before **ovulation**. As the productivity of both beef and dairy herds relies upon cattle having offspring, it is beneficial that cows are fertilised during this short period. Detecting when a cow is in oestrous or 'heat' is therefore of great importance, as it allows the farmer either to introduce a bull or inseminate the cow artificially with maximum chance of **fertilisation**. Clues are provided by certain changes in a cow's behaviour during oestrous. These include:

- increased restlessness, including butting other cows
- mounting other cows
- allowing herself to be mounted by other cows
- feeding less and producing less milk
- mucus discharge from the vagina.

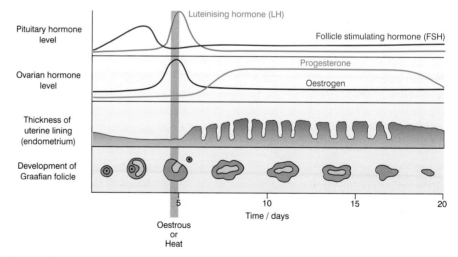

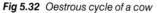

Fig 5.32 *Oestrous cycle of a cow*

5.14.2 Production of animal embryos for transplantation

Both to control the genetic make-up of the offspring and to ensure pregnancy, embryos are frequently transplanted into the females of domesticated animals. In this way, animals with the desired characteristics can be selectively bred. In cattle the process is as follows:

- A cow is treated with follicle stimulating hormone (FSH) to induce it to produce many **oocytes**
- When the cow is in oestrous, it is artificially inseminated with sperm from the selected bull.
- About 8 days after insemination, and before the embryos have implanted in the uterus, they are flushed out of the donor cow.

- Each of the embryos is transplanted into a separate recipient cow, through the vagina and cervix.
- Embryos not immediately transplanted may be stored in liquid nitrogen at −197°C for future use.

Embryo transplantation offers a greater degree of control and selection than artificial insemination. For example, the sex of the embryo can be identified in the laboratory, and only the desired type implanted.

5.14.3 Increasing milk production in cattle

The production of milk by dairy cattle is influenced by many factors, including the breed, food supply, time of calving, state of health, and hormones. Under natural conditions, a cow produces the hormone **bovine somatotrophin (BST)** from the pituitary gland. BST controls milk production (lactation) by increasing the number of cells in the mammary glands. BST can be produced by **genetically engineered** bacteria and, when injected into cattle, can increase milk yields by around 20%. One disadvantage is that cases of mastitis (inflammation of the udder) are more frequent when BST is used, and there is consequently greater need for antibodies to counteract the mastitis, increasing the risk of **antibiotic resistance** in bacteria. Although used in the United States, BST is banned in Europe.

5.14.4 Synchronising breeding behaviour in sheep

It is to a farmer's advantage to ensure that many sheep are fertilised together, so that feeding programmes can be carried out for groups of animals at the same time. There are two ways to control the time of ovulation:

- Maintaining a high progesterone level for many days, by inserting a capsule which releases progesterone into the vagina. The removal of the capsule and the consequent fall in progesterone level triggers follicle development, followed by ovulation a few days later.
- Supplying **prostaglandin F2α**, which is normally secreted from the uterus. If injected at the appropriate time in the oestrous cycle (Fig 5.32), it causes the ewe to come on heat a few days later. Two injections of prostaglandins about 10 days apart are most effective in bringing on oestrous in the majority of a flock.

5.14.5 Synthetic hormones as contraceptives

The contraceptive pill is used to control the menstrual cycle artificially in order to prevent conception. The contraceptive pill contains both oestrogen and progesterone, and if it is taken daily, the levels of both remain high in the blood. This inhibits the production of the **gonadotrophic hormones** from the pituitary gland (unit 5.11). Without these hormones, follicles are not developed, and consequently no ovulation can occur. Typically, the pill is taken for 21 days of the 28-day cycle, allowing menstruation to occur for part of the remaining 7 days.

5.14.6 Synthetic hormones as fertility drugs

Infertility in females may be due to an inability to develop follicles and therefore to ovulate. A fertility drug can overcome this problem in one of three ways:

- Drugs such as **clomiphene** increase the natural production of the gonadotrophic hormones such as follicle stimulating hormone (FSH) and so stimulate ovulation.
- FSH may be provided directly, to induce the ovaries to produce **Graafian follicles** and/or to release the oocytes they contain.
- Drugs may be given which inhibit the production of oestrogen by the ovaries. As oestrogen inhibits the production of FSH, the level of FSH will rise and ovulation will take place.

SUMMARY TEST 5.14

Knowledge of the **(1)** cycles of organisms has allowed humans to manipulate both their own sexual cycles and those of domesticated animals. The contraceptive pill contains progesterone and **(2)** at levels that inhibit the production of **(3)** hormones from the pituitary gland, and so prevent **(4)**. Drugs such as **(5)** can be used to increase fertility by increasing production of **(6)**. Cows can be given follicle stimulating hormone to induce them to produce **(7)**. These cows are then artificially **(8)** when the cow is in oestrus, an event that occurs about every **(9)** days. The embryos that develop are then flushed out of **(10)** and transplanted into other cows immediately or stored in liquid **(11)** at −197°C. Cattle may also be induced to produce more milk using the hormone **(12)**, but this leads to an increased risk of an inflamed udder, a condition called **(13)**.

5.15

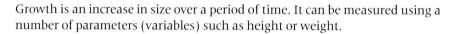

EDEXCEL (Human)

Growth and ageing in humans

Growth is an increase in size over a period of time. It can be measured using a number of parameters (variables) such as height or weight.

5.15.1 Growth patterns

If the size of an organ, an organism or a population is measured over time, the graph produced often has an S-shape, regardless of the species involved. This is known as a **sigmoid curve** (Fig 5.33). It is the result of an early period of slow growth while the number of cells or organisms builds up, followed by rapid growth, and finally a slower growth as maximum size is achieved. In the case of an organism, there is normally a maximum size limit, which is imposed either by its genetic make-up or by external factors such as the availability of food.

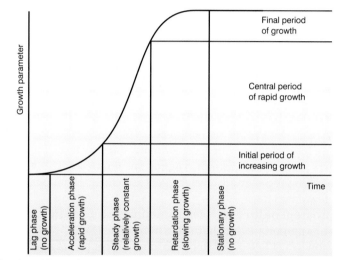

Fig 5.33 *The sigmoid growth curve*

5.15.2 Human growth curves

The usual sigmoid curve is modified in the case of an individual human. Rather than a single curve, it resembles two sigmoid curves, one in early life and the other in adolescence. These are separated by a period of relatively slow growth (Fig 5.34). The growth of a human can therefore be divided into five phases:

- **Prenatal** growth occurs before birth, when the foetus develops rapidly from the zygote over a 9-month period.
- **Infancy** is the first 18 months to 2 years after birth, when there is very rapid growth.
- **Childhood** is the period of slower growth, from 2 years to sexual maturity at around 12 years.
- **Adolescence** is the second period of rapid growth, which usually occurs between the ages of 12 and 19 years.
- **Adulthood** is the rest of a human's life. Although there may be a little growth in the early years of adulthood, this period is largely one of no growth, followed by a decline in later years.

While the human as a whole grows in this fashion, different organs have their own pattern of growth within the overall scheme. Where organs follow the standard pattern, the term used is **isometric growth**, but where they grow at a rate different from the overall one, it is termed **allometric growth**. There are good reasons for the allometric growth of particular organs in humans:

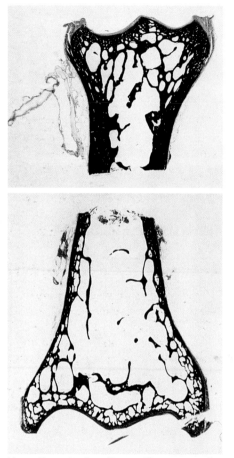

Normal bone (top) and bone with osteoporosis

- **Reproductive organs** grow very little in early life, as the organism is not sufficiently physically or mentally developed to rear offspring successfully. These organs, however, develop rapidly during adolescence.
- **Lymph tissue** produces white blood cells which are used to fight infection. As humans are more open to disease in childhood, because they have yet to develop immunity, the lymph tissue grows rapidly in this period. As immunity is developed, the need for lymph tissue decreases and so this tissue reduces in size during adolescence.
- **The head** grows rapidly in infancy, to allow for the rapid development of the brain, which is essential for early mental development, especially learning.

Figure 5.35 illustrates the relative growth rates of these organs.

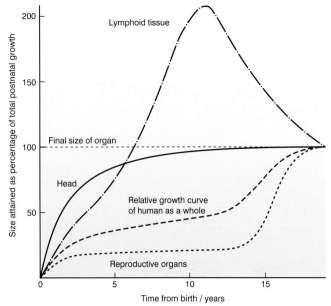

Fig 5.35 Allometric growth as shown by human organs and tissues

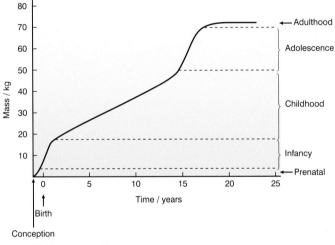

Fig 5.34 Human growth curve

5.15.3 Effects of ageing on the skeletal system

From early adulthood onwards, the functioning of body systems declines. In the skeletal system, this shows in a number of ways:

- The edges of bones become less distinct, and develop extensions and spurs, which may be painful and restrict movement around joints.
- The pattern of bone calcification changes, making bones more porous and more likely to fracture.
- **Osteoporosis** occurs as a result of the loss of calcium and **collagenous** fibres from bone. Bones therefore shorten and weaken, due to increased porosity resulting from loss of the organic bone matrix.
- **Osteoarthritis** occurs, due to the cartilage at the ends of bones degenerating.

5.15.4 Effects of ageing on the cardiovascular system

With age, degenerative changes occur in the heart and blood vessels:

- **Atherosclerosis** is the result of fatty deposits such as cholesterol building up on the walls of blood vessels. This

may lead to heart attacks, due to blockages in the coronary arteries, or strokes due to blockages in the arteries serving the brain.

- **Arteriosclerosis** is the hardening of the arteries as a result of ageing. Being less able to expand, hardened arteries lead to an increase in blood pressure **(hypertension)**, which in turn may cause a heart attack or stroke.
- **Heart muscle weakness**, limiting the activity of older people.

5.15.5 Effects of ageing on the reproductive system

While sexual activity may continue into old age, fertility is nevertheless reduced, due to ageing of the reproductive system. Sperm counts diminish in men and there may be increased difficulty in achieving an erection. In women, the ovaries cease to produce **oocytes** after about 50 years of age. This is known as the **menopause**, and results from a decrease in the production of **gonadotrophic hormones** by the pituitary gland. Changes associated with the menopause include:

- menstrual periods cease
- vaginal wall becomes thinner
- hot flushes may be experienced
- night sweats may occur
- osteoporosis (section 5.15.3) may result from a fall in oestrogen levels.

The effects of the menopause may be offset by the use of **hormone replacement therapy (HRT)**. This supplements the level of oestrogen in the body, using tablets containing hormone substitutes.

1 The diagram shows the main stages of the cell cycle. The letters A to D represent the four stages of mitosis.

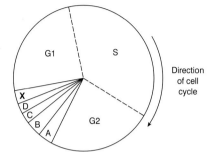

a Identify the stage when each of the following events is taking place.
  (i) DNA replication
  (ii) Individual chromatids from a chromatid pair move to opposite poles of the cell. *(2 marks)*
b What is happening during stage **X**? *(1 mark)*

c *Vinblastine* is an anti-cancer drug that prevents the formation of a spindle.

  (i) What is the function of the spindle? *(1 mark)*
  (ii) How would a drug like vinblastine help prevent the growth of a tumour? (2 marks)
(Total 6 marks)
AQA Jan 2001, HB (A) BYA3, No.6

2 The diagram shows a sperm and an egg from the same species, drawn to the same scale.

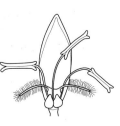

a Assume that both the egg and its nucleus are spheres.

The volume of a sphere is given by $\frac{4\pi r^3}{3}$ where r is the radius of the sphere. Calculate the proportion of the egg cell that is occupied by the nucleus. Show your working. *(2 marks)*

b The nucleus occupies a much smaller proportion of an egg cell than it does of a sperm. Give **one** reason for this difference and explain its importance. *(2 marks)*
(Total 4 marks)
AQA June 2001, B(B) BYB2, No. 2

3 a Explain what is meant by the term **pollination**.
(2 marks)

b The diagram shows the structure of a grass flower.

Describe **two** ways in which this flower is adapted for wind pollination. *(4 marks)*

c This grass flower can be self-pollinated. Suggest how the flowers of other grasses might be adapted to avoid self pollination. *(3 marks)*
(Total 9 marks)
Edexcel 6102/01 June 2001, B AS/A, No.6

4 The graph shows the increase in foetal mass and placental mass during the development of a human embryo.

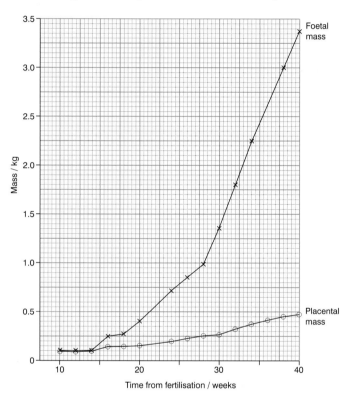

a (i) Determine the placental mass at 30 weeks.
(1 mark)
  (ii) Calculate the percentage increase in placental mass between 10 and 40 weeks. Show your working.
(3 marks)

b Compare the changes in placental mass and fetal mass between 10 and 40 weeks. *(3 marks)*

c A woman at the beginning of labour (the birth process) could be given a mild sedative in the form of a barbiturate-based drug, by injection into a vein. The table below shows the concentration of such a drug in the maternal blood and in the foetal blood over a period of 40 minutes from the time of injection.

Time from Injection / min	Concentration of drug in blood /µg per cm³	
	Mother	Foetus
0	0	0
2	50	2
5	8	5
10	5	3
20	5	3
40	4	3

Explain these results and suggest why you would recommend that such a procedure should **not** be carried out. *(3 marks)*

(Total 10 marks)

Edexcel 6112/01 June 2001, B (H) AS/A, No.8

5 The photographs show stages in mitosis labelled **A–D**. The stages are in sequence with stage **A** being the earliest.

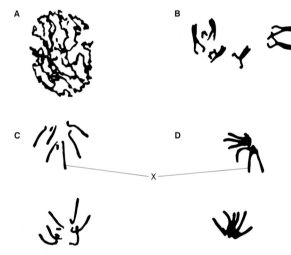

a Describe what occurs between

(i) stage **A** and stage **B**; *(2 marks)*
(ii) stage **B** and stage **C**. *(2 marks)*

b Describe what happens to structure **X** between the end of stage **D** and the next time stage **A** occurs. *(2 marks)*

(Total 6 marks)

AQA Jan 2001, B (B) BYB2, No.1

6 The diagram shows the life cycle of the honey bee.

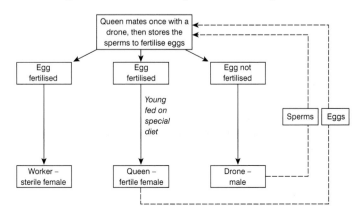

a The queen has body cells containing 32 chromosomes. How many chromosomes are there in the body cells of
(i) the drone;
(ii) the workers? *(2 marks)*

b Name the type of cell division in bees that produces
(i) eggs;
(ii) sperms. *(2 marks)*

c Are all the drones identical? Explain the reason for your answer. *(2 marks)*

(Total 6 marks)

AQA Jan 2001, B (B) BYB2, No.2

7 Fig 7.1 indicates the appearance of a chromosome at early prophase of mitosis.

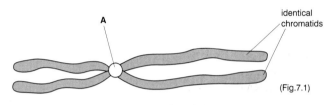

(Fig.7.1)

a With reference to Fig. 7.1,
(i) name the structure labelled **A**; *(1 mark)*
(ii) explain why the two chromatids are identical. *(2 marks)*

Fig. 7.2 represents the nucleus of an animal cell (2n = 6) at early prophase of mitosis.

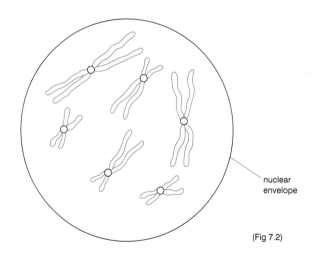

(Fig 7.2)

b On Fig. 7.2, shade **one** pair of homologous chromosomes. *(1 mark)*

c In the space below, draw an annotated diagram to indicate what happens in this cell at anaphase of mitosis. *(4 marks)*

d (i) State the number of chromosomes which would be found in a **haploid** cell in this animal. *(1 mark)*
(ii) Explain why haploid cells need to be produced during a life cycle which includes sexual reproduction. *(3 marks)*

(Total 12 marks)

OCR 2801 June 2001, B (BF), No.6

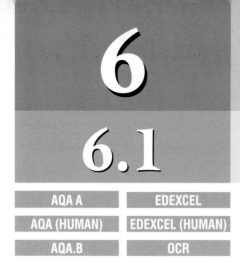

6 Exchange in organisms

6.1 Exchanges between organisms and their environment

Table 6.1 *Showing how the surface area : volume ratio gets smaller as an object becomes larger*

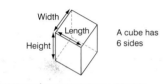

Width
Length
Height
A cube has 6 sides

Length of edge of a cube / cm	Surface area of whole cube (area of one side × 6 sides) / cm²	Volume of cube (length × width × height) / cm³	Ratio of surface area to volume (surface area ÷ volume
1	1 × 6 = 6	1 × 1 × 1 = 1	$\frac{6}{1}$ = 6.0
2	4 × 6 = 24	2 × 2 × 2 = 8	$\frac{24}{8}$ = 3.0
3	9 × 6 = 54	3 × 3 × 3 = 27	$\frac{54}{27}$ = 2.0
4	16 × 6 = 96	4 × 4 × 4 = 64	$\frac{96}{64}$ = 1.5
5	25 × 6 = 150	5 × 5 × 5 = 125	$\frac{150}{125}$ = 1.2
6	36 × 6 = 216	6 × 6 × 6 = 216	$\frac{216}{216}$ = 1.0

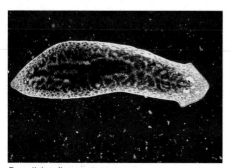

Free-living flatworm

For survival, organisms must transfer materials between themselves and their environment. Examples of things which need to be interchanged include:

- respiratory gases (oxygen and carbon dioxide)
- nutrients (glucose, fatty acids, amino acids)
- excretory products (urea)
- heat.

This exchange can take place in two ways:

- passively (no energy is required), by **diffusion** and **osmosis**
- actively (energy is required), by **active transport**, **pinocytosis** and **phagocytosis**.

6.1.1 Surface area : volume ratio

Exchange takes place at the surface of an organism, but the materials absorbed are used by the cells that mostly make up its volume. For exchange to be effective, therefore, the surface area of the organism must be large compared with its volume.

Small organisms like **protozoa** have a surface area that is large enough, compared with their volume, to allow efficient exchange across their body surface. However, as organisms become larger, their volume increases at a faster rate than their surface area (table 6.1), and so simple diffusion of materials across the surface cannot meet the needs of any but the most sluggish organisms. Even if the surface could supply enough material, it would still take too long for it to reach the middle of the organism if diffusion alone was the method of transport. To overcome this problem, organisms have evolved one or more of the following features:

- a flattened shape, where no cell is ever far from the surface (e.g. flatworm)
- a central region that is hollow or filled with non-metabolising material at its centre (e.g. the gut at the centre of an earthworm)
- Specialised exchange surfaces with large areas which increase the surface area to volume ratio.

6.1.2 Features of specialised exchange surfaces

To allow effective transfer of materials across them by diffusion or active transport, exchange surfaces have the following characteristics:

- **a large surface area : volume ratio** speeds up the rate of exchange
- **very thin** – allows materials to cross rapidly
- **permeable** – allows materials to cross without hindrance
- **moist** – dissolves diffusing substances
- **movement of the environmental medium, e.g. air**, maintains a diffusion gradient
- **movement of the internal medium, e.g. blood**, maintains a diffusion gradient.

The relationship between certain of these factors is described in **Fick's Law**, which is expressed as:

diffusion is proportional to:

$$\frac{\text{surface area} \times \text{difference in concentration}}{\text{distance over which diffusion occurs}}$$

Being thin, specialised exchange surfaces are easily damaged, therefore they are often located inside an organism for protection. This is particularly important in organisms living on land, as the surface would quickly dry out if it were outside the body. Where an exchange surface is located inside the body, the organism needs to have a means of moving the external medium over the surface, e.g. a means of ventilating the lungs in a mammal.

Albino axolotol showing external gills

6.1.3 Epithelial tissues

Exchange often occurs across epithelial tissues that comprise either a single layer of cells **(simple epithelium)** or many layers **(compound epithelium)**. The bottom layer of cells is always attached to a basement membrane of **collagen** fibres, which holds the cells together. The cells are further 'glued' together by small amounts of an intercellular substance.

Simple epithelia are of three types – squamous, cuboidal and columnar (table 6.2).

Table 6.2 *Simple epithelia*

Type of epithelium and where it is found	Description of epithelium	Appearance	Appearance as seen by light microscope
Squamous epithelium Found in: Alveoli of the lungs Bowman's capsule of the kidney Capillary walls	Flat cells with little cytoplasm Irregular in shape, often so thin that the central nucleus forms a bump on the surface	(a) Surface view — Nucleus, Cell membrane, Closely packed cells – no intercellular substance (b) Longitudinal view — Nucleus, Basement membrane, Protoplasmic bridge	
Cuboidal epithelium Found in: Kidney nephrons Salivary glands	Cube-shaped cells with a central nucleus Appear to have flattened sides due to close packing of the cells	Cell membrane, Cytoplasm, Nucleus, Basement membrane	
Columnar epithelium Found in: Stomach lining Ileum lining	Tall, narrow cells with a nucleus near the base Surface area often increased by microvilli Often found alongside mucus-secreting goblet cells	Basement membrane, Goblet cell, Microvilli forming brush border, Nucleus	

Stomata on a rose leaf (SEM) (×500)

Although plants must exchange gases between themselves and their environment, their needs are less than those of mammals of similar size because:

* some of the carbon dioxide needed for photosynthesis is supplied directly by respiring cells
* in light, oxygen for respiration is provided by those cells carrying out photosynthesis
* their energy requirements, and therefore their oxygen needs, are less because they do not move from place to place
* in woody plants, much of the bulk of the plant is non-living tissue, which therefore neither requires, nor produces, gases.

In the day, provided photosynthesis is taking place, flowering plants take in carbon dioxide and remove oxygen. At night, the reverse is normally true, with oxygen being absorbed and carbon dioxide diffused out.

Flowering plants have a very large surface area compared with the volume of living tissue. Also, no living cell is far from the surface, and therefore a source of oxygen and carbon dioxide. Because of this, no specialised transport system is needed for gases, which simply move in and through the plant by diffusion. Most gaseous exchange occurs in the leaves, which show the following adaptations for rapid diffusion:

* a thin, flat shape which provides a large surface area
* many small pores, called **stomata**, in the epidermis (Figs 6.1 and 6.2)
* control over the opening and closing of stomata
* numerous interconnecting air-spaces.

6.2.1 Stomata

Stomata are minute pores which occur mainly, but not exclusively, on the leaves, especially the underside. Each stoma (singular) is surrounded by a pair of **guard cells**, which can open and close the stomatal pore (Fig 6.1). In this way they control the rate of gaseous exchange. The guard cells surrounding each stomatal pore are bean-shaped and, unlike other leaf epidermal cells, they contain chloroplasts. The inner walls of the guard cells are thicker and less elastic than the outer ones.

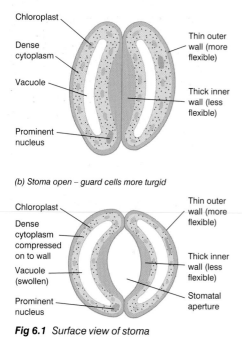

(a) Stoma closed – guard cells less turgid

Chloroplast
Dense cytoplasm
Vacuole
Prominent nucleus
Thin outer wall (more flexible)
Thick inner wall (less flexible)

(b) Stoma open – guard cells more turgid

Chloroplast
Dense cytoplasm compressed on to wall
Vacuole (swollen)
Prominent nucleus
Thin outer wall (more flexible)
Thick inner wall (less flexible)
Stomatal aperture

Fig 6.1 *Surface view of stoma*

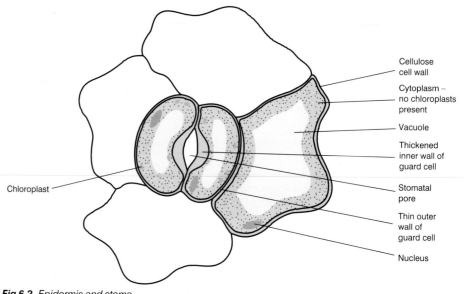

Cellulose cell wall
Cytoplasm – no chloroplasts present
Vacuole
Thickened inner wall of guard cell
Stomatal pore
Thin outer wall of guard cell
Nucleus
Chloroplast

Fig 6.2 *Epidermis and stoma*

6.2.2 Opening and closing of stomata

A number of factors control the opening and closing of stomata (Fig 6.3), including the concentration of carbon dioxide in the leaf air-spaces and the availability of water. The most important factor, however, is light. Stomata tend to open in the light and close in the dark. One suggested mechanism of stomatal movement which explains this observation is:

- light activates ATPase, an enzyme that increases the production of **ATP** by the chloroplasts in the guard cells
- the increased ATP provides more energy for the active transport of potassium (K^+) ions into the guard cells
- the solute concentration of the guard cells increases, due to the increase in K^+ ions drawing more water into them by osmosis
- the extra water causes the guard cells to become more turgid and to swell
- the thinner outer and thicker inner walls of the guard cells cause them to bow outwards, and so widen the stomatal aperture.

Stoma is closed in the dark, but in the presence of light, ATPase is stimulated to convert ADP to ATP, which provides the energy to pump out hydrogen ions (protons) from the guard cells. These protons return on a carrier which also brings chloride ions (Cl^-) with it. At the same time, potassium ions (K^+) also enter the guard cells

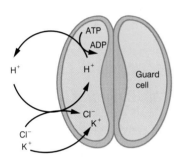

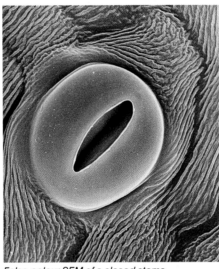

False-colour SEM of a closed stoma

As a result of this influx of ions, the water potential of the guard cells becomes more negative (lower), causing water to pass in by osmosis. The resulting increase in water potential causes the stoma to open

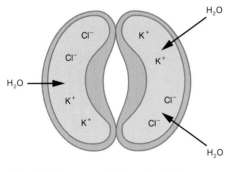

In the dark, the movement of ions and water is reversed

Fig 6.3 *Mechanism of stomatal opening*

SUMMARY TEST 6.2

Stomata are found in leaves and herbaceous stems. Each **(1)** is made up of a pair of **(2)** - shaped guard cells that have an inner wall that is **(3)** and less **(4)** than the outer one. Stomata usually open in the light and close in the dark. One theory to explain this states that in the light the water potential of the guard cells increases due to more **(5)** ions entering the guard cells by the process of **(6)**. This increased water potential causes water to enter the guard cells by **(7)**. Due to the uneven thickening of their walls the guard cells therefore bend more and the stomatal **(8)** widens. Light is thought to stimulate the influx of ions into guard cells by stimulating the enzyme **(9)** to increase the production of **(10)** by the **(11)** of the guard cells. This is then used as the energy source to pump ions into the guard cells.

Gaseous exchange in aquatic organisms

Organisms that exchange gases directly with the water in which they live, typically use two mechanisms:

- Smaller organisms, e.g. **protozoa**, use the body surface to exchange gases.
- Larger organisms, e.g. fish, use specialised organs called gills.

6.3.1 Gaseous exchange in protozoa

Protozoans such as *Amoeba* have a large surface area : volume ratio. Oxygen is therefore absorbed by diffusion over the whole of their body surface, which is covered by a thin cell membrane. In the same way, carbon dioxide from respiration diffuses out into the surrounding water.

6.3.2 Gaseous exchange in a bony fish

The gaseous exchange system in teleost (bony) fish is the gills, which are made up of **gill filaments** (Fig 6.4). The gill filaments are stacked up in a pile, much in the way that pages are in a book. At right angles to the filaments are **gill lamellae**, which increase the surface area of the gills. The gills are located within the body of the fish, behind the head. They are protected by a bony flap called the **operculum**, which has an opening called the **opercular valve**.

6.3.3 Ventilation of the gills

To draw water over the gills, the opercular valve is closed and the mouth is opened. The floor of the buccal cavity (the chamber in which the gills are situated) is then lowered, thus reducing pressure in the buccal cavity and drawing water in through the mouth. As this is happening, the operculum is moved slightly away from the body, drawing some water out of the buccal cavity and across the gills. To remove water, the mouth is closed and the opercular valve is opened. The floor of the buccal cavity is then raised, increasing the water pressure within the buccal cavity and forcing water over the gills and out of the body through the opercular valve. These events are summarised in figure 6.5.

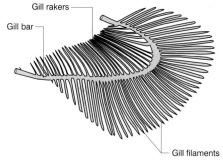

Fig 6.4 *A single fish gill*

Table 6.3 *Adaptations of three organisms to obtaining oxygen from water in which oxygen levels are low*

Worm (*Tubifex*)	Diving beetle (*Dytiscus*)	Mayfly nymph (*Cleon*)
Haemoglobin that is saturated with oxygen at low oxygen levels	Stores air amongst hairs on its abdomen	Has a series of gills along the body
No Bohr effect	Lives amongst pond weed where oxygen levels are highest due to photosynthesis of weeds	Gills vibrate to help bring fresh supplies of oxygen-loaded water into contact with them
Respires sugars anaerobically	Surfaces periodically to collect fresh air	Gills vibrate faster when oxygen levels are low

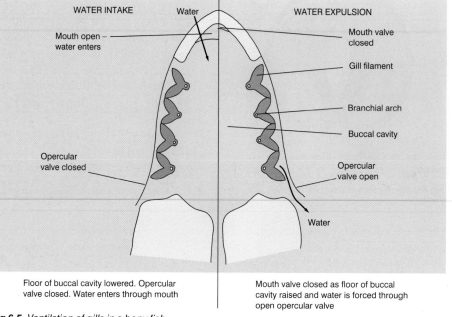

Floor of buccal cavity lowered. Opercular valve closed. Water enters through mouth

Mouth valve closed as floor of buccal cavity raised and water is forced through open opercular valve

Fig 6.5 *Ventilation of gills in a bony fish*

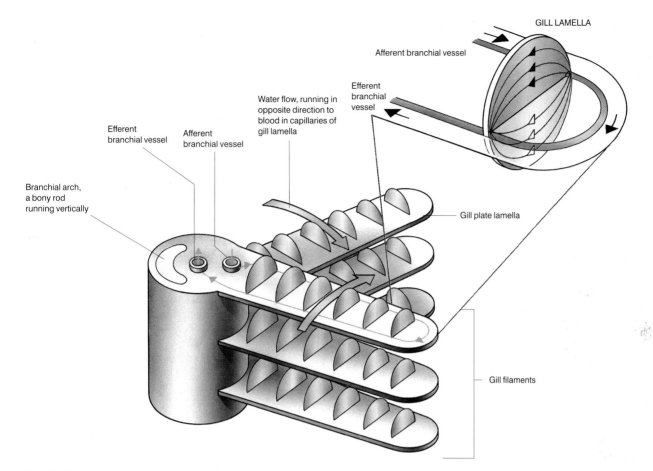

Fig 6.6 *Water flow over the gills of a bony fish*

6.3.4 The counter-current exchange principle

The exchange of gases between water and blood in the gills of fish is a good example of the counter-current exchange principle. The essential feature of this system is that the blood and the water which flow over the gill lamellae do so in opposite directions (counter-current) (Fig 6.6). This arrangement means that blood which is already well loaded with oxygen meets water which has its maximum concentration of oxygen. Diffusion of oxygen from the water to the blood can occur. Similarly, blood with little or no oxygen in it meets water which has had most, but not all, of its oxygen removed. Again, diffusion of oxygen from the water to blood takes place. There is therefore a fairly constant rate of diffusion all the way across the gill lamella. In this way, around 80% of the oxygen available in the water is absorbed into the blood of the fish (Fig 6.7). If the flow of water and blood had been in the same direction (parallel), as it is in the cartilaginous fish, then only 50% of the available oxygen would be absorbed by the blood.

Table 6.4 *Water and air as repiratory media*

Property	Water	Air
Oxygen content	<1%	21%
Oxygen diffusion rate	Low	High
Density	High	Low
Viscosity	High	Low

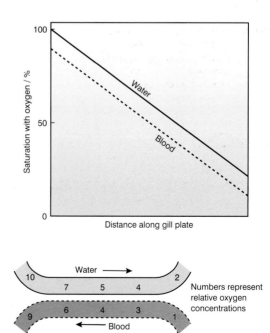

Fig 6.7 *Counter-current exchange in the gills of a bony fish*

109

6.4 Structure of human gaseous exchange system

In mammals, gaseous exchange takes place in the lungs. To avoid excessive water loss, the lungs are situated deep inside the thorax (Fig 6.8). This means that, to be effective, the air must be continually passed in and out of the lungs, i.e. they must be ventilated.

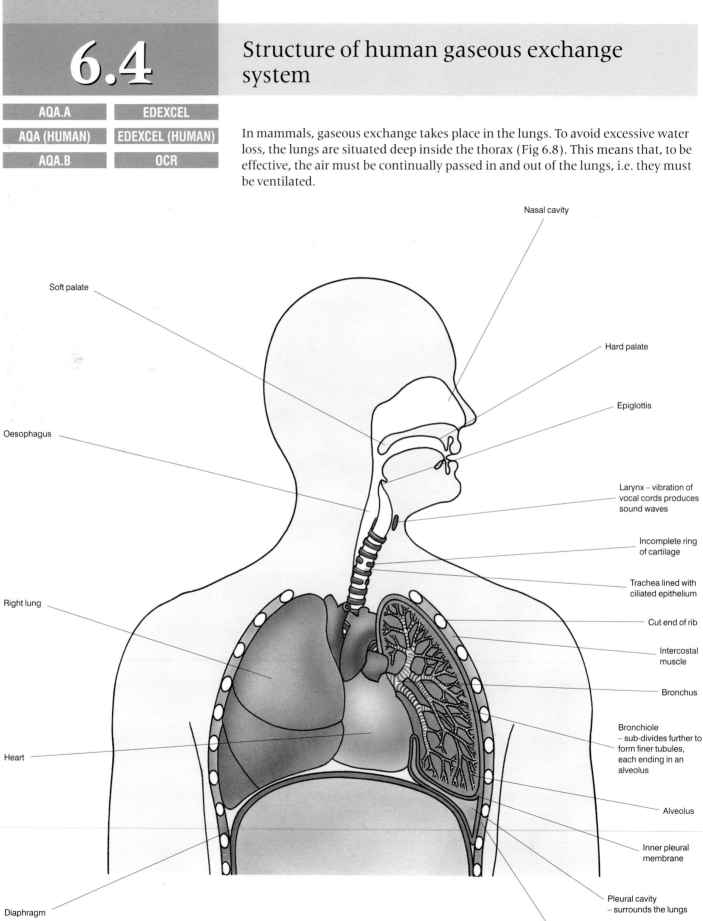

Nasal cavity

Soft palate

Hard palate

Epiglottis

Oesophagus

Larynx – vibration of vocal cords produces sound waves

Incomplete ring of cartilage

Trachea lined with ciliated epithelium

Right lung

Cut end of rib

Intercostal muscle

Bronchus

Bronchiole – sub-divides further to form finer tubules, each ending in an alveolus

Heart

Alveolus

Inner pleural membrane

Pleural cavity – surrounds the lungs

Diaphragm

Outer pleural membrane

Fig 6.8 *Structure of human gas exchange system*

6.4.1 Structure of the human gaseous exchange system

- **Lungs** – a pair of lobed structures made up of highly branched tubules called **bronchioles**, which end in tiny air-sacs called **alveoli**.
- **Ribs** – 12 pairs of curved bones that are attached to the vertebrae at the back. They form a protective cage for the delicate lungs and heart within. The 10 upper pairs of ribs are attached to the sternum at the front, while the lower two pairs are unattached or 'floating'.
- **Intercostal muscles** – two sets of muscles found between the ribs, which move the rib-cage, to help ventilate the lungs.
- **Pleural cavity** – a narrow, air-tight, cavity that is bounded by two membranes which secrete **pleural fluid** into the cavity. This fluid acts as a lubricant, reducing friction between the lungs and the rib-cage during ventilation.
- **Diaphragm** – a sheet of muscle that separates the thorax from the abdomen. Its movements also aid the ventilation of the lungs.
- **Nasal cavity** – a cavity behind the nose that is lined by a mucous membrane which secretes mucus from the **goblet cells** it contains. This mucus helps to clean the inhaled air by trapping dust and other particles on its sticky surface. The incoming air is also warmed as it passes through the nasal cavity.
- **Epiglottis** – a flap of cartilage that closes the trachea during swallowing and so prevents choking caused by food entering the trachea.
- **Larynx** – or **voice box** – is a cartilaginous box that has a number of ligaments stretched across it. These are known as the **vocal cords**, and they vibrate to produce sound when expired air passes over them.
- **Trachea** – a flexible air-way that is supported by horseshoe-shaped pieces of cartilage. The cartilage prevents the trachea collapsing when the air pressure inside it is lowered during inspiration (breathing in). The trachea is made up of **smooth muscle** lined with ciliated epithelium and goblet cells. The goblet cells produce the mucus that traps particles from inspired air, while the cilia move the mucus laden with dirt and microorganisms up to the throat, where it passes down into the stomach, via the oesophagus.
- **Bronchi** – the trachea divides into two bronchi, each leading to one lung. The structure of the bronchi resembles that of the trachea and they too protect the alveoli from damage by having mucus that traps dirt, and cilia that move this dirty mucus away from the alveoli.
- **Bronchioles** – the branching sub-divisions of the bronchi. The larger bronchioles are supported by cartilage, but this gradually disappears as bronchioles get smaller. Their walls are made of smooth muscle lined with cuboidal epithelial cells. The muscle allows them to constrict, and so control the flow of air in and out of the alveoli.
- **Alveoli** – minute air-sacs with a diameter of between 100μm and 300μm. They contain some collagen and elastic fibres, and are lined with squamous epithelium. The elastic fibres allow the alveoli to stretch as they fill with air during inspiration and then recoil during expiration in order to expel the carbon-dioxide-enriched air. The collagen fibres add strength to the thin-walled alveoli and so prevent them bursting during inspiration.

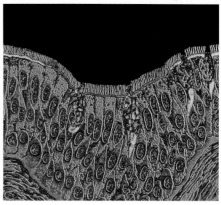

Trachea epithelium (×600)

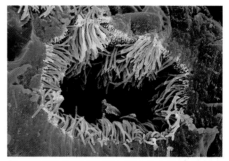

Section of human trachea showing cilated epithelium (×3570 approx.)

SUMMARY TEST 6.4

In humans, air is drawn into the lungs due to the combined movements of the **(1)** muscles found between the ribs and the **(2)**, a sheet of muscle separating the thorax from the abdomen. The air first passes through the nasal cavity which contains mucus secreted by **(3)** cells. The mucus functions to **(4)** the incoming air which is also **(5)** as it passes through the cavity. The air passes the **(6)** that covers the trachea during swallowing, continuing through the **(7)** that contains the vocal cords.

Gaseous exchange in humans

The sites of gaseous exchange in humans are the alveoli, minute air-sacs some 100–300µm in diameter.

6.5.1 Role of alveoli in gaseous exchange

The 300 million alveoli in both lungs of a human have a total surface area of around 50m². Their structure is shown in figures 6.9 and 6.10. Each alveolus is lined mostly with squamous epithelial cells only 0.2µm thick; their structure is shown in section 6.1.3. Around each alveolus is a network of pulmonary capillaries, so narrow (7–10µm) that red blood cells are flattened against the thin capillary walls in order to squeeze through. A review of Fick's Law (section 6.1.1) tells us that diffusion of gases between the alveoli and the blood will be very rapid because:

*   the red blood cells are slowed as they pass through the pulmonary capillaries, allowing more time for diffusion
*   the distance between alveolar air and red blood cells is reduced as the red blood cells are flattened against the capillary walls
*   the walls of both the alveoli and the capillaries are very thin
*   the alveoli and blood capillaries have a very large total surface area
*   breathing movements constantly ventilate the lungs, and the action of the heart constantly circulates blood around the alveoli. Together, these ensure that the concentration gradient of the gases to be exchanged is maintained.

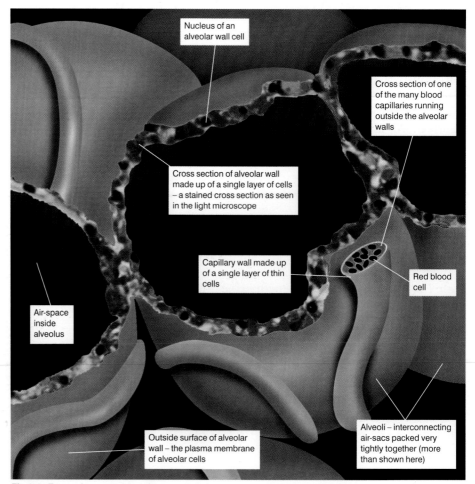

Fig 6.9 *External appearance of a group of alveoli (×300 approx.)*

6.5.2 Lung surfactants and their function

The walls of the alveoli are made up of two types of cell (Fig 6.11):

- **Type I pneumocytes** – these are the squamous epithelial cells referred to in section 6.1.3.
- **Type II pneumocytes** – a special type of epithelial cell which produces a chemical called a **surfactant**.

Surfactants are chemicals that reduce the surface tension of substances, most commonly water. They are best known for their role in detergents. As you are probably aware, detergents such as washing-up liquid make things less sticky and more slippery. The surfactants in the lungs perform exactly the same function, only in this case it is to prevent the surfaces of the alveoli sticking to each other, rather than preventing grease sticking to a plate. Without the phospholipid–protein surfactants of the lung, the moist alveolar surfaces would stick together, making it difficult, if not impossible, to inflate the lungs.

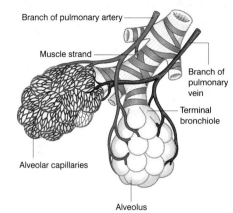

Fig 6.10 *Alveoli*

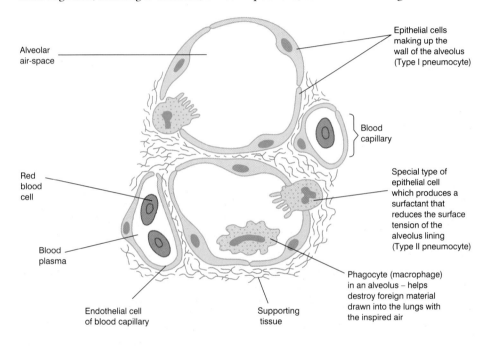

Fig 6.11 *Arrangement of cells and tissues in human alveoli*

SUMMARY TEST 6.5

Each minute alveolus in the lungs is made up of **(1)** fibres to give them strength, **(2)** fibres to allow them to stretch and squamous epithelium also known as **(3)**. The squamous epithelial cells are only **(4)** thick and so allow rapid **(5)** of gases across them. Each alveolus is surrounded by a network of **(6)** capillaries that are around **(7)** in diameter, causing **(8)** within them to be flattened against their surface, thus improving the rate of exchange of gases between themselves and the alveoli. The alveoli also contain **(9)** that produce a chemical, **(10)**, that reduces the surface tension of their walls and therefore makes them less **(11)**. Within the alveoli can be found **(12)** cells that help to destroy foreign material brought into the lungs during breathing.

6.6 Ventilation and its control in humans

To maintain diffusion of gases across the alveolar walls, air must be constantly moved in and out of the lungs. This process, which we call breathing, is also known as **pulmonary ventilation**. When the pressure of the atmosphere is greater than that inside the lungs **(pulmonary pressure)**, air is forced into the alveoli. This is called **inspiration** (inhalation). When the pressure in the lungs is greater than that of the atmosphere, air is forced out of the lungs. This is called **expiration** (exhalation).

6.6.1 Inspiration

Breathing in is an active process (it uses energy) and occurs as follows:
- the external intercostal muscles contract, while the internal muscles relax
- the ribs are pulled upwards and outwards, increasing the volume of the thorax
- the diaphragm muscles contract, causing it to flatten, which also increases the volume of the thorax
- the increased volume of the thorax results in reduction of pressure in the lungs
- atmospheric pressure is now greater than pulmonary pressure, and so air is forced into the lungs.

The pleural membranes and fluid between them are crucial to this process, for two reasons:
- They link the outer wall of the lungs to the inner wall of the thorax; any expansion of the thorax therefore results in expansion of the lungs.
- They lubricate movements between the lung and thorax walls.

6.6.2 Expiration

Breathing out is a largely passive process (it does not require much energy) and occurs as follows:
- the internal intercostal muscles contract, while the external intercostal muscles relax
- the ribs move downwards and inwards, decreasing the volume of the thorax
- the diaphragm muscles relax, making it return to its upwardly domed position, again decreasing the volume of the thorax
- the decreased volume of the thorax increases the pressure in the lungs
- the pulmonary pressure is now greater than that of the atmosphere, and so air is forced out of the lungs.

During normal quiet breathing, the recoil of the elastic lungs is the main cause of air being forced out (like air being expelled from a partly inflated balloon). Only under more strenuous conditions such as exercise do the various muscles play a part.

6.6.3 Control of ventilation

The rate of breathing in humans can be controlled voluntarily, for example when speaking or deliberately sucking or blowing. The basic breathing rate, however, is involuntary and is controlled by a region of the brain called the **medulla oblongata**. The medulla oblongata contains a breathing centre, which has two components:
- an **inspiratory centre** which controls breathing in
- an **expiratory centre** which controls breathing out.

These centres have a basic rhythm that can be influenced by other factors. These factors include:

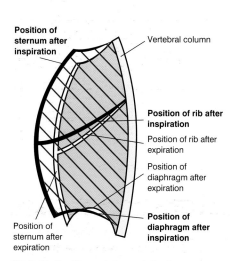

Position of sternum after inspiration
Vertebral column
Position of rib after inspiration
Position of rib after expiration
Position of diaphragm after expiration
Position of sternum after expiration
Position of diaphragm after inspiration

Fig 6.12 *Positions of ribs and diaphragm during inspiration and expiration*

Table 6.5 *Comparison of inspired, alveolar and expired air*

Gas	% Composition by volume		
	Inspired	Alveolar	Expired
Oxygen	20.95	13.8	16.4
Carbon dioxide	0.04	5.5	4.0
Nitrogen	79.01	80.7	79.6

- **carbon dioxide** – any increase in the carbon dioxide level of the blood increases the breathing rate
- **pH** – a fall in the pH of the arterial blood increases the breathing rate
- **oxygen** – a decreased concentration in arterial blood can increase the breathing rate
- **stretch receptors in the lungs** – these are stimulated when the lungs are expanded, and they cause a reduction in breathing rate as protection against over-inflating the lungs.

The breathing centre and the factors that change its base rhythm operate together. If, for example, the carbon dioxide level in arterial blood rises above its normal level, this indicates that it is not being removed by the lungs rapidly enough. An increase in the rate of breathing is required. **Chemoreceptors** which occur in the **carotid and aortic bodies** – found in the carotid artery and the aorta respectively – sense the rise in the level of carbon dioxide. The chemoreceptors send nerve impulses to the inspiratory centre in the medulla oblongata which, in turn, sends impulses along the **phrenic and thoracic nerves** to the diaphragm and external intercostal muscles, making them contract more rapidly. This increases the rate of ventilation, and so more carbon dioxide is removed by the alveoli. When the level of carbon dioxide in the blood falls back to normal, the chemoreceptors stop stimulating the inspiratory centre and the ventilation rate returns to normal.

When the lungs are full of air they become expanded, and this stimulates stretch receptors in their walls to send impulses along the **vagus nerve** to the expiratory centre of the medulla oblongata. The inspiratory centre is then automatically 'switched off', and expiration occurs. When the lungs are deflated, the stretch receptors are no longer stimulated, the expiratory centre is 'switched off', and the inspiratory centre is 'switched on'. Inspiration then takes place, and so the cycle continues.

6.6.4 Lung capacities and their measurement

The total capacity of the human lungs is around $5dm^3$, although this varies between $3.5dm^3$ and $6.0dm^3$ depending on the age and level of fitness of the individual. The composition of inhaled air, alveolar air and exhaled air is given in table 6.5. Figure 6.13 illustrates the various lung capacities, which are explained below. The figures in brackets are average values for humans.

- **Tidal volume** ($0.5dm^3$) – the volume of air exchanged at each breath at rest.
- **Inspiratory reserve volume** ($1.5dm^3$) – the additional volume of air that can be inspired during the maximum breath in.
- **Expiratory reserve volume** ($1.5dm^3$) – the additional volume of air that can be expired during the maximum breath out.
- **Residual volume** ($1.0dm^3$) – the volume of air that cannot be removed from the lungs, even after the

maximum breath out. This residual air is left in the bronchi, bronchioles and alveoli. It is essential that some air is left in the alveoli to prevent their moist walls from sticking together, which would make it impossible to re-inflate the lungs.
- **Vital capacity** ($3.5dm^3$) – the maximum volume of air that can be exchanged between maximum inspiration and maximum expiration.

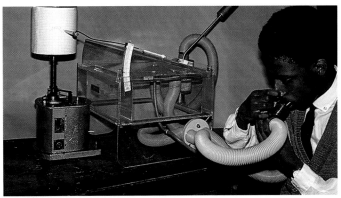

A spirometer

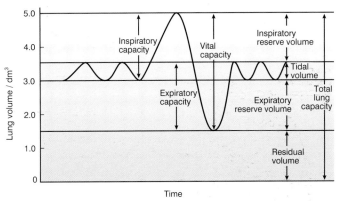

Fig 6.13 *Graph to illustrate lung volume and capacities*

VENTILATION RATE

The **ventilation rate** is equal to the tidal volume multiplied by the number of breaths each minute. Its value changes considerably during exercise.

SUMMARY TEST 6.6

During inspiration the **(1)** intercostal muscles contract and the diaphragm flattens. This **(2)** the volume of the lungs, thereby **(3)** their pressure. Atmospheric pressure is hence greater than **(4)** pressure, causing air to move into the lungs. Ventilation is controlled by the **(5)** of the brain which ensures that an increase in pH will cause the breathing rate to **(6)**. The volume of air normally moved in and out when at rest is called the **(7)** and measures **(8)** dm^3. The maximum volume that can be exchanged in one breath is the **(9)** and measures **(10)** dm^3.

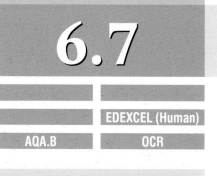

6.7

Effects of exercise and smoking on gaseous exchange

EDEXCEL (Human)

AQA.B   OCR

6.7.1 Energy and exercise

The energy required for the muscular movement that is an integral part of exercise is provided by the breakdown of various substances during cellular respiration. These substances include:

- **Glucose** – this is the usual source of energy, and provides a source of **adenosine triphosphate (ATP)**, whether or not oxygen is available.
- **Glycogen** – this is stored in the liver and muscles, and can be quickly broken down to glucose, which in turn can provide energy as above.
- **Triglycerides** – these are stored as fat in adipose tissue under the skin and around body organs. A little is stored in muscle cells. Triglycerides provide more ATP than an equivalent mass of glucose or glycogen, and are therefore important as long-term, concentrated stores of energy. This breakdown does, however, require 15% more oxygen and so, when oxygen is in short supply, glycogen stores are preferred.

Whatever the original chemical used, it has to be converted to **adenosine triphosphate (ATP)** before it can be used for muscle contraction. ATP is thus the intermediate source of energy for exercise. It can be generated in the body both **aerobically** and **anaerobically**.

6.7.2 Aerobic exercise

In aerobic exercise, the muscles use up oxygen at a rate less than the rate at which it is delivered from the lungs, via the blood. The presence of oxygen means that the muscles produce ATP by the complete oxidation of glucose to carbon dioxide and water. The process releases 40% of the energy available in the glucose molecules. Most moderate exercise such as walking, swimming and jogging is aerobic.

6.7.3 Anaerobic exercise

Anaerobic exercise includes very strenuous exercise that can only be sustained for a short period of time. Running the 100 metres or lifting heavy weights are examples of anaerobic exercise. The muscles use up oxygen faster than it can be supplied, building up an **oxygen debt**. In the absence of oxygen, glucose is broken down, not to carbon dioxide and water, but to lactate (lactic acid). The lactate quickly causes muscle fatigue and cramp if allowed to accumulate. It is therefore rapidly broken down to carbon dioxide and water once the supply of oxygen outstrips its consumption. Although this process releases only 2% of the energy of the glucose molecules, it does allow the muscles to continue to operate despite the lack of an adequate oxygen supply.

6.7.4 Effects of exercise on gaseous exchange

During exercise, the demand for oxygen by muscle cells goes up. To meet the demand, the ventilation rate must be increased by:

- increasing the number of breaths taken each minute, from about 15 at rest to 45 during exercise
- increasing the volume of air exchanged at each breath, from around $0.5dm^3$ at rest to $3.5dm^3$ during exercise.

During moderate exercise such as jogging, the amount of oxygen absorbed by the lungs can increase 13 times, from $0.03dm^3\ min^{-1}$ to $4.0dm^3\ min^{-1}$.

How the increased ventilation rate is triggered is complex, but includes some or all of the following factors:
- **Lowering of the blood pH** due to an increase in lactate (lactic acid) resulting from the anaerobic breakdown of glucose in muscle cells. The presence of more carbon dioxide in the blood also lowers blood pH.
- **Increased production of carbon dioxide** as a result of aerobic breakdown of glucose in muscle cells.
- **Lowering of the oxygen concentration** of the blood as it is used up by muscle cells.
- **Receptors in muscles and joints** stimulating the inspiratory centre in the brain.

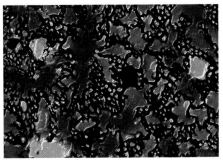

Healthy lung tissue

6.7.5 Long-term consequences of exercise

Through a process of training, individuals can become physically more fit and more able to undertake exercise without ill effects. Good health is more than the mere absence of disease, it is an active state that can be more readily achieved by being physically fit.

The benefits include:
- increased ventilation rate
- larger lung **vital capacity**
- increased diffusion rates in the alveoli
- larger blood volume
- improved **cardiac output** (heart rate)
- stronger and more efficient muscles and **tendons**
- stronger bones and more flexible **ligaments**
- improved endurance and less fatigue
- better mental health.

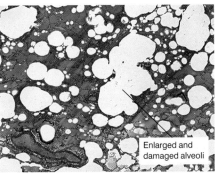

Enlarged and damaged alveoli

Lung tissue damaged by emphysema

6.7.6 Effects of tobacco smoke on gaseous exchange

Tobacco smoke is a complex mixture of chemicals, of which three particularly threaten health:
- **Nicotine** – causes coronary heart disease and increases the risk of strokes and heart attacks.
- **Carbon monoxide** – found also in the exhaust gases of vehicles, it lowers the oxygen-carrying capacity of the blood by combining with haemoglobin more easily than oxygen does.
- **Tars** – cause chronic bronchitis, emphysema and cancer, especially of the lungs.

Tobacco smoking causes 15–20% of all deaths in Britain – around 100 000 extra deaths each year. The three main diseases which affect the lungs, and therefore gaseous exchange, are:
- **Chronic bronchitis** – caused by the tars in tobacco smoke irritating the epithelial lining of the bronchi and bronchioles, which therefore produce more mucus. The cilia lining these tubes are damaged and do not function to remove the mucus, which therefore fills and blocks the bronchioles and alveoli, causing the victim breathlessness and coughing.
- **Lung cancer** – tars may induce the epithelial cells lining the bronchi to divide, forming a bronchial carcinoma, which therefore restricts the passage of air in and out of the lungs. Such cancers may completely disrupt lung function, and spread to other parts of the body.
- **Emphysema** – is a condition in which the walls of the alveoli become less elastic and begin to break down. Not only does the lack of elasticity make breathing out more difficult, but the breakdown of the alveoli reduces the surface area available for gaseous exchange, leading to breathlessness and fatigue.

SMOKING AND PREGNANCY

Tobacco smoke not only affects the health of the mother, but may directly affect the unborn foetus, which therefore has an increased risk of:

- spontaneous abortion
- congenital abnormalities
- being stillborn
- mental retardation
- being physically smaller

6.8 Digestion

EDEXCEL
EDEXCEL (HUMAN)
AQA.B

In holozoic organisms, the breakdown of food usually takes place in two stages:
- mechanical breakdown
- chemical digestion.

6.8.1 Mechanical breakdown (Mastication)

To be ingested, food must be small enough to be taken into the body. If the original source of food is large, it is broken down into smaller pieces, using structures such as **teeth**. Not only does this make it possible to ingest the food, it also provides a larger surface area for chemical digestion by enzymes.

The teeth are found in the mouth or **buccal cavity**. As humans eat food of both plant and animal origin, they are known as **omnivores**. Their teeth are therefore unspecialised, unlike those of herbivores and carnivores (unit 9.2). There are four types of teeth in the human jaw:
- **Incisors** are found at the front of the mouth, and have sharp chisel-shaped edges for biting food. There are 4 incisors in the upper jaw and 4 in the lower jaw.
- **Canines** occur further back in the mouth and, being more pointed, are used to tear food. There are 2 canines in each jaw.
- **Premolars** are situated behind the canines and their flattened tops have ridges or cusps that are used to grind the food. There are 4 premolars in each jaw.
- **Molars** occur at the back of the mouth. Larger than premolars and more ridged, they are used to crush and grind food. There are 6 molars in each jaw.

6.8.2 Chemical digestion

In section 2.1.3, we saw that enzymes are specific. It follows that, in order to break down a large macromolecule, more than one enzyme is needed. Often one enzyme is used to break up a molecule into smaller sections, which are then reduced to their component parts by one or more other enzymes. There are a number of types of digestive enzyme, three of which are particularly important:
- **carbohydrases**, which break down carbohydrates, ultimately to monosaccharides
- **lipases**, which break down lipids (fats and oils) to glycerol and fatty acids
- **proteases**, which break down proteins, ultimately to amino acids.

Carbohydrases

The carbohydrate starch (amylose) is initially **hydrolysed** by the enzyme **amylase** into the disaccharide, maltose. Amylase is produced in two places: by the salivary glands which secrete it into the mouth, and by the pancreas, from where it passes along the pancreatic duct and into the duodenum. In turn, maltose is hydrolysed by the enzyme **maltase** to give the monosaccharide, glucose. Maltase is secreted by the wall of the ileum (small intestine) and acts in the ileum.

Lipases

Fats and oils are broken down by lipases, enzymes that hydrolyse the ester bonds found in triglycerides. Lipases are produced by the pancreas and act in the duodenum.

Proteases

Proteins are large, complex molecules and the protease enzymes that digest them are divided into two groups:
- **endopeptidases**, which hydrolyse the peptide bonds between amino acids in the central region of the protein molecule, reducing it to a series of peptide molecules

- **exopeptidases**, which hydrolyse the peptide bonds on the terminal amino acids of these peptide molecules, progressively reducing them to individual amino acids. The exopeptidases are of two types:
  - **aminopeptidases**, which work at the end of the peptide chain, which has a free amino ($-NH_2$) group
  - **carboxypeptidases**, which work where the free group at the end of the peptide chain is a carboxyl ($-COOH$) group.

6.8.3 Bile

Bile is a complex green fluid produced by the liver. While it contains no enzymes, it is nevertheless made up of two groups of substances important in digestion:

- **Mineral salts** such as sodium hydrogen carbonate neutralise the acid material from the stomach. This provides a more neutral pH for the enzymes of the small intestine to work in.
- **Bile salts** mechanically break down lipids into tiny droplets. The process, called **emulsification**, produces a much larger surface area for lipase from the pancreas to work on.

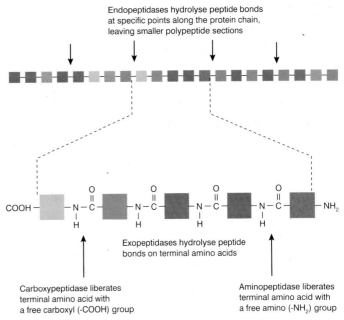

Fig 6.14 Action of endo- and exopeptidases

Table 6.6 *Summary of chemical digestion*

Enzyme group	Enzyme name	Source of enzyme	Site of action	Effect
Carbohydrases	Amylase	Salivary glands	Mouth	Starch (amylose) → maltose
		Pancreas	Ileum	Starch (amylose) → maltose
	Maltase	Ileum wall	Ileum	Maltose → glucose
	Lactase	Ileum wall	Ileum	Lactose → glucose and galactose
	Sucrase	Ileum wall	Ileum	Sucrose → glucose and fructose
Lipases	Lipase	Pancreas	Duodenum	Fats → fatty acids and glycerol
Proteases	Endopeptidases e.g. pepsin	Stomach wall	Stomach	Protein → peptides
	Exopeptidases e.g. aminopeptidases	Ileum wall	Ileum	Peptides → amino acids

SUMMARY TEST 6.8

Mechanical digestion is also known as **(1)** and is performed by the teeth located in the mouth, which is also known as the **(2)**. From the back of the mouth, towards the front, the four types of teeth, in order, are **(3)**, **(4)**, **(5)**, **(6)**. Chemical digestion is carried out by **(7)**, of which the three main groups are carbohydrases, proteases and **(8)**. An example of a carbohydrase is lactase, which is produced by the **(9)** and breaks down lactose into **(10)** and **(11)**. Of the two main types of protease, exopeptidases work on the **(12)** bonds at the end of an amino acid chain. Exopeptidases are of two types, namely **(13)** and **(14)**, depending on at which end of the chain they operate. The breakdown of lipids into fatty acids and **(15)** by the process of **(16)** is aided by bile salts, which increase the surface area of the lipids by a process called **(17)**.

6.9

The alimentary canal

EDEXCEL

EDEXCEL (HUMAN)

AQA.B

The alimentary canal is a long muscular tube which is around 10 metres in length in adult humans. It extends from the mouth to the anus, and a variety of secretory glands are associated with it (Fig 6.15).

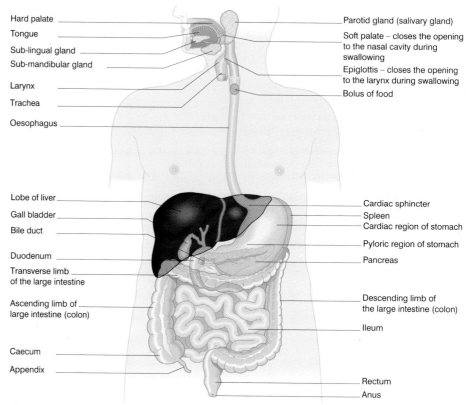

Hard palate
Tongue
Sub-lingual gland
Sub-mandibular gland
Larynx
Trachea
Oesophagus

Parotid gland (salivary gland)
Soft palate – closes the opening to the nasal cavity during swallowing
Epiglottis – closes the opening to the larynx during swallowing
Bolus of food

Lobe of liver
Gall bladder
Bile duct
Duodenum
Transverse limb of the large intestine
Ascending limb of large intestine (colon)
Caecum
Appendix

Cardiac sphincter
Spleen
Cardiac region of stomach
Pyloric region of stomach
Pancreas
Descending limb of the large intestine (colon)
Ileum
Rectum
Anus

Fig 6.15 *Human digestive system*

6.9.1 The oesophagus

The oesophagus carries food from the mouth to the stomach. It is therefore adapted for transport rather than for digestion or absorption, and lacks many of the features characteristic of the rest of the alimentary canal. The oesophagus is characterised by:
- having no villi or microvilli to increase its surface area
- lacking any secretory glands
- a thick wall made up of circular and longitudinal muscle.

6.9.2 The stomach

The stomach is a muscular sac which has an inner folded layer, known as the **gastric mucosa**, which produces enzymes. The role of the stomach is to store and digest food, especially proteins; it is adapted to this function by:
- possessing gastric glands that produce enzymes to digest proteins
- producing hydrochloric acid to provide a low pH for gastric enzymes and to kill microorganisms that might cause diseases such as food poisoning
- having a muscular wall to churn and mix the food with the gastric juice
- producing mucus to line the wall, thus protecting it from being digested by its own enzymes
- lacking features associated with absorption, such as villi and microvilli
- releasing food into the rest of the alimentary canal gradually over many hours,

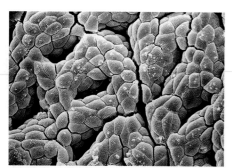

Gastric mucosa of the stomach (SEM) (×450 approx.)

in order to provide a steady supply of nutrients and avoid the need for continuous eating.

6.9.3 The duodenum

From the stomach, food enters the duodenum, which is adapted for the further digestion of the food and for some absorption of the products of this breakdown. These adaptations include:

- the presence of Brunner's glands, which produce mineral salts to provide the more neutral pH suited to the digestive enzymes in this region
- a muscular wall to move food along and mix it with the secretions of its wall, the pancreas and the liver
- a wall folded into villi and containing epithelial cells with microvilli, both of which increase the surface area and speed up absorption.

6.9.4 The ileum

The final stages of digestion take place in the ileum, but its main function is to absorb the soluble products of digestion, such as sugars, amino acids, fatty acids, glycerol and minerals. The structure of the ileum wall is illustrated in figure 6.16 and the photograph. Its adaptations include:

- digestive enzymes such as maltase and sucrase are bound to the membranes of the epithelial cells
- the villi contain fibres of **smooth muscle** that contract and relax rhythmically to keep food in contact with enzymes on the ileum wall
- an immense surface area is provided both for the attachment of enzymes and the absorption of the soluble products of digestion. This is achieved by:
  - the length of the ileum – some 5 metres in humans and 45 metres in cattle
  - multiple folding of the ileum wall
  - large numbers of finger-like projections called **villi**
  - minute projections called **microvilli** on the epithelial cells, which collectively form a **brush border**
- the villi are well supplied with blood vessels, to carry away the absorbed products and so ensure that a concentration gradient for diffusion is maintained
- there is a very thin epithelial layer, which speeds absorption by reducing the distance across which diffusion occurs
- the fatty acids and glycerol are absorbed separately into lacteals, rather than blood capillaries
- epithelial cells are rich in mitochondria, which are required to generate the energy needed for absorption by active transport.

6.9.5 Movement of food along the alimentary canal

In figure 6.16 you can see that the wall of the alimentary canal is made up of two layers of smooth muscle:

- An outer layer of **longitudinal muscle**, fibres running along the length of the canal. When these contract, the length of the canal becomes shorter.
- An inner layer of **circular muscle**, fibres running around the circumference of the canal. When these contract, the diameter of the canal becomes smaller.

Food is moved along the alimentary canal by waves of contraction of these muscles moving down from the mouth towards the anus. Food is formed into lumps, each known as a **bolus**. In front of the bolus, circular muscles relax and longitudinal muscles contract, opening the cavity (the **lumen**) of the canal, so that the bolus meets little resistance from the walls of the canal. Behind the bolus, the circular muscle contracts and the longitudinal muscle relaxes. This constricts the lumen behind the bolus, and so moves the food along the alimentary canal. The process is called **peristalsis**.

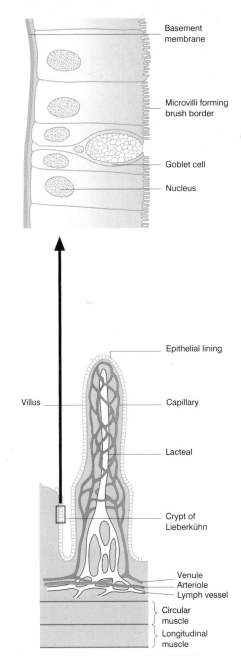

Fig 6.16 Intestinal wall, showing villi (LS)

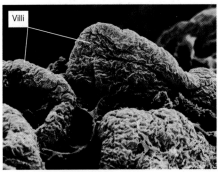

Lining of the ileum (SEM) (×250)

121

1 The graph shows some of the changes which take place in a man during breathing.

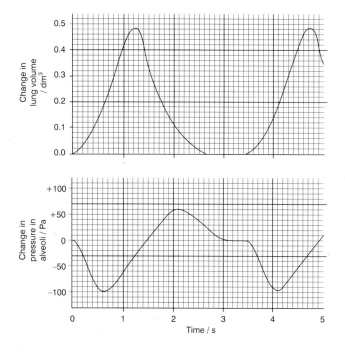

a (i) Use the information in the graph to calculate the man's rate of breathing in breaths per minute. Show your working. *(2 marks)*

(ii) The volume of air in the man's lungs after he had exhaled was 2400 cm³. What is the volume of air in his lungs immediately after he had inhaled? *(1 mark)*

b Explain how muscles cause the change in alveolar pressure between 0 and 0.5s *(2 marks)*

The table shows the composition of atmospheric air and gas samples obtained while this man was breathing normally.

Sample	Percentage of total volume	
	Oxygen	Carbon dioxide
Atmospheric air	21.0	0.0
Air from alveoli	13.8	5.5
Exhaled air	16.9	4.1

c Explain why the percentage of carbon dioxide in atmospheric air is given in the table as zero. *(1 mark)*

d (i) Not all the air that a person breathes out has been in the alveoli. Explain why. *(1 mark)*

(ii) Use your answer to part (i) to explain the difference in carbon dioxide concentration between exhaled air and air from alveoli. *(2 marks)*

e Describe how the structure of the lungs and the red blood cells enable efficient diffusion and transport of oxygen. *(6 marks)*
(Total 15 marks)
AQA June 2001, B/HB (A) BYA1, No.8

2 The diagram shows the structure of part of the ileum as seen in transverse section.

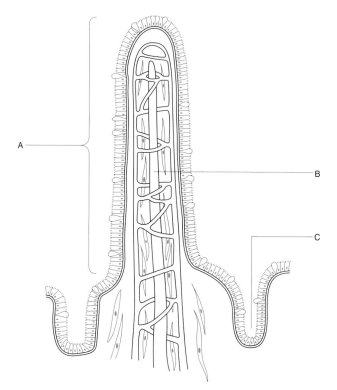

a Name the parts labelled A, B and C. *(3 marks)*

b Describe **two** ways in which the structure of part A is adapted for the absorption of the products of digestion. *(4 marks)*

c The table below lists some enzymes associated with carbohydrate digestion, their site of secretion and the products of their action.

Complete the table by filling in the blank spaces.

Enzyme	Site of secretion	Products
	Pancreas	Maltose
Lactase		
Sucrase	Lining (mucosa) of ileum	

(4 marks)
(Total 11 marks)
Edexcel 6102/01 June 2001, B AS/A, No.3

3 The diagram shows a section through the stomach wall.

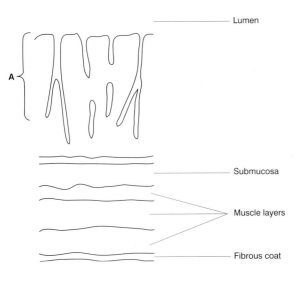

a Name the part labelled **A**. *(1 mark)*

b Give **two** ways in which the structure of the wall of the stomach differs from the wall of the oesophagus. *(2 marks)*

c Give **one** way in which the stomach wall is adapted
  (i) to churn food; *(1 mark)*
  (ii) to make food poisoning by bacteria less likely; *(1 mark)*
  (iii) to prevent enzymes produced in the stomach wall digesting the surface of the stomach. *(1 mark)*
  (Total 6 marks)
  AQA June 2001, B (B) BYB1, No.1

4 The diagram shows some of the nerve pathways involved in the control of ventilation in humans.

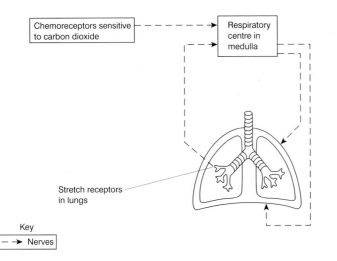

a Give **one** place where chemoreceptors sensitive to carbon dioxide are found. *(1 mark)*

b Describe the role of the stretch receptors in the lungs in the maintenance of breathing. *(2 marks)*

c A person was asked to hyperventilate while his pattern of breathing was being recorded. The trace shows the changes in the pattern of breathing before, during and after hyperventilation.

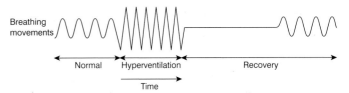

  (i) Describe how the pattern of breathing changed when the person was hyperventilating. *(1 mark)*
  (ii) Suggest an explanation for the long delay after hyperventilation before breathing started again. *(2 marks)*
  (Total 6 marks)
  AQA June 2001, B (B) BYB3, No.5

5 The figure shows a transverse section of a bronchus from the lung of a mammal.

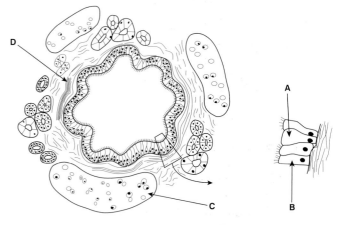

a Name **A** to **D**. *(4 marks)*

b Describe how the cells lining the bronchus protect the alveoli from damage. *(4 marks)*

There are elastic fibres between the cells lining the gaseous exchange surface in the alveoli.

c Describe the function of the elastic fibres in the alveoli. *(3 marks)*

Table 1.1

tidal volume at rest	500 cm^3
vital capacity	4600 cm^3
breathing rate at rest	12 breaths per minute
ventilation rate during exercise	20 000 cm^{3}min^{-1}

d (i) Calculate the ventilation rate at rest.
  (ii) Explain the meaning of the term *vital capacity*.
  (iii) State how the person increased their ventilation rate even though their breathing rate remained constant. *(4 marks)*
  (Total 15 marks)
  OCR 2802 June 2001, B(HHD), No.1

7

7.1

EDEXCEL

AQA.B   OCR

Transport in plants

Transport systems in organisms

Transport over short distances, such as between adjacent cells, is adequately achieved by processes such as **diffusion** (unit 4.9), **osmosis** (unit 4.10) or **active transport** (unit 4.11). The situation in larger organisms is, however, altogether different.

7.1.1 Why large organisms need a transport system

All organisms need to exchange materials between themselves and their environment (unit 6.1). In small organisms, this exchange takes place satisfactorily over the surface of the body. With increasing size, however, the surface area : volume ratio decreases to a point where the needs of the organism cannot be met (section 6.1.1). A specialist exchange surface is therefore needed, to absorb nutrients and respiratory gases and to remove excretory products. These exchange surfaces are located in specific regions of the organism. A transport system is needed to take materials from the cells that require them or produce them, to or from the exchange surfaces that absorb or remove them. However, not only do materials need to be transported between these organs and the environment, they also need to be transported between the different parts of the organism. As organisms have evolved into larger and more complex structures, the tissues and organs of which they are made have thus become more specialised and dependent upon one another.

7.1.2 Features of transport systems

Whether plant or animal, any large organism has the same problems in transporting materials within itself. Not surprisingly, therefore, the transport systems of organisms have many common features:

* A suitable medium in which to carry materials, e.g. blood. This is usually a liquid based on water, because water easily dissolves substances, and can be moved around relatively easily.
* A form of mass flow transport in which the transport medium is moved around in bulk over large distances.
* A closed system of tubular vessels, which contains the transport medium and forms a branching network to distribute it along specific routes, to all parts of the organism.
* A mechanism for moving the transport medium within the vessels. This requires a pressure difference between one part of the system and another. It is achieved in two main ways:
  - Animals use muscular contraction either of the body muscles or of a specialised pumping organ such as the heart (unit 8.6).
  - Plants do not possess muscles and so rely on passive natural physical processes such as the evaporation of water or differences in solute concentrations.
* A mechanism to maintain the mass flow movement in one direction, e.g. valves.
* A means of controlling the flow of the transport medium to suit the changing needs of different parts of the organism.

Giant redwoods need to transport water up to a height of over 100m without the expenditure of energy

7.1.3 Transport systems in plants

The range of materials transported by plants is more limited than that moved by animals. Plants, for example, have no need to transport respiratory gases because:

- most gases are produced or required by leaves, which in any case have a large surface area for the capture of light, and gases can therefore diffuse directly in and out of them
- plants do not move from place to place and their energy requirements are low, which means a reduced need for respiratory gases
- in light, the oxygen that plants require for respiration can mostly be supplied from photosynthesis, while the carbon dioxide they produce is used up in photosynthesis
- no photosynthetic or respiratory tissue is far from the surface, and so diffusion can meet the needs of cells. The trunk of any tree is composed mainly of dead non-respiratory tissue at its centre.

Plants therefore need mainly to transport water and mineral salts from the roots to aerial parts and the products of photosynthesis – sugars – in the opposite direction.

Plants do not possess contractile cells, and therefore depend on passive modes of transport. Evaporation through the stomata in the leaves creates an osmotic gradient, which draws water across the leaf from the xylem. The xylem forms an uninterrupted column of water, and so water is drawn up this column in much the way that it can be sucked up through a straw. Water likewise moves across the root by means of an osmotic gradient. Only in the passage of water from the root tissues into the xylem is energy expended. The movement of sugars is less clearly understood, but it probably combines elements of active and passive transport through a different tissue, the phloem.

7.1.4 Transport systems in animals

Animals have circulatory systems that move the transport medium, the blood, around the body. Smaller animals such as insects use an **open blood system**, in which blood flows freely over the cells and tissues. Large organisms have a **closed blood system**, in which blood is confined to vessels. A muscular pump, the heart, circulates the blood around the body. A greater diversity of substances is transported in animals than in plants. These include nutrients such as glucose, amino acids, minerals and vitamins, respiratory gases such as oxygen and carbon dioxide, waste metabolites such as urea, and hormones. The blood system of animals also carries white blood cells, which are involved in providing immunity and protecting against disease. The ability to clot, and so prevent leakage that results from blood being under pressure, is another feature of blood. Mammals and birds have a **double circulatory system** (Fig 7.1) in which the blood, having had its pressure reduced as it is forced over the lung capillaries, is returned to the heart to boost its pressure, before being circulated to the rest of the body. This assists the rapid delivery of material, which is necessary because birds and mammals have a higher rate of metabolism, due to their high body temperature.

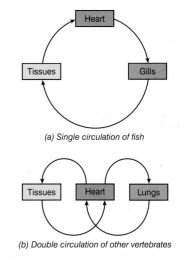

(a) Single circulation of fish

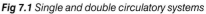

(b) Double circulation of other vertebrates

Fig 7.1 *Single and double circulatory systems*

SUMMARY TEST 7.1

Small organisms such as protozoa exchange materials using their (1), but as large organisms evolved, more specialist surfaces developed and so too did an efficient transport system within the organisms. In some animal groups such as (2), the blood flows freely over the internal organs. This is called (3) blood system. Other organisms confine the blood to vessels – this is (4) blood system. Materials are circulated using the muscular contractions of a heart. Such materials include nutrients, respiratory gases and waste metabolites such as (5). Some animals pass the blood through the heart twice in each complete circulatory cycle. This is called a (6) system and occurs in two groups of animals, (7) and (8). Plants do not use internal sources of energy to transport water and the mechanism is therefore said to be (9). Water evaporating from (10) creates an (11) gradient that draws water up the plant through a tissue called (12).

Structure of vascular tissues in plants

A flowering plant can be thought of as having two main functional areas: the leaves, which manufacture sugars by photosynthesis at one end, and the roots, which absorb water and minerals at the opposite end. Each relies on the other – the leaves needing water and minerals to photosynthesise, and the roots requiring sugar to respire and keep alive. No less important, therefore, are the communication channels between the two, namely the **vascular tissue**, of which there are two types:

- **xylem** – which carries water from the roots, up the plant to the aerial parts
- **phloem** – which carries sugars produced by leaves to other parts of the plant.

7.2.1 Structure of xylem

Xylem performs the functions both of supporting the plant and transporting water and minerals within it. **Parenchyma** cells and **sclerenchyma fibres** in the xylem all contribute to support, whereas the **vessels** and **tracheids** have both support and transport roles.

- **Xylem parenchyma** is composed of unspecialised cells that act as packing tissue around the other components of the xylem. They are roughly spherical in shape, but when they are turgid they press upon and flatten each other in places. In this way they provide support.
- **Xylem fibres** are elongated sclerenchyma cells with walls that are thickened with ligin; these features suit them to their role of support.
- **Vessels** (Fig 7.2) vary in structure depending on the type and amount of thickening of their cell walls, but all are hollow and elongated. As they mature their walls become impregnated with **lignin**, which causes them to die. The end walls break down to form a perforated plate which allows the cells to form a continuous tube. (The word 'element' is sometimes used rather than 'cell', because a cell is a living structure, whereas mature xylem vessels are dead.) Sometimes the lignin forms rings around the vessel; in other cases it forms a spiral. This arrangement is better than a continuous thickening, because it allows elongation of the vessels as the plant grows. Pits in the lignified walls allow for lateral (sideways) movement of water.
- **Tracheids** (Fig 7.2) have a structure similar to vessels, except that they are longer and thinner, and have tapering ends. They, too, are thickened with lignin and therefore die when mature. As with vessels, the end walls break down, and their side walls possess bordered pits which allow lateral movement of water between adjacent cells.

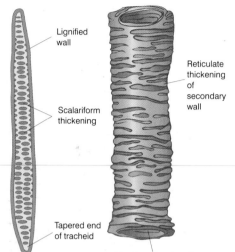

Lignified wall

Reticulate thickening of secondary wall

Scalariform thickening

Tapered end of tracheid

Simple perforation plate

Fig 7.2 Tracheid (left) and xylem vessel (right)

Xylem squash as seen under a light microscope (×200 approx.)

7.2.2 How xylem's structure is related to its functions

The vessels and tracheids of which xylem is mostly made up are structurally suited to the transport of water in a number of ways:

- The cells are long and arranged end to end to form a continuous column.
- The cell contents die when mature, which means that
  - there is no cytoplasm or nucleus to impede water flow
  - the end walls can break down, so that there is no barrier to water flow between adjacent cells.
- Cell walls are thickened with lignin, which
  - makes them more rigid and therefore less likely to collapse under the tension created by transpiration pull (section 7.3)
  - increases the **adhesion** of water molecules, enabling them to rise by capillarity (section 7.3.3).
- There are pits throughout the cells, allowing lateral movement of water.

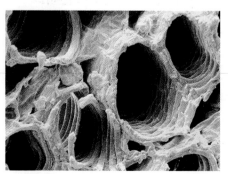

Leaf of tobacco showing xylem vessels (SEM) (×500 approx.)

- The narrow **lumen** of the cells increases the height to which water rises by capillarity.

7.2.3 Structure of phloem

Phloem carries organic material, such as sugars and amino acids, from the leaves to other parts of the plant. It is composed of a number of cell types:

- **Phloem parenchyma** is an unspecialised tissue of thin-walled cells which provide support when they are turgid.
- **Phloem fibres** are elongated sclerenchyma cells which are adapted to their role of support by having thickened, lignified walls.
- **Sieve tube elements** are elongated structures which are joined end to end to form long tubes. Unlike xylem, at maturity the cytoplasm remains, along with mitochondria and a modified form of endoplasmic reticulum. The Golgi, nucleus and ribosomes are, however, broken down. The loss of these structures makes the sieve tubes more hollow, reducing resistance to the flow of liquid within them. Their end walls are perforated by large pores, 2–6μm in diameter, giving rise to their name of **sieve plates**. Within the lumen of the tubes are strands of cytoplasm 1–7μm wide, made up of **phloem-protein**. These stretch along the cell and on through the sieve plates.
- **Companion cells** are always associated with sieve tube elements and both come from the same cell division. They have a nucleus, dense cytoplasm, a thin cell wall, a small vacuole and many mitochondria. They carry out metabolic activities for the sieve tube, to which they are linked by **plasmodesmata** in their walls. The two function intimately, to the extent that, if one dies, so does the other. At the tips of veins in the leaf, these cells have very folded cell walls and cell surface membranes. Known as **transfer cells**, they are thought to actively transport sucrose into the sieve tubes.

7.2.4 How phloem's structure is related to its function

As with xylem, the structure of sieve tubes has evolved to suit their function of transporting organic materials in solution.

- The elements are elongated and arranged end to end to form a continuous column.
- The nucleus and many of the organelles are located in the companion cells, leaving the lumen of the sieve tube elements more open and lessening obstruction of the flow of liquid.
- The end walls are perforated, so give less resistance to liquid flow.
- The companion cells have many mitochondria that release the energy needed for translocation of organic materials (unit 7.8).
- The walls contain cellulose microfibrils that run around the cells, giving strength and preventing the tubes bursting under pressure.
- Cytoplasmic strands run along the tubes and are continuous through the sieve plates, which aids movement of organic materials.

Phloem as seen under a light microscope (×400 approx.)

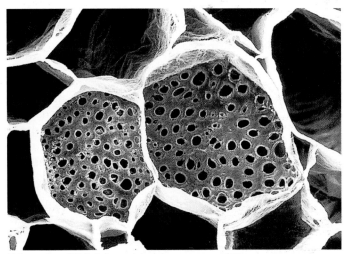

False-colour SEM of sieve plates

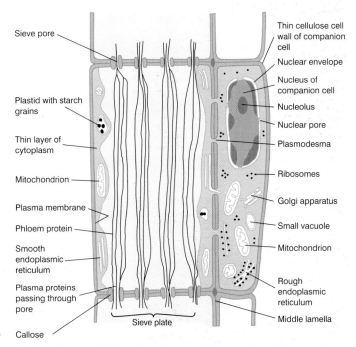

Fig 7.3 *Structure of sieve tube element and companion cell as revealed by electron microscope*

Transpiration

Transpiration is the evaporation of water from plants. It takes place at three sites:
- **Stomata** (section 6.2.1) occur in leaves and **herbaceous** stems, and account for 90% of water loss.
- **Cuticle** is a waxy external layer on plant surfaces which serves to limit water loss through cell walls, although up to 10% of water nevertheless escapes by this route.
- **Lenticels** are areas of loosely packed cells on the surface of woody stems through which gas exchange, and therefore water loss, take place.

7.3.1 Role of transpiration

Although transpiration is universal in flowering plants, it is the unavoidable result of plants having leaves designed for photosynthesis. To absorb light effectively, leaves have a large surface area. To allow adequate inward diffusion of carbon dioxide, they have stomata. Both features result in an immense loss of water – up to 700 litres a day in a large tree. Transpiration is not essential as a means of bringing water to the leaves – osmotic processes could achieve this. It does help, however, although less than 1% of water moved in the transpiration stream is used by a plant. What then, if any, are the benefits of transpiration?
- It contributes to the supply of water to aerial parts.
- On hot days, it cools the plant.
- It speeds the process of mineral absorption and transport.

7.3.2 Measurement of water loss

The rate of water loss in a plant can be measured using a **potometer** (Fig 7.4). The experiment is carried out in the following stages:
- A leafy shoot is cut under water, to prevent air entering the xylem.
- The potometer is filled completely with water, making sure there are no air bubbles.
- The leafy shoot is fitted to the potometer under water, using a rubber tube.
- The potometer is removed from under the water and all joints are sealed with vaseline.
- Using a syringe, an air bubble is introduced into the capillary tube.
- The distance moved by this air bubble in a given time is measured, and from this the volume of water lost can be calculated ($\pi r^2 l$).

The experiment can be repeated to compare the rates of water loss under different conditions, e.g. at different temperatures, humidities, light intensity.

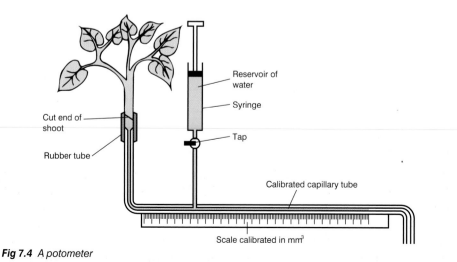

Reservoir of water

Syringe

Cut end of shoot

Tap

Rubber tube

Calibrated capillary tube

Scale calibrated in mm³

Fig 7.4 *A potometer*

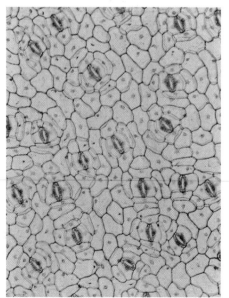

Surface view of stomata

7.3.3 The transpiration stream

The flow of water from the root to the aerial parts of plants is known as the **transpiration stream**. The process of movement up the plant is passive, i.e. there is no expenditure of energy by the plant. Instead, the energy needed to overcome the gravitational forces acting on the water comes ultimately from the sun. It is heat that causes water to evaporate from aerial parts of the plant. This creates a more negative water potential in those cells losing water, which in turn causes water to move in by osmosis (unit 4.10) from adjacent cells (section 7.4.2). Water is ultimately drawn from the xylem vessels and tracheids, which have the least negative water potential of all. How, then, is water drawn up the xylem? The mechanism is known as the **cohesion–tension theory**:

- Water forms a continuous column from the roots to the leaves.
- As water evaporates from leaves, a water potential gradient pulls water into them from the xylem.
- Water molecules 'stick' to one another by **cohesion** (section 1.1.5).
- As one water molecule leaves the xylem, another is pulled up behind it, and so along the whole length of the water column in the xylem, by the **transpiration pull**.

- The transpiration pull creates a negative pressure in the xylem – **tension** – if the xylem is cut, it draws in air rather than leaking water.

Such are the cohesive forces between the water molecules that the transpiration pull can draw water to heights in excess of 100m – enough to supply the tops of Californian Redwoods, the world's tallest trees. **Adhesion** makes a small contribution to the transpiration pull. It is the force created by the attraction of different molecules to each other (cohesion = attraction of the **same** molecules). In this case, the water molecules are attracted to the walls of the xylem vessels and are therefore drawn up by a phenomenon called **capillarity**. It is really only effective in smaller plants – indeed, the 'drag' it creates over large distances may even be detrimental to a tall tree.

7.3.4 Factors affecting transpiration

Both internal and external factors influence the rate of transpiration (table 7.1). Plants living in conditions that lead to high transpiration rates often have adaptations which allow them to reduce water loss. Such plants are called **xerophytes**; their features are considered in section 11.1.1.

Table 7.1 Summary of factors affecting transpiration

Type	Factor	How factor affects transpiration	Increase in transpiration caused by	Decrease in transpiration caused by
External	Light	Stomata open in the light and close in the dark	Higher light intensity	Lower light intensity
	Humidity	Affects the diffusion gradient between the air-spaces in the leaf and in the atmosphere	Lower humidity	Higher humidity
	Temperature	Alters the kinetic energy of the water molecules and the relative humidity of the air	Higher temperatures	Lower temperatures
	Wind speed	Changes the diffusion gradient by altering the rate at which moist air is removed from around the leaf	Higher speeds	Lower speeds
	Water availability	Influences water potential gradient between soil and leaf	Wetter soils	Drier soils
Internal	Leaf area	Some water is lost over the whole surface of the leaf	Larger leaf area	Smaller leaf area
	Cuticle	Forms a waterproofing layer over the leaf surface	Thinner cuticle	Thicker cuticle
	Number of stomata	Most water is lost by evaporation through stomata	More stomata	Fewer stomata
	Distribution of stomata	Upper surface is more exposed to environmental factors that increase the rate of transpiration.	Greater proportion of stomata on upper surface of leaf.	Greater proportion of stomata on lower surface of leaf

SUMMARY TEST 7.3

In plants, water evaporating through the stomata and the **(1)** of the leaves creates a **(2)** water potential in the cells. A water potential gradient therefore extends across the leaf and water moves from the xylem to the surface of the leaf by the process of **(3)**. The ability of molecules to stick to one another is called **(4)** and accounts for water being pulled up through the continuous column of xylem vessels. In daylight the rate of transpiration is **(5)** than in the dark. High humidity **(6)** the transpiration rate. The rate is decreased when the temperature is **(7)**.

Leaves carry out photosynthesis and therefore are adapted not only for capturing light, but also for ensuring a constant supply of another raw material – water.

7.4.1 Structure of the leaf

The leaf is adapted to bring together the three raw materials of photosynthesis (light, water and carbon dioxide), and to remove its products (oxygen and glucose). Figure 7.5 illustrates the structure of a leaf of a typical **dicotyledonous plant**. The various components include:

- **cuticle** – a waxy outer waterproofing layer
- **epidermis** – a layer, usually one cell thick, of close-fitting cells which protect the leaf from disease and physical damage
- **palisade mesophyll** – layer of closely fitted, columnar cells one or two cells deep, which contain many chloroplasts for photosynthesis
- **spongy mesophyll** – layer of more spherical cells with fewer chloroplasts than the palisade layers. There are many air-spaces, allowing the rapid diffusion of gases, including water vapour
- **stomata** (section 6.2.1) – pores surrounded by guard cells which control the entry and exit of gases to and from the leaf
- **main vein** – large central area of vascular tissue made up of xylem and phloem, which carries liquids to and from the leaf
- **side veins** – small branches of the main vein, which radiate out, forming a network throughout the leaf

7.4.2 Movement of water across the leaf

The humidity of the atmosphere is usually less than that of the sub-stomatal air-space and so, provided that the stomata are open, water diffuses out of the air-spaces into the surrounding air. Water lost from the air-spaces is replaced by water evaporating from the surrounding spongy mesophyll cells. These cells therefore have a lower (more negative) **water potential** (section 4.10.2) and so draw water by **osmosis** (unit 4.10) from the neighbouring cells. The loss of water from these adjacent cells causes them to have a lower (more negative) water potential and so they, in turn, draw water from their neighbours by osmosis. In this way, a water potential gradient is established that pulls water from the xylem, across the leaf mesophyll, and finally out into the atmosphere. It is this process that drives the **transpiration** stream.

7.4.3 Adaptations of palisade mesophyll to its function

The palisade mesophyll tissue performs the function of photosynthesis and is adapted in two main ways:

- **To absorb the maximum light possible, it**
  - is packed with chloroplasts – the site of photosynthesis
  - is made of long, thin cells with their long axis perpendicular to the leaf surface
  - has chloroplasts that can move to arrange themselves in a way that captures as much light as possible
  - is arranged in a continuous layer near the surface of the leaf.
- **To obtain certain gases (e.g. carbon dioxide) and remove others (e.g. oxygen), it has**
  - interconnecting air-spaces around cells to allow rapid diffusion of gases between different parts of the plant
  - cellulose cell walls that are thin and not impregnated with substances that might slow diffusion
  - chloroplasts that are forced close to the cell surface by a large central vacuole.

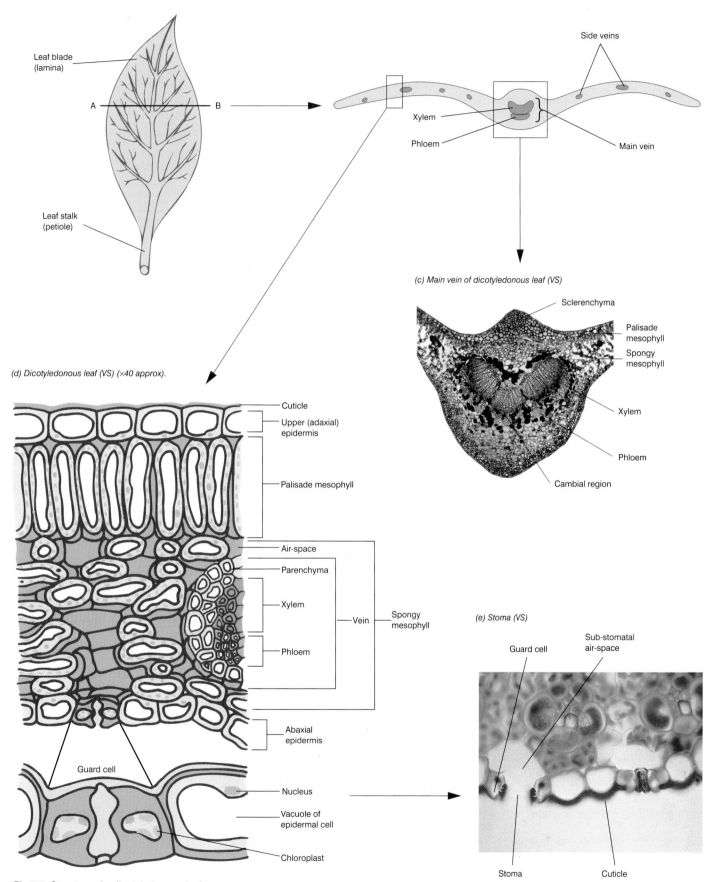

(a) Whole leaf

Leaf blade (lamina)

Leaf stalk (petiole)

A — B

(b) Section A/B through leaf

Side veins

Xylem

Phloem

Main vein

(c) Main vein of dicotyledonous leaf (VS)

Sclerenchyma

Palisade mesophyll

Spongy mesophyll

Xylem

Phloem

Cambial region

(d) Dicotyledonous leaf (VS) (×40 approx).

Cuticle

Upper (adaxial) epidermis

Palisade mesophyll

Air-space

Parenchyma

Xylem

Phloem

Vein

Spongy mesophyll

Abaxial epidermis

Guard cell

Nucleus

Vacuole of epidermal cell

Chloroplast

(e) Stoma (VS)

Guard cell

Sub-stomatal air-space

Stoma

Cuticle

Fig 7.5 *Structure of a dicotyledonous leaf*

131

7.5 Root structure and water uptake

EDEXCEL

AQA.B   OCR

The loss of water by plants during **transpiration** is unavoidable, but if they are to survive, it must be replaced. This might mean absorbing some 700 litres a day in the case of a large tree. This water is drawn from the soil by tiny extensions of epidermal cells known as **root hairs**. These are situated only on the youngest parts of roots, near their tips, and they remain functional for only a few weeks before being replaced by others near the growing points.

7.5.1 Structure of the root

The **vascular tissue** in a root is concentrated at the centre, rather than in the outer layers as in stems. This is because xylem has a supporting function as well as a transport one, and roots are subject to vertical pulling forces. Such forces are best resisted by a central column of supporting tissue. The root has a number of component parts (Fig 7.6):

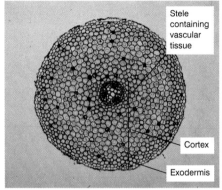

Root of Ranunculus *(buttercup) (TS) (×20)*

- **Epidermis** – a layer of closely fitting cells around the outside, which protects the root from physical damage. There is no waxy cuticle; water loss is not a problem, because the root is surrounded by soil. Where root hairs are present, the epidermis is known as the **piliferous layer**.
- **Cortex** – a large layer of **parenchyma** cells, where starch is stored. Many air-spaces in this layer enable the diffusion of gases between root tissues and air-spaces in the soil.
- **Endodermis** – a one-cell-thick ring of cells around the central vascular tissue. The cells are elongated and have a distinctive band known as the **Casparian strip**, which is made up of **suberin**. This strip has an important role in the movement of water into the xylem (section 7.5.4).
- **Pericycle** is a layer of parenchyma cells inside the endodermis, from which lateral roots arise.
- **Vascular tissue**, made up of xylem, phloem and cambium and sometimes known, along with the pericycle, as the **stele**.

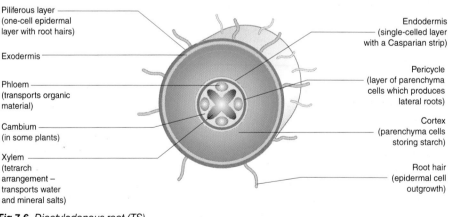

Fig 7.6 Dicotyledonous root (TS)

7.5.2 Uptake of water by roots

The root hairs of the piliferous layer grow into the spaces around the soil particles. In damp conditions, these hairs are surrounded by a soil solution which contains small quantities of mineral ions but is mostly water and therefore has a water potential only slightly less than zero. In contrast, the root hairs and other cells of the root have a much lower water potential; the root hairs readily absorb water by osmosis.

7.5.3 Movement of water across the root cortex

Having been absorbed into the epidermal cell, water continues its journey to the cortex in three ways, which are illustrated in figure 7.7:

- **The apoplast pathway** accounts for most of the transport of water that moves along the water-filled spaces of the cell walls. As water is drawn into the endodermal cells, it pulls more water along behind it due to the cohesive properties of the water molecules. This creates a tension that draws water along the adjacent cell walls of the cortical cells, in exactly the same way as water is drawn up through the xylem (section 7.3.3).

- **The symplast pathway** is where water passes across tiny strands of cytoplasm in the cell walls, known as plasmodesmata. Movement occurs because, water having been absorbed from the soil solution, the water potential of the epidermal cell is raised to a level greater than that of the adjacent cortical cell. Water therefore moves from the epidermal cell with its less negative (higher) water potential. Having absorbed water, this first cortical cell now has a higher water potential than its neighbour to the inside, and so water moves into it by osmosis. In this way, a water potential gradient is set up across the cortex, which carries water along the cytoplasm from the epidermis to the endodermis.

- **The vacuolar pathway** involves the movement of water through the vacuoles of cells by osmosis. The same water potential gradient found in the symplast pathway is the mechanism by which water moves, but, instead of being confined to the cytoplasm, it also passes across the vacuoles of cells.

7.5.4 Movement of water into the xylem

Water reaching the endodermis by the apoplast pathway now comes up against the **Casparian strip** (section 7.5.1), a waterproof band of suberin, which prevents its further passage along the cell wall. The water is therefore forced into the living protoplast of the endodermal cell, to join the rest of the water which has reached there by the symplast and vacuolar pathways. It is thought that salts are then actively pumped into the xylem, thus creating a much lower (more negative) water potential in the xylem next to the endodermis. This lower water potential causes water to be drawn from the endodermis. It is this active pumping of mineral salts followed by water that causes water to leak from a newly cut stem. It is known as **root pressure**, and contributes to the movement of water up small plants, but is inadequate to account for its transport up large trees.

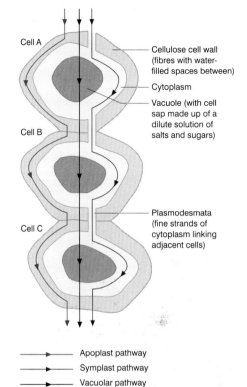

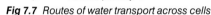

Apoplast pathway
Symplast pathway
Vacuolar pathway

Fig 7.7 *Routes of water transport across cells*

SUMMARY TEST 7.5

The outer layer of a root is called the **(1)** and where it bears root hairs it is known as the **(2)**. Beneath this outer layer is a cortex made of **(3)** cells. At the centre of the root is the vascular tissue made up of xylem, phloem and **(4)**. Around this vascular tissue is a one-cell-thick ring of cells called the **(5)**, the cells of which have a band of **(6)** known as the **(7)**. Inside the endodermis is a layer of parenchyma cells called the **(8)**. Water absorbed by a root reaches the xylem in three ways: via the strands of cytoplasm called **(9)** that extend across cell walls – the **(10)** pathway; via the cell walls themselves – the **(11)** pathway; or through the vacuoles of the cells – the **(12)** pathway.

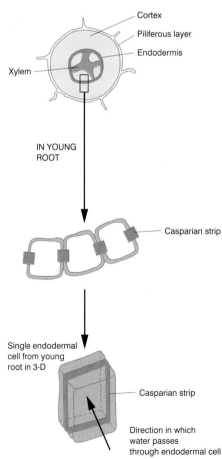

Fig 7.8 *Water transport in the root*

Stem structure and water transport

Stems serve a number of functions in plants: they store food and water, some are used as a means of asexual reproduction, and others photosynthesise. The two main functions are, however, to support the aerial parts of the plant and to transport water and organic materials between the leaves and the roots.

7.6.1 Structure of the stem

The vascular tissue in the stem is found towards the outside, rather than centrally as it is in roots. This is because xylem and its associated sclerenchyma provide support. Stems are subject to lateral forces from sources such as the wind blowing them from side to side. These lateral forces are best resisted by an outer supporting cylinder. The stem is made up of the following parts:

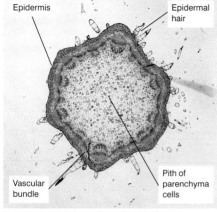

Young stem of Helianthus (TS) (×3)

- **Epidermis** – in young herbaceous plants, this is a single-celled layer of close-fitting cells. The epidermis usually has stomata for gas exchange, and a waxy cuticle that reduces water loss. Hairs may also be present. In woody stems it is a multilayered structure known as **bark**.
- **Cortex** – mostly parenchyma cells, which provide support when turgid and also allow gaseous diffusion through the air-spaces between them. Sometimes a layer of **collenchyma** is present, which provides additional support.
- **Vascular bundles** are mostly of xylem and phloem, which transport water and organic materials, respectively. **Sclerenchyma** is often associated with the bundles and this provides support, with its heavily thickened and lignified walls. **Cambium** is a dividing tissue found between the xylem and phloem.
- **Pith** is the large central region, which is made up of parenchyma cells that are mainly used for storage of starch.

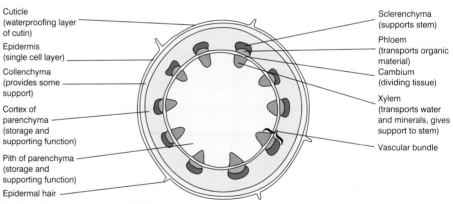

Fig 7.9 Dicotyledonous stem (TS)

7.6.2 Movement of water up the stem

The vessels and tracheids of xylem are responsible for the movement of water up the stem. There is a variety of evidence that supports this view:

- If a leafy shoot is cut under water which contains a dye, the xylem is found to be the only tissue that is coloured by the dye when the stem is cut across and is inspected some hours later.
- A ring of tissue can be removed from around a woody stem. Where only the bark, including the phloem, is removed, the flow of water is unaffected. Cut deeper, into the xylem, the plant wilts.
- Radioactive **isotopes** of minerals can be added to water given to a plant. Later examination shows the xylem to be radioactive. This evidence assumes that water and minerals are carried together.

The movement of water is driven by the process of transpiration (unit 7.3), whereby the evaporation of water from the stomata (section 7.4) causes water to be drawn across the leaves by **apoplast**, **symplast** and **vacuolar** pathways (section 7.5.3). Due to the cohesion–tension process (section 7.3.3), water is pulled up the xylem from the roots. Helping the process to operate effectively, the xylem is specially adapted to the role of water transport (section 7.2.2). The whole process of water transport from roots to leaves is illustrated in Figure 7.10.

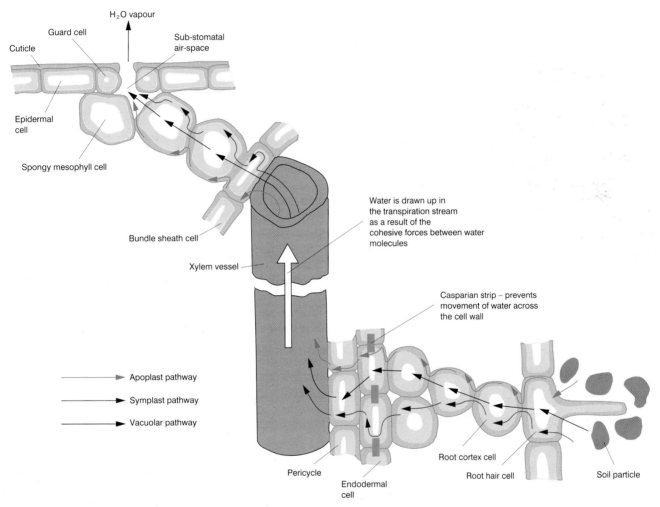

Fig 7.10 *Summary of water transport through a plant*

SUMMARY TEST 7.6

The vascular bundles of a dicotyledonous stem form a ring towards the outside. The area outside this ring is called the **(1)** and it contains many air-spaces to allow **(2)** to take place. The area inside the ring is known as the **(3)** and its main fuction is to store **(4)**. A line from the outside of a dicotyledonous stem to the centre, passing through a vascular bundle, would cross in turn seven types of cell. In order these would be **(5)**, collenchyma, **(6)**, **(7)**, phloem, **(8)**, **(9)** and **(10)**.

7.7

Uptake and translocation of minerals

EDEXCEL

AQA.B   **OCR**

Minerals are needed in only tiny amounts, but they are essential for the healthy functioning of all plants. Some of the more important ones are listed in section 1.1.8. These minerals are absorbed almost entirely by root hairs, which have thin walls and a large surface area that make them ideal for this purpose. A small quantity of minerals may be absorbed through the leaves.

7.7.1 Mineral uptake

Minerals are absorbed by plants as ions, in a number of ways:
- **Absorption with water uptake** – some mineral ions are passively absorbed, dissolved in the water that is drawn into the plant. They pass along the apoplast pathway (section 7.5.3) to the root endodermis. The process is not selective, and the ions are absorbed in the proportions in which they are present in the soil solution. They may include ions such as gold or tin, which are of no functional use to the plant.
- **Diffusion** – any mineral ion that is present in a greater concentration in the soil solution than in the root hairs will diffuse into the plant. This process, also, is non-selective.
- **Active transport** – most minerals are absorbed against a concentration gradient, by active transport (unit 4.11). The minerals are absorbed selectively, with the plant taking up only those ions it requires. There is much evidence to support the view that active transport is involved in mineral absorption:
  - The addition of mineral salts to plants growing in deionised water causes a rise in the respiratory rate (Fig 7.11); this provides the energy needed for active transport.
  - An increase in temperature of the roots leads to an increase in respiratory rate, with an increase in mineral ion uptake that is more than that expected by diffusion (Fig 7.12).
  - The addition of a respiratory inhibitor, e.g. cyanide, causes a decrease in the rate of mineral ion absorption (Fig 7.13).

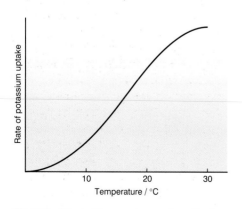

Fig 7.11 *Relationship between rate of respiration and mineral uptake*

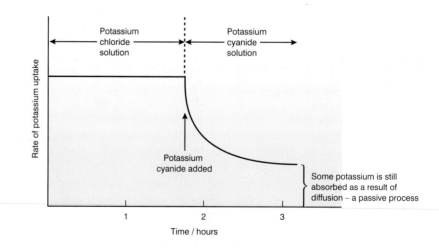

Fig 7.13 *Effect of a respiratory inhibitor on mineral uptake*

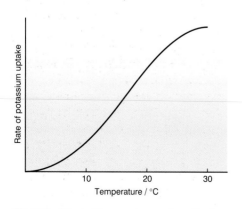

Fig 7.12 *Effect of temperature on the rate of potassium uptake in plants*

136

7.7.2 Translocation of minerals

Absorbed mineral ions move along the root cortex via the apoplast pathway (section 7.5.3) in two ways:
- **diffusion** along a concentration gradient
- **mass flow** with water drawn along in the transpiration stream.

Once the minerals reach the endodermis, they encounter the Casparian strip, which prevents further movement along the cell wall (section 7.5.4). At this point the ions have to enter the cytoplasm of the endodermal cell, and from there they either diffuse or are actively transported into the xylem. From the xylem, they are drawn up the plant by the mass flow of the transpiration stream.

7.7.3 Evidence for translocation of minerals in the xylem

The view that mineral ions are carried throughout the plant by the xylem is supported by the following (Fig 7.14):
- mineral ions are present in the contents of the xylem vessels and tracheids
- the rate of transport of mineral ions in the plant matches the rate of flow of the transpiration stream
- experiments using radioactive tracers, carried out as follows:
  - A section of stem of an actively growing plant has its xylem and phloem separated by waxed paper, which is impervious to water and minerals.
  - The plant is watered with a solution containing radioactively labelled potassium ions (^{42}K), for a number of hours.
  - The rate of transpiration is increased by blowing air over the leaves.
  - After 5 hours, the amount of radioactive potassium in the phloem and xylem of the separated region of the stem is measured.
  - The radioactive potassium occurs almost exclusively in the xylem.

The mineral ions carried up the plant in the xylem either diffuse or are actively transported into the cells of the leaves, flowers, fruits etc which require them.

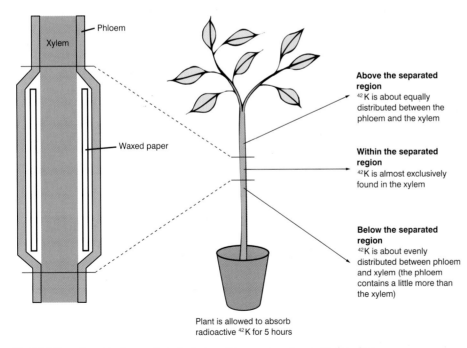

Phloem

Xylem

Waxed paper

Above the separated region
^{42}K is about equally distributed between the phloem and the xylem

Within the separated region
^{42}K is almost exclusively found in the xylem

Below the separated region
^{42}K is about evenly distributed between phloem and xylem (the phloem contains a little more than the xylem)

Plant is allowed to absorb radioactive ^{42}K for 5 hours

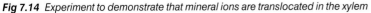

Fig 7.14 Experiment to demonstrate that mineral ions are translocated in the xylem

---

SUMMARY TEST 7.7

Minerals required by plants are absorbed from the soil either passively by (**1**) or against a concentration gradient by (**2**). The minerals are absorbed as (**3**) and include magnesium and (**4**), both of which are needed to form (**5**) which is essential for photosynthesis. The minerals pass to the endodermis by the apoplast pathway, which involves movement through the (**6**) of cells. On reaching the endodermis, the band of suberin called the (**7**) forces the minerals to enter the (**8**) of the cell, from where they either enter the xylem by either (**9**) or (**10**). If the temperature of the roots of a plant is raised to 70 °C, the rate of absorption of minerals is likely to (**11**).

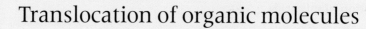

7.8 Translocation of organic molecules

EDEXCEL

AQA.B   OCR

Having produced sugars during photosynthesis, the plant needs to transport them from the sites of production, known as **sources**, to the places where they will be used directly or stored for future use – known as **sinks**. Other organic materials to be transported include amino acids.

7.8.1 Mechanism of translocation

It is accepted that organic materials are transported in the phloem (section 7.8.2) and that the rate of movement is too fast to be explained by diffusion. What is in doubt is the precise mechanism of how translocation is achieved. Current thinking favours the **pressure flow hypothesis**, a theory which can be divided into three phases:

* **Transfer of sucrose into sieve tube elements from photosynthesising tissue**, which takes place by two possible routes:
  - **The apoplast route** occurs in those plants that have few, if any, cytoplasmic connections between the sieve element–companion cell complex and the photosynthesising cell. In this case, the sucrose is actively pumped into the apoplasts (spaces in the cell wall) by the photosynthesising cell, and is then actively taken up into the phloem using transporter proteins in the cell surface membranes.
  - **The symplast route** occurs in those plants that have branched plasmodesmata between the photosynthesising cells and the sieve element–companion cell complex. The sucrose moves freely along these plasmodesmata into the companion cells, which then convert it into larger sugars which cannot flow back, but easily move into the sieve tube. The plasmodesmata therefore act as a type of valve. As the conversion into larger sugars needs energy, the overall process is an active one.
* **Mass flow of sucrose through sieve tube elements**. The sucrose produced by photosynthesising cells (source) causes them to have a more negative (lower) water potential. Water therefore moves into them from the xylem, which has a very much less negative (higher) water potential. At the sink, sucrose is used up or converted to starch for storage. These cells therefore have a low sucrose content, giving them a less negative (higher) water potential. As a result of water being drawn into cells at the source and leaving at the sink, there is a mass flow along the xylem. This is a passive process. A model of this theory is shown in figure 7.15.
* **Transfer of sucrose from the sieve tube elements into storage or other sink cells**. The sucrose is actively transported by companion cells, out of the sieve tubes and into the sink cells.

Some of the evidence for and against the pressure flow theory is listed in table 7.2. Alternative theories for the translocation of organic molecules include:

* **cytoplasmic streaming** – solutes move along in the stream of cytoplasm, which can be observed moving through sieve tube elements
* **contractile proteins** similar to the **actin** and **myosin** in mammals have been found in phloem. It is suggested that these might be responsible for solute movement.

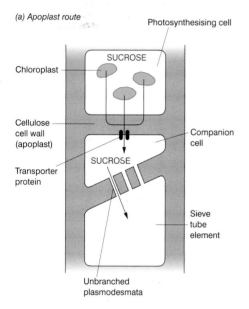

(a) Apoplast route

Photosynthesising cell

SUCROSE

Chloroplast

Cellulose
cell wall
(apoplast)

Companion
cell

SUCROSE

Transporter
protein

Sieve
tube
element

Unbranched
plasmodesmata

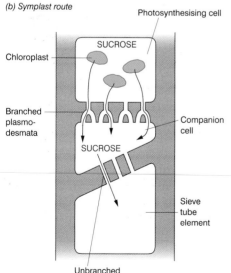

(b) Symplast route

Photosynthesising cell

SUCROSE

Chloroplast

Branched
plasmo-
desmata

Companion
cell

SUCROSE

Sieve
tube
element

Unbranched
plasmodesmata

Fig 7.15 *Loading of phloem with sucrose*

Table 7.2 *Evidence for and against the pressure flow theory*

Evidence supporting the pressure flow theory	Evidence questioning the pressure flow theory
• There is a pressure within sieve tubes, as shown by sap being released when they are cut. • The concentration of sucrose is higher in leaves (source) than in roots (sink). • Downward flow in the phloem occurs in daylight, but ceases when leaves are shaded, or at night. • Increases in sucrose levels in the leaf are followed by similar increases in sucrose levels in the phloem a little later. • Metabolic poisons and/or lack of oxygen inhibit translocation of sucrose in the phloem. • Companion cells possess many mitochondria and readily produce ATP.	• The function of the **sieve plates** is unclear, as they would seem to hinder mass flow (it has been suggested that they may have a structural function, helping to prevent the tubes from bursting under pressure). • Not all solutes move at the same speed – they should do so if movement is by pressure flow. • Sucrose is delivered at more or less the same rate to all regions, rather than going more quickly to the ones with the lowest sucrose levels, which the pressure flow theory would suggest.

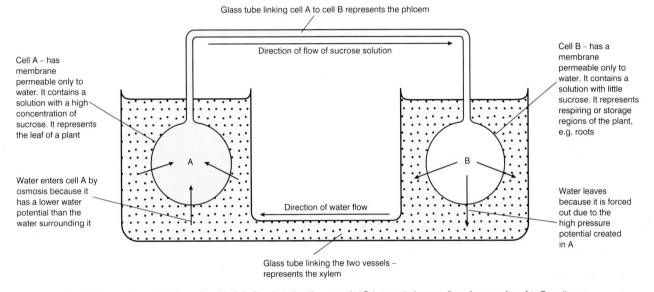

Glass tube linking cell A to cell B represents the phloem

Direction of flow of sucrose solution

Cell A – has membrane permeable only to water. It contains a solution with a high concentration of sucrose. It represents the leaf of a plant

Cell B – has a membrane permeable only to water. It contains a solution with little sucrose. It represents respiring or storage regions of the plant, e.g. roots

Water enters cell A by osmosis because it has a lower water potential than the water surrounding it

Direction of water flow

Water leaves because it is forced out due to the high pressure potential created in A

Glass tube linking the two vessels – represents the xylem

Provided sucrose is continually produced in A (leaf) and continually removed at B (e.g. root), the mass flow of sucrose from A to B continues

Fig 7.16 *Model illustrating the movement of sucrose solute by mass flow in phloem*

7.8.2 Evidence that translocation of organic molecules occurs in phloem

- When phloem is cut, a solution of organic molecules is exuded.
- Plants provided with radioactive carbon dioxide can be shown to have radioactively labelled carbon in phloem after a short time.
- Aphids that have penetrated the phloem with their needle-like mouth parts can be used to extract the contents of the sieve tubes. These contents show **diurnal** variations in the sucrose content of leaves that are mirrored a little later by identical changes in the sucrose content of the phloem (Fig 7.17).
- The removal of a ring of phloem from around the whole circumference of a stem leads to the accumulation of sugars above the ring, and their disappearance from below it.

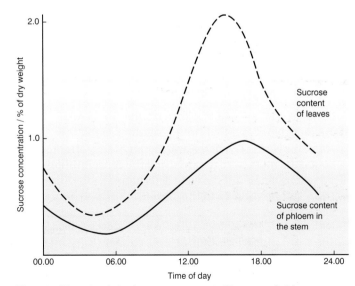

Sucrose content of leaves

Sucrose content of phloem in the stem

Fig 7.17 *Diurnal variation in sucrose content of leaves and phloem*

1 The photomicrograph below shows some cells viewed using a light microscope.

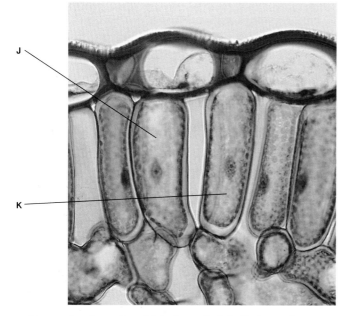

a Name the tissue in which the cells labelled J and K are found, giving **one** reason for your identification.
(2 marks)

b Give the function of this tissue. *(1 mark)*

c In the space below make an accurate drawing, enlarged × 1.5, of the cells labelled J and K. Do not label your drawing.
(5 marks)
(Total 8 marks)
Edexcel 6101/01 Jan 2001, B B(H) AS, No.5

2 Experiments were carried out to investigate the uptake of mineral ions by barley roots.

In the first investigation, isolated barley roots were immersed in an aerated culture solution containing potassium ions (K^+) and nitrate ions (NO_3^-). After ten hours, the roots were removed and the concentrations of these ions in the cell sap were determined.
The results are shown in the table.

Ion	Concentration in culture solution / mmol per dm³	Concentration in cell sap / mmol per dm³
Potassium	7.98	97.8
Nitrate	7.29	38.1

a Suggest why the culture solution was aerated.
(2 marks)

b These results show that the concentration of potassium ions in the cell sap is 12.3 times greater than that in the culture solution. This is referred to as the **accumulation ratio**.

Calculate the accumulation ratio for nitrate ions. Show your working. *(2 marks)*

c What do these results suggest about the mechanism for the uptake of potassium and nitrate ions? Explain your answer. *(2 marks)*

d In a further experiment, the effect of temperature on the uptake of potassium ions was investigated. Isolated barley roots were kept in aerated nutrient solutions at a range of temperatures, and the concentrations of potassium ions in the cell sap were measured after ten hours. The results are shown in the table below.

Temperature / °C	Concentration of potassium ions in cell sap / mmol per dm³
6	35
12	42
18	70
24	95
30	110

(i) What effect does temperature have on the concentration of potassium ions in the cell sap?
(ii) Suggest an explanation for these results. *(4 marks)*
(Total 10 marks)
Edexcel 6102/01 June 2001, B AS/A, No.8

3 The diagram shows part of a cross-section through a primary root.

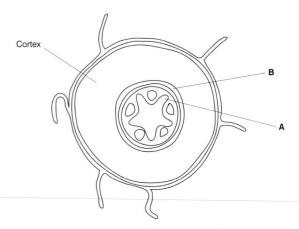

a Name the tissues labelled **A** and **B**. *(2 marks)*

b Water enters root hair cells and moves across the cortex through both apoplast and symplast pathways.

(i) Which part of the cortex cells forms the apoplast pathway? *(1 mark)*

(ii) Explain in terms of water potential how water enters root hair cells from the soil. *(2 marks)*

(Total 5 marks)

AQA June 2001, B (B) BYB3, No.1

4 Radioactive carbon dioxide, $^{14}CO_2$ was used in an investigation of the transport of organic compounds in plants. One leaf on each of two plants was supplied with air containing $^{14}CO_2$. The stem of one plant was ringed below the treated leaf by removing phloem tissue; the other was not ringed. The figures on the diagram show the concentration of radioactive compounds in the treated leaf and in the roots after two hours in sunlight.

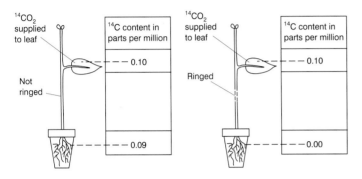

a Name **one** organic compound transported in plants. *(1 mark)*

b Explain what the results show about the transport of organic compounds in plants. *(2 marks)*

c Name the process by which organic compounds are transported in plants. *(1 mark)*

(Total 4 marks)

AQA June 2001, B (B) BYB3, No.6

5 a Describe how the structure of xylem is related to its function. *(4 marks)*

b Describe the roles of root pressure and cohesion-tension in moving water through the xylem.

(i) root pressure
(ii) cohesion-tension *(8 marks)*

c Describe and explain how **three** structural features reduce the rate of transpiration in xerophytic plants. *(3 marks)*

(Total 15 marks)

AQA June 2001, B (B) BYB3, No.8

6 The cells shown in the figure are adapted for transport in flowering plants.

a Name the tissue in which these cells are found. *(1 mark)*

b Identify and explain **two** features of these cells that adapt them to their role in transport. *(4 marks)*

In plants, sucrose is transported in sieve tubes.

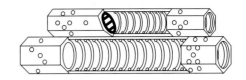

c Describe a possible method for the transport of sugars through sieve tubes. *(In this question, 1 mark is awarded for the quality of written communication).* *(9 marks)*

(Total 14 marks)

OCR 2803/1 June 2001, B(T), No.2

7 Up to 99% of the water that plants take up through their roots may be lost by transpiration.

a Define the term *transpiration*. *(2 marks)*

b Explain briefly why so much water is lost by transpiration. *(2 marks)*

The rates of transpiration for two different species of flowering plant, **A** and **B**, were measured over several hours. One of the plants, **B**, is adapted to survive in very dry conditions. The figure shows the transpiration rate measured in μ**g per cm² of leaf surface** for the two different species.

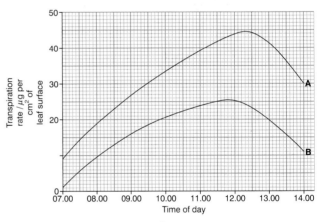

c With reference to this figure, calculate:

(i) the difference in rate between species **A** and **B** at 10.30;
(ii) the increase in rate for species **A** between 8.00 and 11.00. *(2 marks)*

d State and explain **two** possible reasons for the change in the rate of transpiration seen in both species between 8.00 and 11.00 *(4 marks)*

Species **B** is adapted to living in dry conditions.

e (i) State the general name given to plants which can live sucessfully in dry areas. *(1 mark)*
(ii) State **two** features that such a plant may possess and explain how each of these may contribute to its success in dry areas. *(4 marks)*

(Total 15 marks)

OCR 2803/1 June 2001, B(T), No.4

AQA.A	EDEXCEL
AQA.A (Human)	EDEXCEL (Human)
	OCR

Blood is the medium by which materials are transported between different parts of the body. It is made up of a liquid – the **plasma** (55%) – and two types of cells (45%) – **red cells** and **white cells**.

8.1.1 The plasma

Blood plasma is 90% water and 10% chemicals, which are either dissolved or suspended in it. These chemicals include:
- **nutrients**, e.g. glucose, amino acids and vitamins
- **waste products**, e.g. urea
- **mineral salts**, e.g. calcium, iron
- **hormones**, e.g. insulin, adrenaline
- **plasma proteins**, e.g. fibrinogen, prothrombin, albumin
- **respiratory gases**, e.g. oxygen, carbon dioxide.

Between them, these chemicals make the plasma slightly alkaline – around pH 7.4. The removal from the plasma of the proteins involved in clotting (prothrombin and fibrinogen) results in a liquid called **serum**, which does not clot.

8.1.2 Red blood cells

Red blood cells, or **erythrocytes**, are bi-concave discs around 8µm in diameter (Fig 8.1). There are 5 million in each mm^3 of blood and each lives for around 120 days. This means that, to maintain their numbers, the bone marrow of certain bones (cranium, sternum, vertebrae and ribs) needs to make over 2 million red blood cells each second in humans. Red blood cells are unusual in having no nucleus when mature – a feature which, although it gives them a shorter life-span, makes them more efficient in their role of transporting oxygen because:
- they have a larger surface area : volume ratio
- without the nucleus, there is more room for the pigment which carries oxygen.

Oxygen is carried in combination with haemoglobin, the pigment which gives erythrocytes their characteristic colour. The structure of haemoglobin is illustrated in section 1.7.2 and its role in oxygen transport is described in unit 8.2.

8.1.3 White blood cells

White blood cells, or **leucocytes**, exist in a variety of forms, which are illustrated in Fig 8.2. Made in the thymus gland and the marrow of the limb bones, the leucocytes function to protect the body against infection.

8.1.4 Functions of the blood

The blood has two main functions:
- **Transport** – the blood is the body's delivery service, distributing a wide range of materials between different parts (table 8.1).
- **Defence** – the leucocytes protect the body in various ways, including engulfing foreign material such as bacteria by **phagocytosis**, and producing antibodies and other chemicals which stimulate defensive reactions to disease and provide immunity against future infection.

Surface view

8µm

Transverse section

Fig 8.1 *Red blood cell*

In addition, the blood is capable of clotting. This prevents dangerous loss of blood when a vessel is ruptured. Cell fragments known as **platelets (thrombocytes)** have an important role in the clotting of blood.

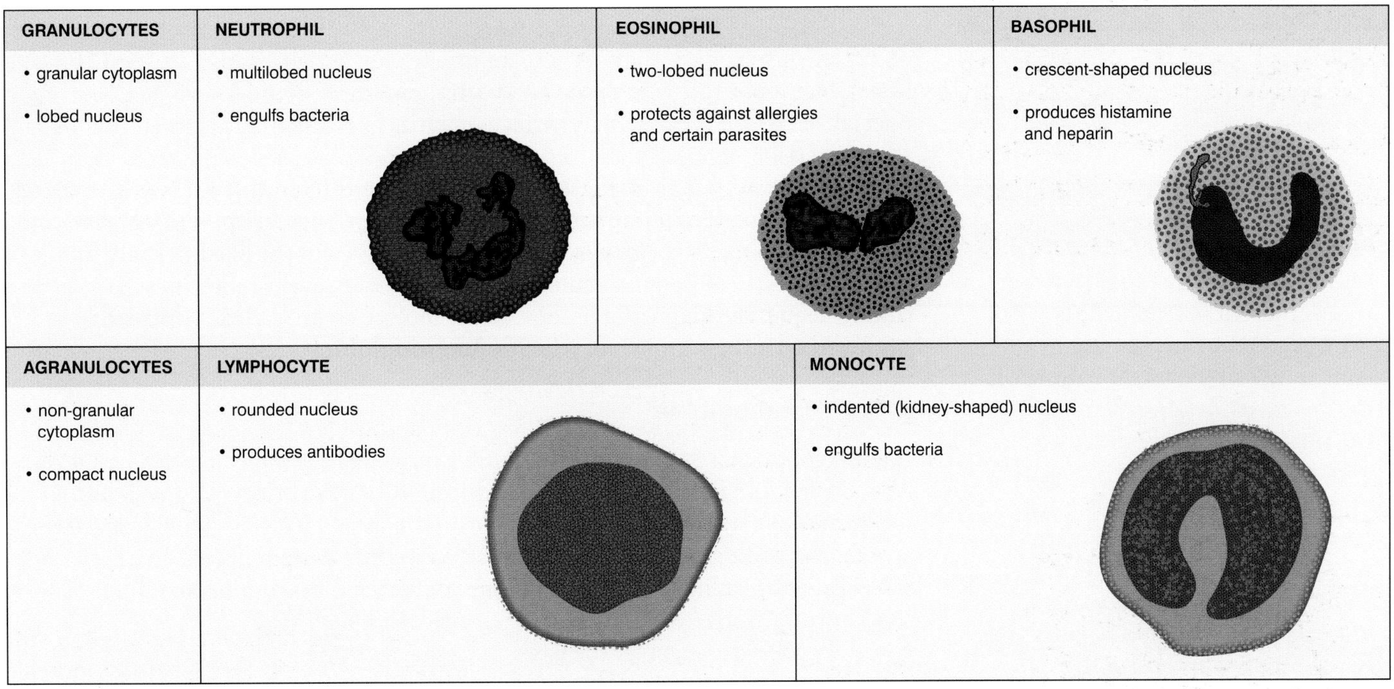

GRANULOCYTES	NEUTROPHIL	EOSINOPHIL	BASOPHIL
• granular cytoplasm • lobed nucleus	• multilobed nucleus • engulfs bacteria	• two-lobed nucleus • protects against allergies and certain parasites	• crescent-shaped nucleus • produces histamine and heparin

AGRANULOCYTES	LYMPHOCYTE	MONOCYTE	
• non-granular cytoplasm • compact nucleus	• rounded nucleus • produces antibodies	• indented (kidney-shaped) nucleus • engulfs bacteria	

Fig 8.2 *Types of white blood cell*

Table 8.1 *Summary of the transport functions of blood*

Materials transported	Examples	Transported from	Transported to	Transported in
Respiratory gases	Oxygen	Lungs	Respiring tissues	Haemoglobin in red blood cells
	Carbon dioxide	Respiring tissues	Lungs	Haemoglobin in red blood cells Hydrogen carbonate ions in plasma
Organic digestive products	Glucose	Intestines	Respiring tissues/liver	Plasma
	Amino acids	Intestines	Liver/body tissues	Plasma
	Vitamins	Intestines	Liver/body tissues	Plasma
Mineral salts	Calcium	Intestines	Bones/teeth	Plasma
	Iodine	Intestines	Thyroid gland	Plasma
	Iron	Intestines/liver	Bone marrow	Plasma
Excretory products	Urea	Liver	Kidney	Plasma
Hormones	Insulin	Pancreas	Liver	Plasma
	Antidiuretic hormone	Pituitary gland	Kidney	Plasma
Heat	Metabolic heat	Liver and muscle	All parts of the body	All parts of the blood

SUMMARY TEST 8.1

Blood is made up of a watery liquid called the **(1)**, in which lie a variety of cells. The most numerous are the **(2)** that are bi-concave discs about **(3)** in diameter. They live for about **(4)** and their role is to transport **(5)**, which is carried in combination with the red pigment, **(6)**, that they contain. The second type of cells are the **(7)**, which have a variety of types. The **(8)** have granular cytoplasm, a crescent-shaped nucleus and produce histamine. The **(9)** have a rounded nucleus and produce antibodies, while the cytoplasm of **(10)** is granular and they protect against allergies. The two types that engulf bacteria by **(11)** are **(12)** and **(13)**. In the plasma are two proteins important in clotting blood. These are **(14)** and **(15)**. Also important in clotting are cell fragments known as **(16)**.

8.2

Transport of oxygen

| | EDEXCEL |
| EDEXCEL (Human) |
| AQA.B | OCR |

MEASURING OXYGEN CONCENTRATION

The amount of a gas such as oxygen that is present in a mixture of gases is measured by the pressure it contributes to the total pressure of the gas mixture. This is known as the **partial pressure** of the gas, and in the case of oxygen is written as pO_2. It is also known as **oxygen tension** and is measured in the usual unit for pressure, namely **kiloPascals (kPa)**. Normal atmosphere pressure is 100kPa. As oxygen makes up 21% of the atmosphere, its partial pressure is normally 21kPa

As organisms evolved, they became more complex and, in many cases, much larger. Metabolic rates increased and with them the demand for oxygen. Specialised gaseous exchanges surfaces such as gills and lungs evolved that met this need (Chapter 6). This solved the problem only if there was a mechanism to transport the oxygen from these surfaces to the cells requiring it. Even with blood vessels and a heart to pump the blood around them, the transport of oxygen would be totally inadequate if the gas were simply dissolved in the plasma. Only the evolution of specialised molecules capable of carrying large quantities of oxygen could adequately supply the tissues. These molecules are called **respiratory pigments**, the best known of which is **haemoglobin**.

8.2.1 Haemoglobin

Haemoglobin is a red pigment which has a large relative molecular mass of 68 000. As such, it could be lost from the body during **ultrafiltration** (section 8.4.2) in the kidneys. It is therefore contained within the red blood cells which carry it around the body, separated from the plasma. Its structure is described in section 1.7.2. One oxygen molecule can combine with each of its four haem groups to form **oxyhaemoglobin**. To be efficient, haemoglobin must
- readily pick up oxygen at the gaseous exchange surface
- readily release oxygen at those tissues requiring it.

These two requirements may appear contradictory, but are achieved by the remarkable property of haemoglobin changing its affinity for oxygen under different conditions (table 8.2).

Table 8.2 Affinity of haemoglobin for oxygen under different conditions

Region of body	Oxygen tension (concentration)	Carbon dioxide tension (concentration)	Affinity of haemoglobin for oxygen	Result
Gaseous exchange surface	High	Low	High	Oxygen is absorbed
Respiring tissues	Low	High	Low	Oxygen is released

Haemoglobin has a greater affinity for carbon monoxide than for oxygen – a feature which makes this gas potentially lethal, especially because, once it is attached to haemoglobin, the carbon monoxide molecule remains there permanently, and so prevents oxygen molecules being loaded.

8.2.2 Oxygen dissociation curves

When haemoglobin is exposed to different partial pressures of oxygen (see margin), it does not absorb the oxygen evenly. At very low concentrations of oxygen, the four polypeptides of the haemoglobin molecule are closely united, and so it is difficult to absorb the first oxygen molecule. However, once loaded, this oxygen molecule causes the polypeptides to bind more loosely, and so the remaining three oxygen molecules are loaded very easily. The graph of this relationship is known as the **oxygen dissociation curve**; it is illustrated in figure 8.3. The graph tails off at very high oxygen concentrations, simply because the haemoglobin is almost saturated with oxygen. There are many different oxygen dissociation curves because:

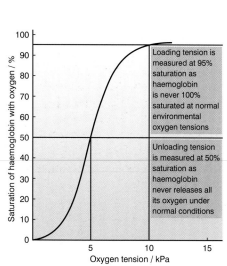

Fig 8.3 Oxygen dissociation curve for adult human haemoglobin

- there are a number of different respiratory pigments
- haemoglobin exists in a number of different forms
- the characteristics of each pigment change under different conditions.

The many different oxygen dissociation curves are better understood if two facts are always kept in mind:

- the more to the left the graph is, the more readily the pigment picks up oxygen but the less easily it releases it
- the more to the right the graph is, the less readily the pigment picks up oxygen but the more easily it releases it.

8.2.3 Foetal haemoglobin

Even within a single species, haemoglobin can exist in more than one form at different stages of life. Consider the situation in mammals when the foetus is being carried by the mother. At the placenta (section 5.13.1) the two blood systems are very close together, without the blood within them actually mixing. If the mother and foetus had the same haemoglobin then, under the same conditions, there would be no reason for the mother's haemoglobin to give up its oxygen to the foetal haemoglobin. In practice, foetal mammals produce their red blood cells in the liver, and the haemoglobin in these cells has two polypeptide chains that differ from the adult's. This gives foetal haemoglobin a higher affinity for oxygen than maternal haemoglobin (Fig 8.4). Oxygen therefore more readily transfers from the mother to the foetus at the placenta. After birth, red blood cell production moves from the liver to the marrow of bones such as the ribs. This produces the adult haemoglobin, which has a lower affinity for oxygen then foetal haemoglobin.

8.2.4 Myoglobin

Myoglobin is a different respiratory pigment whose main role is to store, rather than transport, oxygen. Found in muscles of all vertebrates, it is used to provide an emergency supply of oxygen at times when the rate at which muscles are using oxygen exceeds the rate of supply by the blood. Myoglobin has a higher affinity for oxygen than haemoglobin and so its dissociation curve is to the left of haemoglobin's (Fig 8.5). This ensures that oxygen for muscle action is taken from haemoglobin rather than from myoglobin – the latter being used only when the haemoglobin supply is exhausted. It also ensures that myoglobin is rapidly re-loaded with oxygen after exercise has ended and supply again exceeds demand. The muscles of diving mammals such as whales and seals have much myoglobin. This acts as a store of oxygen to sustain them during long periods of submersion.

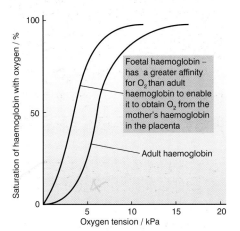

Fig 8.4 *Comparison of oxygen dissociation curves of foetal and adult haemoglobin*

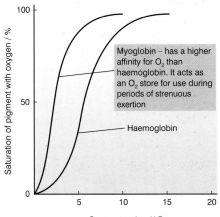

Fig 8.5 *Comparison of oxygen dissociation curves of human haemoglobin and myoglobin*

SUMMARY TEST 8.2

Haemoglobin in humans is an example of a (**1**). One molecule has a relative molecular mass of (**2**) and possesses a total of (**3**) haem groups, each of which possesses a single atom of (**4**). The 'globin' part of the molecule comprises four (**5**) that are of two different types, (**6**) and (**7**). Each haemoglobin molecule can carry a total of (**8**) molecules of oxygen. The amount of oxygen present in a mixture of gases is known as the oxygen (**9**) and the graph of the relationship between this and the amount of oxygen taken up by haemoglobin is called the (**10**). The haemoglobin in a foetus is different from that in an adult, in that it has a (**11**) affinity for oxygen and it is produced in the (**12**) rather than the (**13**) as it is in an adult. Another oxygen carrying substance is (**14**), which is found in muscles. It has a (**15**) affinity for oxygen than haemoglobin and is found in large quantities in mammals such as a (**16**), where it stores oxygen.

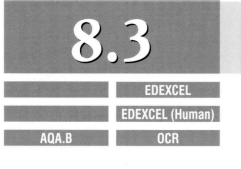
The carbon dioxide produced by respiratory tissues must be carried back to the gaseous exchange surface for removal from the body, because its accumulation is harmful. This carbon dioxide is, however, essential to the efficient release of oxygen from haemoglobin to the tissues. This is the **Bohr effect**, named after its discoverer in 1904, Christian Bohr.

8.3.1 The Bohr effect

Haemoglobin has a reduced affinity for oxygen in the presence of carbon dioxide. The greater the concentration of carbon dioxide, the more readily it releases its oxygen. This is the Bohr effect, and explains the differing behaviour of haemoglobin in different regions of the body:

- At the gaseous exchange surface (e.g. lungs), the level of carbon dioxide is low, because it diffuses across the exchange surface and is expelled from the organism. Haemoglobin's affinity for oxygen is increased which, coupled with the high concentration of oxygen in the lungs, means that oxygen is readily absorbed by haemoglobin. The reduced carbon dioxide level has shifted the oxygen dissociation curve to the left (Fig 8.6).

- In the respiratory tissues (e.g. muscles), the level of carbon dioxide is high in the blood because of its production during respiration. Haemoglobin's affinity for oxygen is reduced which, coupled with the low concentration of oxygen in the muscles, means that oxygen is readily released from the haemoglobin to the muscle cells. The increased carbon dioxide level has shifted the oxygen dissociation curve to the right (Fig 8.6). This is especially important during exercise because the more carbon dioxide that is produced, the more readily oxygen is supplied from the haemoglobin to meet extra energy demands due to exercise.

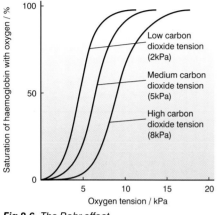

Fig 8.6 *The Bohr effect*

The Bohr effect is a consequence of the acidic nature of dissolved carbon dioxide: it forms hydrogen ions and hydrogen carbonate ions. It is the hydrogen ions that lower the affinity of haemoglobin for oxygen and lower the pH. Low pH caused by other chemicals, e.g. lactic acid, therefore also reduces haemoglobin's affinity for oxygen in the same way.

8.3.2 Transport of carbon dioxide

Carbon dioxide is carried from the tissues to the gaseous exchange surface in three ways:

- **In solution in the plasma** – although carbon dioxide is more soluble in the aqueous plasma than oxygen is , only 5% of the total carbon dioxide is carried in this way.

- **In combination with haemoglobin** – carbon dioxide can combine with amino groups in the protein part of the haemoglobin molecule:

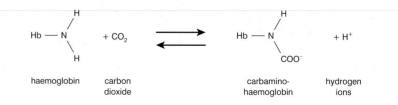

About 10% of the total carbon dioxide is carried in this way.

- **As hydrogen carbonate ions** – the remaining 85% of the total carbon dioxide is transported in this form. The carbon dioxide combines with water to form carbonic acid, which then dissociates (splits) into hydrogen ions (H^+) and hydrogen carbonate ions (HCO_3^-). The reaction is catalysed by the enzyme **carbonic anhydrase** and is summarised as:

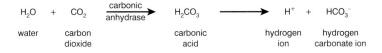

$$H_2O \quad + \quad CO_2 \xrightarrow{\text{carbonic anhydrase}} H_2CO_3 \longrightarrow H^+ \quad + \quad HCO_3^-$$

water   carbon   carbonic   hydrogen   hydrogen
          dioxide   acid   ion   carbonate ion

This reaction takes place in red blood cells. The hydrogen ions produced combine with haemoglobin to form **haemoglobinic acid** and so cause it to release its oxygen, which diffuses out of the cell into the nearby respiratory tissue. In this way, haemoglobin acts as a buffer, helping to keep the pH of the blood around neutral.

The loss of the negatively charged hydrogen carbonate ions would upset the ionic balance of the red blood cells, were it not for the **chloride shift**.

8.3.3 The chloride shift

The chloride shift involves the replacement of hydrogen carbonate ions lost from red blood cells with chloride ions from the dissociation of sodium chloride. The process is shown in figure 8.7 and can be summarised as:
- hydrogen carbonate ions (HCO_3^-) diffuse out of red blood cells into the plasma
- they combine with sodium ions (Na^+) from the dissociation of sodium chloride molecules (NaCl)
- the chloride ions (Cl^-) from the dissociation of sodium chloride diffuse into the red blood cells, replacing the hydrogen carbonate ions (HCO_3^-) that were lost, and so maintain the correct ionic balance.

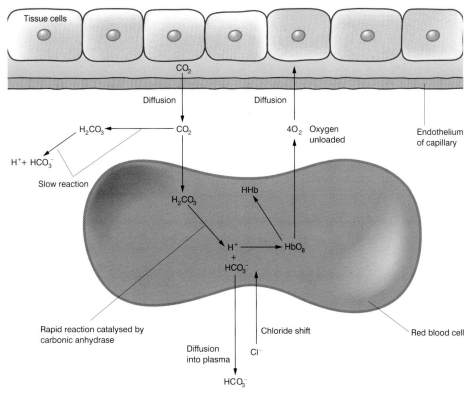

Fig 8.7 *The chloride shift*

SUMMARY TEST 8.3

Carbon dioxide produced by tissues during the process of (**1**) is transported to the gaseous exchange surface in three ways. Firstly, it combines with haemoglobin to form (**2**) and (**3**) ions; this accounts for about (**4**)% of the total carbon dioxide carried. Secondly, around (**5**)% is transported in solution in the (**6**). The remaining (**7**)% is carried in the form of (**8**) that are formed from the dissociation of (**9**) by the enzyme (**10**). One product of this reaction is hydrogen ions, which then combine with haemoglobin to form (**11**) which acts as a (**12**) by helping to keep the blood pH neutral. The ionic balance of the red blood cells could be upset by this reaction were it not maintained through a process called the (**13**). The affinity of haemoglobin for oxygen is reduced in the presence of carbon dioxide. This change is known as the (**14**).

147

8.4 — Tissue fluid and lymph

AQA.A	EDEXCEL
AQA.A (Human)	EDEXCEL (Human)
AQA.B	OCR

Blood supplies nutrients to the tissues of the body via tiny vessels called capillaries. Small though they are, these cannot serve every single cell directly, and therefore the final stage of the nutrient's journey is made in solution in a liquid that bathes the tissues – **tissue fluid**.

8.4.1 Tissue fluid

Formed from the plasma of the blood, tissue fluid is a watery liquid which contains glucose, amino acids, fatty acids, salts and oxygen, all of which it supplies to the tissues. In return, it receives carbon dioxide and other waste materials. Tissue fluid is therefore the means by which materials are exchanged between blood and cells and, as such, it bathes all cells of the body.

8.4.2 Formation of tissue fluid

Blood pumped by the heart passes along arteries, then the narrower arterioles and, finally, the even narrower capillaries. This creates a pressure, called **hydrostatic pressure**, of around 4.8 kPa at the arterial end of the capillaries, which tends to force liquid out of the blood. The outward pressure is, however, opposed by two other forces:

- hydrostatic pressure of the tissue fluid outside the capillaries, which prevents outward movement of liquid
- **osmotic** forces due to the plasma proteins, which tend to pull water back into the capillaries.

The combined effect of all these forces is to create an overall pressure of 1.7 kPa, which pushes tissue fluid out of the capillaries. This pressure is only enough to force small molecules out of the capillaries, leaving all cells and proteins in the blood. This type of filtration under pressure is called **ultrafiltration**. The loss of the tissue fluid reduces the pressure in the capillaries and so, by the time the blood has reached the venous end of the network, its hydrostatic pressure is less than that of the tissue fluid outside it. Along with the osmotic forces due to the proteins in the blood plasma that pull water back into the capillary, there is also an overall negative pressure of 1.5 kPa drawing tissue fluid back into the capillaries. This fluid has lost much of its oxygen and nutrients by diffusion to the cells it bathed, but has gained carbon dioxide and excretory products in return. These events are summarised in figure 8.9. Not all the tissue fluid can return to the capillaries; the remainder is carried back via the lymphatic system.

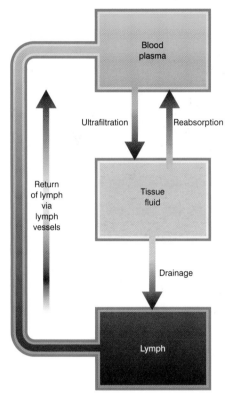

Fig 8.8 *Relationship between plasma, tissue fluid and lymph*

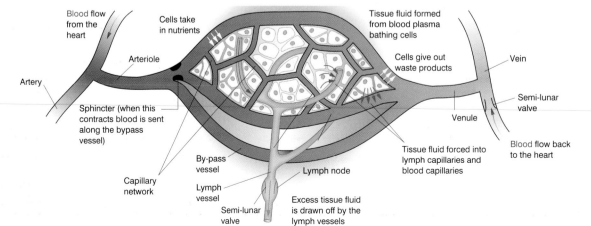

Fig 8.9 *Formation of tissue fluid*

8.4.3 Lymph and the lymphatic system

Lymph is a milky liquid made up of material from three sources:

- **tissue fluid** that has not been reabsorbed at the venous end of the capillary network
- **fatty substances** absorbed by lacteals in the ileum (section 6.9.4).
- **lymphocytes** (section 8.1.3) which have either been produced by the lymph nodes or have migrated from capillaries to fight infection.

Lymph is carried in the **lymphatic system**, which is made up of capillaries resembling blood capillaries, which merge into larger vessels that form a network around the body (Fig 8.10). The lymph vessels drain their contents back into the blood stream via two ducts:

- **the right lymphatic duct** drains the thorax, right side of the head and right arm into the right subclavian vein near the heart
- **the thoracic duct** draws the rest of the body's lymph into the left subclavian vein.

At points along the lymph vessels are a series of **lymph nodes**, which produce and store lymphocytes. Lymph nodes filter from the blood any bacteria and other foreign material, which are then engulfed by lymphocytes. This causes the nodes to swell with dead cells, and is the reason for the tenderness often felt in the groin, armpits and neck during an infection.

Lymph is moved along lymph vessels in three ways:

- **hydrostatic pressure** of the tissue fluid leaving the capillaries
- **contraction of body muscles** squeezes the lymph vessels. Valves ensure that the fluid inside them moves away from the tissues in the direction of the heart
- **enlargement of the thorax during breathing in** reduces pressure in the thorax, drawing lymph into this region and away from the tissues.

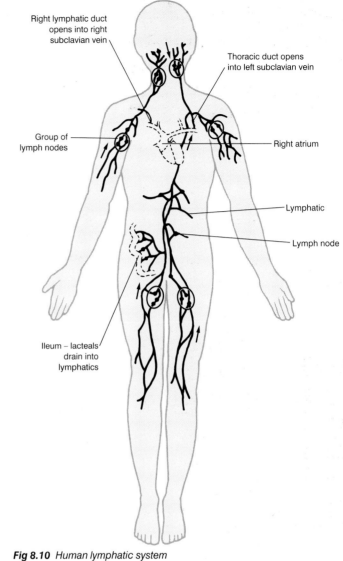

Right lymphatic duct opens into right subclavian vein

Thoracic duct opens into left subclavian vein

Group of lymph nodes

Right atrium

Lymphatic

Lymph node

Ileum – lacteals drain into lymphatics

Fig 8.10 *Human lymphatic system*

Table 8.3 *Comparison of various body fluids*

	Blood	Plasma	Tissue fluid	Lymph
Cells	Erythrocytes, leucocytes and **platelets** (cell debris)	None	None	Lymphocytes
Location	Within heart, arteries, veins and capillaries	Within heart, arteries, veins and capillaries	Outside vessels, bathing cells	Within lymph vessels
Moved by	Pumping of heart, muscle contraction and breathing action	Pumping of heart, muscle contraction and breathing action	Hydrostatic and osmotic forces	Hydrostatic forces, muscle contraction and breathing action
Direction of flow	Circulates around the body – to and from heart	Circulates around the body – to and from heart	Out of arterial end and into venous end of capillaries	Towards the heart from the tissues
Derived from	Bone marrow, thymus gland and lymph nodes	Water and substances absorbed by alimentary canal	Plasma	Tissue fluid, fatty material absorbed by ileum and lymph nodes
Function	Transport and defence	Transport over long distances	Transport over short distances	Transport and defence

8.5

The circulatory system

AQA.A	EDEXCEL
AQA.A (Human)	EDEXCEL (Human)
AQA.B	OCR

NAMING ARTERIES AND VEINS

There is often confusion over naming arteries and veins. The absolute rule is that arteries carry blood **away** from the heart (remembered by the fact that 'artery' and 'away' both start with the letter 'a'). Veins therefore always carry blood **towards** the heart. Although arteries usually carry oxygenated blood and veins deoxygenated blood, the pulmonary artery and vein are both exceptions to this rule

To permit both rapid transport of blood and control of its distribution, larger, more active organisms have evolved a **closed circulatory system** in which blood is retained within **vessels**. In mammals and birds, the blood loses pressure as it is pumped through the capillaries of the lungs. It is then returned to the heart to boost its pressure and ensure its rapid circulation to the rest of the body. Blood therefore passes twice through the heart in each complete circuit of the body (Fig 8.11). This is known as a **double circulatory system**.

8.5.1 Blood vessels

There are three types of blood vessel in a closed circulatory system:
- **Arteries** carry blood away from the heart. Smaller arteries are called **arterioles**.
- **Veins** carry blood towards the heart. Smaller veins are called **venules**.
- **Capillaries** are smaller vessels which link arteries to veins.

At any time, 75% of the blood is in veins, 20% is in arteries and 5% in capillaries. The three types of vessel are illustrated and described in table 8.4. From these diagrams, it can be seen that arteries and veins are good examples of an organ (section 4.3.3), as they are made up of different tissues such as epithelial tissue (endothelium), connective tissue (elastic and **collagen** fibres) and muscular tissue (**smooth muscle**), which are grouped to perform the major function of distributing blood around the body.

Table 8.4 *Comparison of arteries, veins and capillaries*

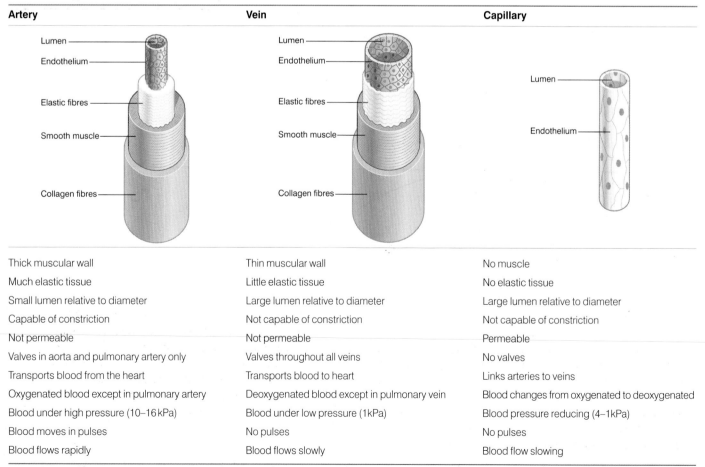

Artery	Vein	Capillary
Thick muscular wall	Thin muscular wall	No muscle
Much elastic tissue	Little elastic tissue	No elastic tissue
Small lumen relative to diameter	Large lumen relative to diameter	Large lumen relative to diameter
Capable of constriction	Not capable of constriction	Not capable of constriction
Not permeable	Not permeable	Permeable
Valves in aorta and pulmonary artery only	Valves throughout all veins	No valves
Transports blood from the heart	Transports blood to heart	Links arteries to veins
Oxygenated blood except in pulmonary artery	Deoxygenated blood except in pulmonary vein	Blood changes from oxygenated to deoxygenated
Blood under high pressure (10–16 kPa)	Blood under low pressure (1 kPa)	Blood pressure reducing (4–1 kPa)
Blood moves in pulses	No pulses	No pulses
Blood flows rapidly	Blood flows slowly	Blood flow slowing

8.5.2 The circulatory system

The circulatory system is a vast network of arteries, veins and capillaries, the total length of which in a human would stretch for 160 000km – enough to go around the world four times. The circulatory system in mammals carries blood between different parts of the body. Its role is to transport respiratory gases, water, nutrients, metabolic wastes, hormones and heat from their origins to their destinations (section 8.1.4). A plan of the mammalian circulatory system is given in figure 8.11. The circulation of blood within the system is maintained in three ways:

- **By the heart** – a muscular organ that pumps blood around the body (unit 8.6).
- **By contraction of skeletal muscle** during movement – this squeezes veins, which have valves to ensure that the displaced blood moves away from the tissues towards the heart.
- **Enlargement of the thorax during breathing in** reduces the pressure in the thorax and helps to draw venous blood back towards the heart.

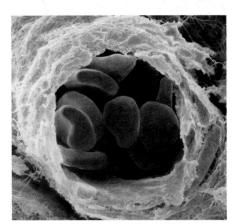

Red blood cells in an arteriole (SEM)

Table 8.5 *Showing the rate of blood flow to different parts of the body at rest and during strenuous exercise*

Part of the body	Rate of blood flow / cm³ min⁻¹	
	At rest	During exercise
Liver and intestines	2500	90
Body muscle	1000	16000
Kidneys	1000	300
Brain	750	750
Skin	500	1000
Heart muscle	250	1200

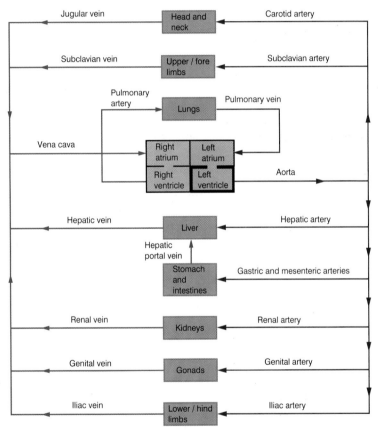

Fig 8.11 *Plan of the mammalian circulatory system*

SUMMARY TEST 8.5

Where blood is circulated within vessels it is called a **(1)** circulatory system. The vessels are of three types – arteries, veins and capillaries. Arteries carry blood **(2)** the heart and they have more **(3)** and **(4)** in their walls compared with the others. With the exception of the **(5)** and **(6)**, other arteries do not possess valves because the **(7)** prevents blood moving backwards. The only vessels that are permeable are the **(8)**, while only **(9)** are capable of constricting. The space at the centre of all blood vessels is called the **(10)** and is smallest relative to the vessel's diameter in **(11)**. The inner lining of blood vessels is called the **(12)**. Blood going to the brain leaves the **(13)** chamber of the heart, passes along the **(14)** and completes its journey via the **(15)**. Blood from the small intestine passes first to the organ called the **(16)** via the vessel known as **(17)**, and then passes along the **(18)** and the **(19)** before entering the **(20)** chamber of the heart.

Structure of the heart

AQA.A   EDEXCEL

AQA.A (Human)   EDEXCEL (Human)

AQA.B   OCR

The heart is a muscular organ which, in humans, pumps 13 000 litres of blood each day – enough to fill a small road tanker. It operates continuously and tirelessly throughout the life of an organism – a period of up to 100 years.

8.6.1 Structure of the human heart

Lying in the thoracic cavity between the two lungs, the human heart is made up of a unique type of muscle, **cardiac muscle**, which is capable of rhythmical contraction and relaxation over a long period, without tiring. It also contains connective tissue, which provides strength and prevents tearing. The heart is covered by a tough membrane called the **pericardium**, which encloses the **pericardial fluid**. This surrounds the heart and lubricates its movement relative to the pericardium. The human heart is really two separate pumps lying side by side. The left-hand pump deals with oxygenated blood from the lungs, while the right-hand one deals with deoxygenated blood from the body. Each pump has two chambers:

- **the atrium** is thin-walled and elastic and distends as it collects blood
- **the ventricle** has a much thicker wall, to pump blood away.

The right ventricle pumps blood to the lungs, a distance of only a few centimetres, and has a relatively thin muscular wall. The left ventricle, in contrast, has a thick muscular wall, enabling it to create enough pressure to pump blood to the extremities of the body, a distance of about 1.5 metres. Although the two sides of the heart are separate pumps and there is no mixing of the blood in each after birth, they nevertheless pump in time with each other: both atria pump together, followed by both ventricles together.

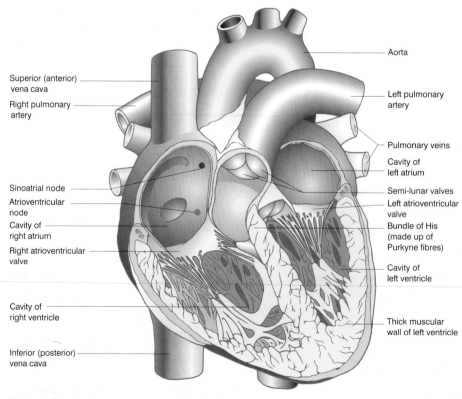

Fig 8.12 *Section through the human heart (VS)*

Between the atrium and ventricle are valves, which prevent the back flow of blood into the atria when the ventricles contract. There are two sets of valves.

- **Left atrioventricular (bicuspid) valves** are formed of two cup-shaped flaps on the left side of the heart.
- **Right atrioventricular (tricuspid) valves** or **mitral valves** are formed of three cup-shaped flaps on the right side of the heart.

To prevent these valves inverting under pressure, they are attached to special pillars of muscle on the heart wall by fibres called the **tendinous chords**. Each of the four chambers of the heart is served by large blood vessels that carry blood into or away from it. The ventricles are thicker than the atria because they pump blood away from the heart, and are therefore always connected to arteries. The atria are thinner because they receive blood and are therefore connected to veins (remember A/V: Atria/Veins and Arteries/Ventricles). Vessels connecting the heart to the lungs are called **pulmonary** vessels. The vessels connected to the four chambers are therefore as follows:

- **the aorta** is connected to the left ventricle and carries oxygenated blood to the all parts of the body except the lungs
- **the vena cava** is connected to the right atrium and brings deoxygenated blood back from the tissues of the body
- **the pulmonary artery** is connected to the right ventricle and carries deoxygenated blood to the lungs, where its oxygen is replenished and its carbon dioxide is removed. Unusually for an artery, it carries deoxygenated blood
- **the pulmonary vein** is connected to the left atrium and brings oxygenated blood back from the lungs. Unusually for a vein, it carries oxygenated blood.

The structure of the heart and its associated blood vessels is shown in figure 8.12.

8.6.2 Supplying the cardiac muscle with oxygen

Although oxygenated blood passes through the left side of the heart in vast quantities, the heart does not use this oxygen to meet its own great respiratory needs. Instead, the heart muscle is supplied by its own blood vessels, called the **coronary arteries**, which branch off the aorta shortly after it leaves the heart. Blockage of these arteries, e.g. by a blood clot, leads to **myocardial infarction** or a **heart attack**, because an area of the heart is deprived of oxygen and so ceases to function effectively.

8.6.3 Benefits of aerobic exercise to the heart

- increase in the amount of heart muscle, leading it to pump more blood at each beat and with greater force
- increase in **cardiac output** (section 8.8.1) due to a greater stroke volume and maximum heart rate
- decrease in resting heart rate.

Other effects of exercise on the heart are given in section 8.8.5.

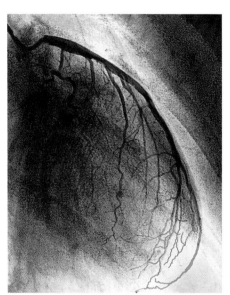

Human heart showing coronary arteries

SUMMARY TEST 8.6

The mammalian heart is made up of **(1)** muscle and is covered by a tough layer called the **(2)**. It is made up of four chambers, a pair of thin-walled elastic ones called **(3)** and a pair of thick muscular ones called **(4)**. Between the chambers on the left side of the heart are the **(5)** valves, while those on the right side are called **(6)** valves. These valves are prevented from turning inside out by **(7)**. Blood from the lungs passes to the heart by the **(8)** and into the chamber called **(9)**; it will leave the heart via the vessel called the **(10)**. Glucose absorbed from the intestines will first enter the **(11)** chamber of the heart. If this glucose is destined for the heart muscle itself, the next blood vessel it will pass along will be the **(12)**. Oxygen is essential to heart muscle and is supplied via the **(13)** that branch off the aorta. A blockage of these vessels can lead to a heart attack that is also known as a **(14)**.

8.7

AQA.A   EDEXCEL
AQA.A (Human)   EDEXCEL (Human)
AQA.B   OCR

Cardiac cycle

The heart undergoes a sequence of events which is repeated in humans around 70 times each minute, and is known as the **cardiac cycle**.

8.7.1 Action of the heart

There are two basic components to the beating of the heart – contraction or **systole** and relaxation or **diastole**. Diastole occurs simultaneously in the atria and ventricles, but systole takes place independently in each. There are therefore three stages to a single cardiac cycle:

- **diastole**, when atria and ventricles relax and both fill with blood
- **atrial systole**, when the atria contract and push blood into the ventricles
- **ventricular systole**, when the ventricles contract and push blood into major arteries.

These events are summarised in figure 8.13.

8.7.2 Pressure and volume changes of the heart

As with any fluid, blood will always move from a region where its pressure is greater to a region where it is lower – a fact which helps to explain the changes in pressure illustrated in figure 8.14.

8.7.3 Control of the cardiac cycle

The sequence of events and their precise timing is vital for the efficient functioning of the heart. Cardiac muscle is **myogenic** – its contraction is started from within the muscle itself, rather than by nervous impulses from outside (neurogenic) as is the case with other muscle. Within the wall of the right atrium of the heart is a distinct group of cells known as the **sinoatrial node**. It is from here that the stimulus for contraction originates. The sinoatrial node has a basic rhythm of stimulation that determines the beat of the heart. For this reason it is often referred to as the **pacemaker**. The sequence of events that controls the cardiac cycle is:

- A wave of excitation spreads out from the sinoatrial node across both atria, causing them to contract.
- A layer of non-conductive tissue (atrioventricular septum) prevents the wave crossing to the ventricles.
- The wave of excitation is, however, picked up by a second group of cells called the **atrioventricular node**, which lies between the atria.
- The atrioventricular node, after a short delay, conveys a wave of excitation between the ventricles along a series of specialised muscle fibres called the **bundle of His**.
- The bundle of His conducts the wave through the atrioventricular septum to the base of the ventricles, where the bundle branches into smaller fibres known as **Purkyne fibres**.
- The wave of excitation is released from the Purkyne fibres, causing both ventricles to contract at the same time, from the apex of the heart upwards.

8.7.4 Artificial pacemakers

Ageing or disease can damage the heart's natural pacemaker, the sinoatrial node. Although, in these circumstances, the atrioventricular node takes over, its rhythm is slower and leads to an abnormally slow heart beat. To overcome the problem, artificial pacemakers have been developed which can be implanted under the skin

1.
Blood enters atria and ventricles from pulmonary veins and venae cavae

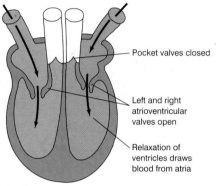

Pocket valves closed

Left and right atrioventricular valves open

Relaxation of ventricles draws blood from atria

Diastole
Atria are relaxed and fill with blood. Ventricles are also relaxed.

2.

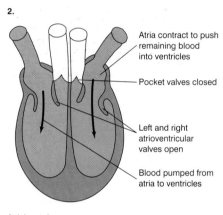

Atria contract to push remaining blood into ventricles

Pocket valves closed

Left and right atrioventricular valves open

Blood pumped from atria to ventricles

Atrial systole
Atria contract, pushing blood into the ventricles. Ventricles remain relaxed.

3.
Blood pumped into pulmonary arteries and the aorta

Pocket valves open

Left and right atrioventricular valves closed

Ventricles contract

Ventricular systole
Atria relax. Ventricles contract, pushing blood away from heart through pulmonary arteries and the aorta.

Fig 8.13 *The cardiac cycle*

in a relatively simple operation. The size of a thin matchbox and weighing 20–60g, they are powered by a lithium battery and generate rhythmic electrical pulses that mimic those of the sinoatrial node. These impulses pass to tiny electrodes which are inserted through veins into the right atrium and right ventricle, where they stimulate normal heart muscle contraction.

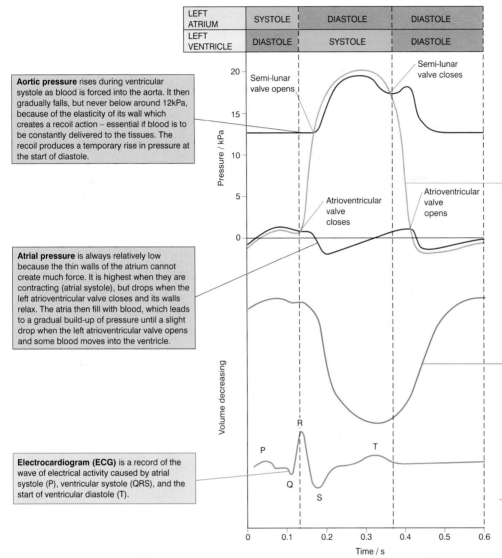

Aortic pressure rises during ventricular systole as blood is forced into the aorta. It then gradually falls, but never below around 12kPa, because of the elasticity of its wall which creates a recoil action – essential if blood is to be constantly delivered to the tissues. The recoil produces a temporary rise in pressure at the start of diastole.

Atrial pressure is always relatively low because the thin walls of the atrium cannot create much force. It is highest when they are contracting (atrial systole), but drops when the left atrioventricular valve closes and its walls relax. The atria then fill with blood, which leads to a gradual build-up of pressure until a slight drop when the left atrioventricular valve opens and some blood moves into the ventricle.

Electrocardiogram (ECG) is a record of the wave of electrical activity caused by atrial systole (P), ventricular systole (QRS), and the start of ventricular diastole (T).

Ventricular pressure is low at first, but gradually increases as the ventricles fill with blood during atrial systole. The left atrioventricular valves close and pressure rises dramatically as the thick muscular walls of the ventricle contract. As pressure rises above that of the aorta, blood is forced into the aorta past the semi-lunar valves. Pressure falls as the ventricles empty and the walls relax at the start of diastole.

Ventricular volume is almost the mirror image of ventricular pressure, but with a short time-delay. Volume rises during atrial systole as ventricles fill with blood, and then drops suddenly as blood is forced out into the aorta when ventricular pressure exceeds aortic pressure. Volume increases again as the ventricles fill with blood during systole.

Fig 8.14 Pressure, volume and ECG changes in the left side of the heart during the cardiac cycle

8.7.5 The electrocardiogram (ECG)

During the cardiac cycle, the heart undergoes a series of electrical current changes related to the waves of excitation created by the sinoatrial node and the heart's response to these. Picked up by a **cathode ray oscilloscope**, these changes can produce a trace known as an **electrocardiogram**. An example, related to the stages of the cardiac cycle, is shown as part of figure 8.14.

SUMMARY TEST 8.7

The mammalian heart beat is initiated from within the heart muscle itself and is therefore termed **(1)**. The pacemaker of the heart is the **(2)** that lies in the wall of the **(3)** of the heart. A wave of excitation from this pacemaker causes both **(4)** to contract, pushing blood into the **(5)**. This phase of the cardiac cycle is called **(6)**. The wave is picked up by another group of specialised cells called the **(7)**, which lies between the atria of the heart.

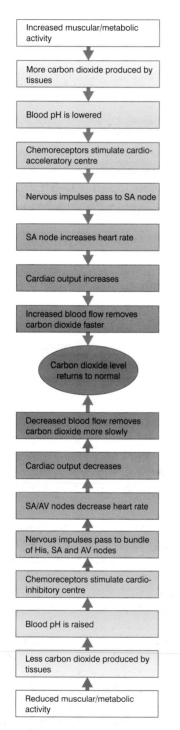

Fig 8.15 *Effects of exercise on cardiac output. SA node = sinoatrial node; AV node = atrioventricular node*

The usual resting heart rate (pulse rate) is 70 beats min^{-1} and normal blood pressure is 140mmHg (systolic) and 90mmHg (diastolic). (In SI units, these would be 18.7kPa and 12kPa respectively, but mmHg continues to be used in medical practice.) It is, however, essential that these levels are modified to meet an organism's varying demands for oxygen.

8.8.1 Cardiac output

Cardiac output is the volume of blood pumped by the heart in a given time. It is usually measured in dm^3 min^{-1} and depends upon two factors:
- the **heart rate** (the rate at which the heart beats)
- the **stroke volume** (how much blood is pumped out at each beat):

$$\text{Cardiac output } = \text{ heart rate } \times \text{ stroke volume}$$

Changes to the cardiac output are controlled by a region of the brain called the **medulla oblongata**, which has two centres:
- the **cardio-acceleratory centre,** which is linked to the **sinoatrial node** by the **sympathetic nervous system** and increases cardiac output
- the **cardio-inhibitory centre**, which is linked to the sinoatrial node by **parasympathetic fibres** within the vagus nerve. It decreases cardiac output.

Which of these centres is stimulated depends upon the information they receive from two types of receptor:
- **chemoreceptors**, which respond to chemicals in the blood
- **baroreceptors**, which respond to pressure changes in the blood.

8.8.2 Control by chemoreceptors

Chemoreceptors occur within a swelling of the carotid artery (which serves the brain) called the **carotid sinus**. They are sensitive to changes in the pH of the blood which result from changes in the concentration of carbon dioxide, which forms an acid in solution and therefore lowers pH. The sequences of events that follow changes in activity levels are given in figure 8.15.

8.8.3 Control by baroreceptors

Baroreceptors (pressure receptors) occur within the walls of the carotid sinuses and the aorta. If stimulation of these receptors indicates a higher than normal blood pressure, they convey a nervous impulse to the cardio-inhibitory centre. This, in turn, stimulates the parasympathetic nervous system to decrease the heart rate and so reduce cardiac output. At the same time, the parasympathetic nervous system also causes the arterioles of the body to dilate (vasodilation). This increase in the diameter of the arterioles means that there is a larger volume within the blood vessels for the blood to occupy, and therefore the pressure within them is reduced. This mechanism prevents blood pressure reaching a level that might burst the arteries. If blood pressure drops below a certain level, baroreceptors in the carotid sinuses and aorta stimulate the cardio-acceleratory centre, which increases the heart rate via the sympathetic nerves. At the same time, these nerves cause smooth muscle in the walls of the arterioles to contract. This **vasoconstriction** also increases blood pressure. This prevents blood pressure falling below a level at which the tissues, especially the brain, are not receiving enough oxygen.

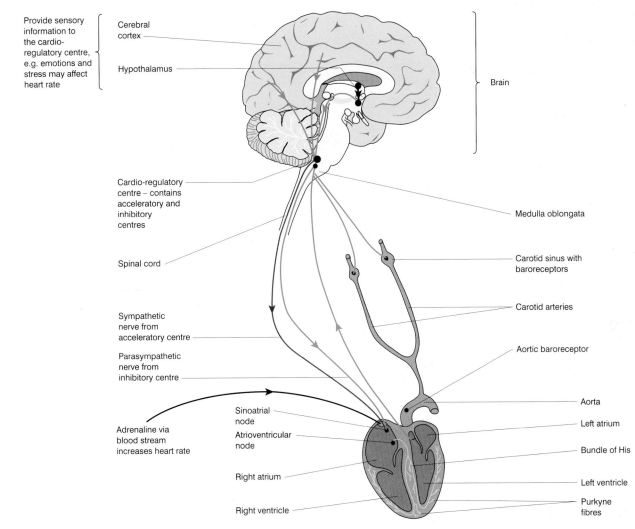

Provide sensory information to the cardio-regulatory centre, e.g. emotions and stress may affect heart rate

Cerebral cortex

Hypothalamus

Brain

Cardio-regulatory centre – contains acceleratory and inhibitory centres

Medulla oblongata

Spinal cord

Carotid sinus with baroreceptors

Carotid arteries

Sympathetic nerve from acceleratory centre

Parasympathetic nerve from inhibitory centre

Aortic baroreceptor

Aorta

Adrenaline via blood stream increases heart rate

Sinoatrial node

Atrioventricular node

Left atrium

Bundle of His

Right atrium

Left ventricle

Purkyne fibres

Right ventricle

Fig 8.16 *Summary of factors controlling heart rate*

8.8.4 Control by hormones

Certain hormones, such as adrenaline, affect the heart rate by stimulating the sympathetic nervous system, thereby increasing cardiac output. Under conditions of excitement or danger, more adrenaline is produced from the adrenal medulla. It travels in the blood to increase both the stroke volume and the heart rate, and so produces a marked increase in cardiac output.

8.8.5 Effects of exercise on the heart

During exercise, the body's demand for oxygen increases, and the rate of ventilation goes up in order to absorb more oxygen at the gaseous exchange surface (section 6.7.4). To be effective, the rate at which blood carries this oxygen to the muscles must likewise be increased. Accordingly, the cardiac output can rise from $5dm^3 min^{-1}$ at rest to a maximum of $30dm^3 min^{-1}$ during very strenuous exertion, although $20dm^3 min^{-1}$ is more normal. This is achieved by increasing the heart rate from 70 beats min^{-1} to 190 beats min^{-1} and the stroke volume from $80cm^3$ at each beat to $110cm^3$. As a result of training, an athlete can increase both the thickness of the ventricle walls and the size of the heart chambers. This

means that, even at rest, they require fewer beats than an untrained individual to pump the same volume of blood. A fit athlete may have a resting heart rate of 55 beats min^{-1}, compared with the normal 70 beats min^{-1}. Heart muscle is also stronger and the coronary arteries wider, to supply more blood to the heart muscle itself.

One major effect of exercise on the vascular system is the redistribution of blood during exertion. **Vasodilation** occurs in the arterioles serving the muscles involved in exercise, to increase their oxygen supply. The percentage of blood to these muscles can increase from 13% at rest to 66% during exercise. This extra blood must be taken from parts of the body that can temporarily exist with a more limited supply. These parts include the liver and intestines. It is important, however, to maintain the blood supply to other vital organs such as the heart, kidneys and brain, as they need to continue operating normally with no loss of function, however temporarily. The skin must also have an increased supply of blood, to allow the heat generated during exertion to be lost and thereby prevent a dangerous rise in temperature. The percentage of the blood supplied to the skin may double from 5% to 10% during exercise.

157

1 The table shows the rate of blood flow to various parts of the body at rest and during vigorous exercise.

Part of body	Rate of blood flow	
	at rest	during vigorous exercise
Skeletal muscles	1000	16 000
Heart muscle	250	1 200
Skin	500	750
Kidneys	1000	300
Liver and gut	1250	375
Brain	750	

a (i) Suggest suitable units for the rate of blood flow.
(1 mark)

(ii) Suggest a suitable value for the rate of blood flow to the brain during vigorous exercise. Give a reason for your answer. *(2 marks)*

b The blood flow to the skeletal muscles changes during vigorous exercise. Describe the part played by each of the following in bringing about this change:

(i) the nerves going to be the sinoatrial node in the heart; *(2 marks)*

(ii) the arterioles taking blood to the skeletal muscles. *(2 marks)*

c Use information from the table to suggest why it is recommended that vigorous exercise should not be undertaken until at least two hours after a meal.
(1 mark)
(Total 8 marks)
AQA June 2001, B/HB (A) BYA1, No.6

2 The table refers to features of the respiratory pigments haemoglobin and myoglobin. If the statement is correct, place a tick (✔), and if the statement is incorrect, place a cross (✘) in the appropriate box.

Feature	Haemoglobin	Myoglobin
Is carried in the blood		
Transports oxygen		
Acts as an oxygen store in muscle		
Transports carbon dioxide		

(Total 4 marks)
Edexcel 6112/01 June 2001, B (H) AS/A, No.1

3 The diagram shows a section through the heart of a mammal.

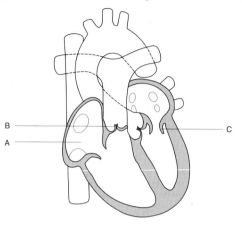

a Name the parts labelled A, B and C. *(3 marks)*

b Each time the heart beats, the atria contract first and then the ventricles contract. Explain how this sequence of events is coordinated. *(4 marks)*
(Total 7 marks)
Edexcel June 2001, B (H) AS/A, No.2

4 The diagram shows the relationship between different body fluids.

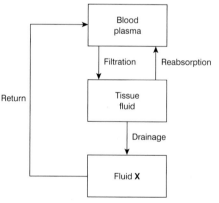

a Name fluid **X**. *(1 mark)*

b Give **one** way in which the composition of tissue fluid differs from that of plasma. *(1 mark)*

c Filtration of blood plasma is partly due to high blood hydrostatic pressure. How is the high blood hydrostatic pressure produced? *(1 mark)*

d Children who have insufficient protein in their diet develop a swollen abdomen. This is due to the accumulation of tissue fluid. Explain the link between insufficient protein in the diet and the accumulation of tissue fluid. *(3 marks)*
(Total 6 marks)
AQA June 2001, B (B) BYB3, No.2

5 The diagrams show the left side of the heart at two stages in a cardiac cycle.

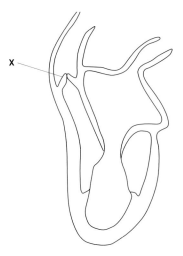

Diagram **A**

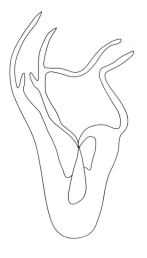

Diagram **B**

a Name the structure labelled **X**. *(1 mark)*

b Describe **two** pieces of evidence in diagram **B** which indicate that the ventricle is emptying. *(2 marks)*

c The cardiac output is the volume of blood that one ventricle pumps out per minute. The resting heart rate of an athlete often decreases as a result of training, even though the cardiac output remains the same. Suggest an explanation for this. *(1 mark)*

d During exercise the rate of blood flow to heart muscle increases from 270cm³ per minute to 750cm³ per minute.

  (i) Calculate the percentage increase in rate of blood flow to heart muscle during exercise. Show your working. *(2 marks)*

  (ii) Explain the advantage of the increase in the rate of blood flow to heart muscle during exercise. *(2 marks)*

(Total 8 marks)

AQA June 2001, B (B) BYB3, No.4

6 The figure shows cross sections of two types of vessel from the mammalian blood system.

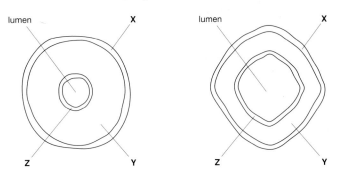

a With reference to the figure,

  (i) name the blood vessels **A** and **B**; *(1 mark)*
  (ii) name **X** to **Z**; *(3 marks)*
  (iii) describe briefly the structure of **X** to **Z**. *(3 marks)*

(Total 7 marks)

OCR 2803/1 Jan 2001, B(T), No.2

7 a Explain the difference between **systolic** and **diastolic blood pressure**. *(3 marks)*

An athlete's blood pressure was monitored for a period of twenty minutes. For the first three minutes, the athlete rested. The athlete then ran on a treadmill at a steady speed for ten minutes before resting again. The results are shown in the figure.

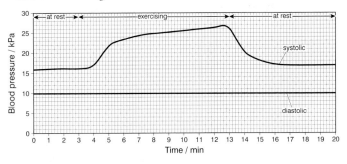

b With reference to the figure,

  (i) state the resting values for systolic and diastolic blood pressure; *(2 marks)*
  (ii) explain the change in systolic blood pressure during the period of exercise; *(4 marks)*
  (iii) explain why the athlete's systolic blood pressure did not return to the resting value immediately the exercise stopped. *(3 marks)*

c State **three** beneficial effects of regular aerobic exercise on the heart. *(3 marks)*

(Total 15 marks)

OCR 2802 Jan 2001, B(HHD), No.4

Energy and environment

Modes of nutrition

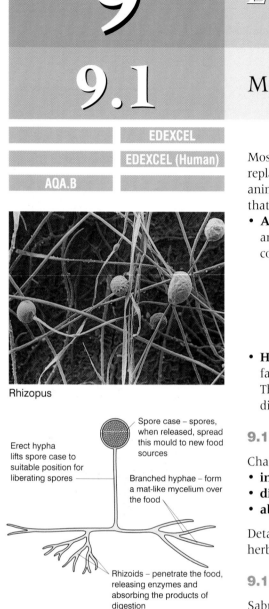

Rhizopus

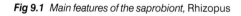

Erect hypha lifts spore case to suitable position for liberating spores

Spore case – spores, when released, spread this mould to new food sources

Branched hyphae – form a mat-like mycelium over the food

Rhizoids – penetrate the food, releasing enzymes and absorbing the products of digestion

Fig 9.1 *Main features of the saprobiont,* Rhizopus

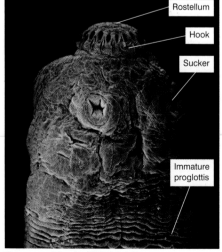

Rostellum

Hook

Sucker

Immature proglottis

Tapeworm Taenia *(SEM) (×28)*

Most chemical processes inside every living organism use energy, and this must be replaced from the environment. Plants obtain their energy from the sun, and animals and fungi obtain theirs by breaking down the complex organic compounds that make up other living organisms. These are the two basic types of nutrition:

- **Autotrophic nutrition** ('auto' = self, 'trophic' = feeding). Most autotrophs are green plants that use light energy, carbon dioxide and water to build up complex organic compounds by photosynthesis; they are **photoautotrophs**.

$$\text{carbon dioxide} + \text{water} \xrightarrow[\text{chlorophyll}]{\text{light energy}} \text{glucose} + \text{oxygen}$$

$$6CO_2 + 6H_2O \rightarrow C_6H_{12}O_6 + 6O_2$$

- **Heterotrophic nutrition** ('hetero' = varied, 'trophic' = feeding). The most familiar heterotrophs are animals, but others include some bacteria and fungi. They all consume complex organic food material. There are a number of different forms of heterotrophic nutrition.

9.1.1 Holozoic nutrition

Characteristic of most animals, this includes:
- **ingestion** of complex organic matter from the bodies of other organisms
- **digestion** to simpler, soluble molecules
- **absorption** of soluble materials for use in body cells.

Details of holozoic nutrition in humans are in unit 6.8 and adaptations of herbivores and carnivores in unit 9.2.

9.1.2 Saprobiontic nutrition

Sabrobionts are heterotrophic organisms which obtain carbon by absorption from dead organisms or organic wastes. The majority of saprobionts are bacteria, or fungi such as *Rhizopus* (Fig 9.1). These organisms do not 'eat' food in the way that typical holozoic animals do. Instead, they secrete enzymes directly onto the food source **(extracellular digestion)** and then absorb the soluble products of digestion. *Rhizopus*, a typical saprobiont found as a white mould growing on damp bread, feeds in the following way:
- **spores** land on a suitable, moist food source and germinate
- **hyphae**, branched thread-like structures, grow over the surface of the substrate (bread)
- **rhizoids**, which are specialised hyphae, penetrate the substrate
- **enzymes** (carbohydrates, lipases and proteases) are secreted by the rhizoids
- **extracellular digestion** of the complex molecules in the food takes place outside the structure of the fungus
- **absorption** of the soluble products of digestion into the rhizoids occurs.

9.1.3 Parasitic nutrition

Parasitism is a close association between two organisms in which one, the **parasite**, is nutritionally dependent on the other, the **host**. A parasite is different from a predator or scavenger, because it spends significantly longer feeding on its host and may induce an immune reaction in it. Parasites such as fleas live as **ectoparasites** on the outside of the host, but others, like tapeworms, are **endoparasites** living on the inside. Adults of the tapeworm, *Taenia*, live in the small intestine of humans. As they are surrounded by digested food, they do not require a digestive system of their own. *Taenia* (Fig 9.2) is adapted for parasitic nutrition as follows:

- **hooks and suckers** attach the tapeworm to the wall of the host's intestine, so it is not dislodged during **peristalsis**
- **long, thin shape** provides a large surface area over which the absorption of pre-digested food can occur. The adult worm is about 3 metres long and 6mm wide. It is made up of numerous segments, or proglottids, each only 1.5mm thick
- **thick outer tegument** made of protein and chitin prevents digestion of the worm by the host's enzymes
- **absence of mouth and digestive system** because food has already been digested by the host.

In addition, tapeworms must be adapted to survive in a low oxygen environment, and they must produce numerous eggs, to ensure that some successfully reach a new host. Tapeworms rarely cause severe harm to healthy adult hosts – it is not in the parasite's interest to kill its host – but children and infirm adults may suffer abdominal pain, loss of appetite, constipation and vomiting.

9.1.4 Mutualistic nutrition

Mutualism involves a close association between members of two species, in which both derive some benefit from the relationship. Examples of mutualism include the cellulose-digesting organisms found in the guts of **ruminants** (section 9.2.2) and **nitrogen-fixing bacteria** found in the root nodules of Papilionaceae (members of the bean family commonly known as legumes). Nitrogen fixation involves the transfer of hydrogen ions from carbohydrates, such as glucose, to nitrogen. The reaction is catalysed by the enzyme, **nitrogenase**, and the first product is ammonia. The ammonia is combined with glutamate to form glutamine, from which other amino acids are synthesised. Many nitrogen-fixing bacteria are found in the soil, but *Rhizobium* is found in swellings, called **root nodules** (Fig 9.3), that develop in the roots of leguminous plants such as peas, beans and alfalfa. The development of a root nodule involves:

- secretion of a hormone by the root of the legume. This attracts *Rhizobium*
- entry of *Rhizobium* into root hair
- *Rhizobium* stimulates the production of plant growth substances such as auxins and **cytokinins**
- these substances cause increased cell division in the root cortex, to form a swelling or **nodule**
- **bacteria** inside the nodules are Y-shaped and are known as **bacteroids**. These contain the nitrogenase required for nitrogen fixation
- a pigment called **leghaemoglobin** surrounds the bacteroids and absorbs oxygen. This provides the anaerobic conditions required by nitrogenase.

In this relationship, both *Rhizobium* and the leguminous plant benefit, as shown in table 9.1.

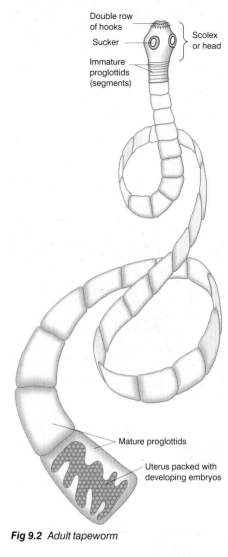

Fig 9.2 *Adult tapeworm*

Table 9.1 *Mutualistic benefits to* Rhizobium *and a legume*

Benefits to *Rhizobium*

- Supplied with carbohydrate by photosynthesis of legume
- Wall of nodule protects *Rhizobium*
- Anaerobic conditions provided for nitrogenase activity

Benefits to leguminous plant

- Receives a supply of amino acids from *Rhizobium*
- Legumes able to grow in nitrogen-deficient soils

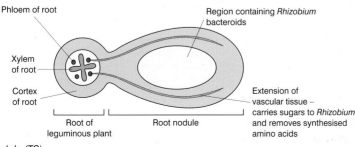

Fig 9.3 *Root nodule (TS)*

Adaptations of herbivores and carnivores

EDEXCEL

EDEXCEL (Human)

Skull of cow

Animals such as cattle and sheep, which eat mostly plant material, are called **herbivores**. They have a number of adaptations to this tough and largely indigestible food source. **Carnivores**, such as a dog, eat meat, which is mainly the muscle of another animal and therefore is rich in nutrients and relatively easily digested.

9.2.1 Herbivore dentition, e.g. of a cow

Vegetation is coarse, and so teeth are specialised to grind it up, disrupting the tissue and increasing its surface area. Adaptations of herbivores, such as a cow, include:

- **toughened gum pad**, which replaces upper incisors and canines
- **small sharp incisors** on lower jaw, which bite against the gum pad and tear up grass
- **broad molars and premolars**, which provide a large surface area for grinding
- **ridges of enamel** on molars and premolars, which help grind the food
- **teeth grow continuously** (also known as 'open pulp'), which compensates for their being worn away
- **loose jaw articulation**, which allows the narrower lower jaw to move from side to side during chewing
- **diastema** – a gap between incisors and premolars, in which newly nibbled food can be kept separate from that being chewed at the back of the mouth. This allows collection of food to continue even during chewing.

Plant food is so low in accessible nutrients that herbivores require large quantities and so spend a long time feeding. They have therefore evolved eyes placed laterally on the head, which enables them to detect the movement of approaching predators; many of them also have acute hearing.

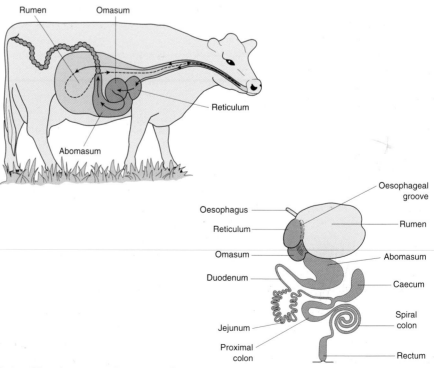

Fig 9.4 *Digestive system of a cow*

9.2.2 Ruminants

All herbivores need a long alimentary canal in relation to their body size, to give time for the digestion of vegetation to take place. Herbivorous mammals are unable to produce their own cellulase to digest the cellulose cell wall of plants. **Ruminants** therefore contain microorganisms that can produce cellulase, located in their complex, four-chambered stomach. The compartments of the stomach are the **rumen, reticulum, omasum** and then the true stomach, or **abomasum** (Fig 9.4):

- **The reticulum** – this is concerned with the movement of food between the rumen and oesophagus, and between the rumen and the omasum.
- **The rumen** – after being swallowed, food may spend 30 hours here, being fermented by huge populations of microorganisms. Coarse food may be returned to the mouth for re-chewing, called **chewing the cud**. This has the advantage of allowing the now partly fermented food, the cud, to be mixed with saliva, and the plant fibres it contains to be further broken up. The rumen of a cow has a capacity of 150 litres and the **microflora** within it:
  - convert polysaccharides, including cellulose, to short-chain organic acids, carbon dioxide and methane:

$$\underset{\text{(e.g. cellulose)}}{\text{polysaccharides}} \xrightarrow{\text{fermentation}} \text{glucose} \rightarrow \underset{\text{organic acids}}{\text{short chain}} + \underset{\text{dioxide}}{\text{carbon}} + \text{methane}$$

  - break down protein:

$$\text{protein} \rightarrow \underset{\text{acids}}{\text{amino}} \xrightarrow{\text{deamination}} \text{ammonia} \nearrow \underset{\searrow}{\overset{\text{proteins in microbes}}{}} \begin{array}{l} \text{proteins in} \\ \text{microbes} \\ \\ \text{blood of} \\ \text{ruminant} \end{array}$$

  - synthesise vitamins of the B complex.
- **The omasum** – food passes here from the rumen. It has no digestive secretions, but absorbs water and organic acids from the digested material. Its capacity is only one-tenth that of the rumen.
- **The abomasum** – this is the site of gastric secretions, similar to those produced by the human stomach (unit 6.8). Conditions are acidic, and protein digestion takes place. Ruminants often have a low-protein diet, but this is compensated for by the protein-synthesising bacteria in the rumen. These protein-rich bacteria are engulfed by single-celled organisms (**protoctists**) which are in turn digested by the ruminant, further along the digestive system.

9.2.3 Carnivore dentition e.g. of a dog

Being meat eaters, carnivores must have teeth adapted for holding and killing prey, as well as for the mechanical digestion of the flesh. Adaptations of a typical carnivore, such as a dog, include:

- **sharp incisor**s used for holding prey and removing muscle from the bones
- **long, curved canines**, which pierce and hold the prey as well as tear flesh
- **sharp, pointed molars and premolars**, which separate muscle fibres
- **carnassial teeth**, which are particularly large and effective at slicing flesh. They are the last upper premolar and the first lower molar on each side
- **tight jaw articulation**, which prevents sideways movement and possible dislocation caused by struggling prey
- **well-developed jaw muscles**, providing a powerful grip on prey.

The alimentary canal of carnivores is short, giving a clue to the relative ease with which meat can be chemically digested. Many carnivores are hunters; they can run fast, may be camouflaged, and have sharp claws to hold their prey. Their **stereoscopic vision**, provided by forwardly directed eyes, enables them to judge distances accurately, and so enables them to capture prey.

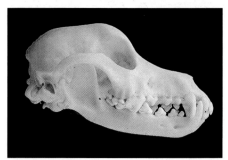

Skull of dog

SUMMARY TEST 9.2

Herbivores have small sharp (1) teeth to bite off vegetation, and broad (2) and (3) with ridges to grind food. There may be a gap, called the (4), between these two types of teeth. Compared with carnivores, herbivores have a (5) alimentary canal. In cattle, the stomach has four chambers, the (6) which houses microorganisms that produce the enzyme (7), the (8) that absorbs water and organic acids, the (9) that secretes gastric juices and the (10). Carnivores, by contrast, have sharp, pointed teeth, some of which are enlarged and slide past one another to slice the food. These are called (11) teeth.

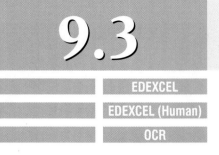
Ecosystems

Bluebells in a woodland habitat

Ecology is the study of the interrelationships between organisms and their environment. The term **environment** refers to the conditions that surround an organism. This includes both non-living **(abiotic)** components and living **(biotic)** components. Obviously, it is never possible to study every organism in every part of the world, but the area of study does need to be defined. It is important to be able to describe and give examples of the ecological terms in most common use.

9.3.1 Biosphere

The term biosphere refers to the relatively thin film of land, air and water around the Earth's surface that supports life. It extends about 8km above sea level and 10km below it. It consists of two major divisions:
• the aquatic environment
• the terrestrial environment.

Local climatic conditions lead to the development of particular kinds of dominant plants, so that the terrestrial environment may be sub-divided into **biomes** such as tropical rainforests, tundra and hot deserts. Within these sometimes vast areas a series of smaller zones, known as **habitats**, can be recognised.

9.3.2 Habitats

Examples of habitats include a rocky shore, a fresh-water pond and an oak woodland. The term refers not just to a single, specific rocky shore, pond or wood, but to all areas where a particular organism or group of organisms lives. The organisms within a habitat form populations and communities:
• **A population** is a group of organisms of the **same species** occupying the **same place** at the **same time**. In the oak woodland, there will be separate populations of, e.g. bluebells, oak trees, worms, woodpeckers, etc. The boundaries of populations are often difficult to define, except perhaps within the habitat of a small pond. All members of a particular (sexually reproducing) population have the chance of interbreeding.
• **A community** is defined as **all the populations** of different organisms living and interacting in a **particular place** at the **same time**. In our oak woodland, this would involve a very large number of organisms: the bluebells, blackbirds, worms, woodpeckers etc would all form part of the community.

In any environment it is unrealistic to consider only the interrelationships between living organisms, without looking at the effects of the non-living, abiotic factors.

9.3.3 Ecosystems

An ecosystem consists of all the interacting biotic and abiotic elements in a particular area. Ecosystems are recognised as relatively self-contained functional units. Within the oak woodland ecosystem, leaves fall and decay, and their minerals are returned to the soil. The plants then absorb them and use them for growth (units 9.6 and 9.7). The communities in the woodland interact with each other, especially through feeding relationships, and they are all affected by the abiotic conditions around them (Fig 9.5). These abiotic factors include:
• temperature
• availability of sunlight
• availability of water
• availability of carbon dioxide and oxygen
• nature of the soil.

SUMMARY TEST 9.3

The band of land, water and air around the Earth's surface that supports life is the **(1)**. The area of land within this band is called the **(2)** environment and is divided, according to the different types of vegetation, into **(3)** such as deserts, tundra and tropical rainforests. Smaller zones within these areas are called habitats, examples of which include **(4)** and **(5)**. Within a habitat are groups of organisms of the **(6)** that occupy the same space and time, called populations. All these populations interact to form a **(7)**. The interrelationship of the living, or **(8)**, components and the non-living, or **(9)**, components of a biological system is called an **(10)** and its study is referred to as **(11)**.

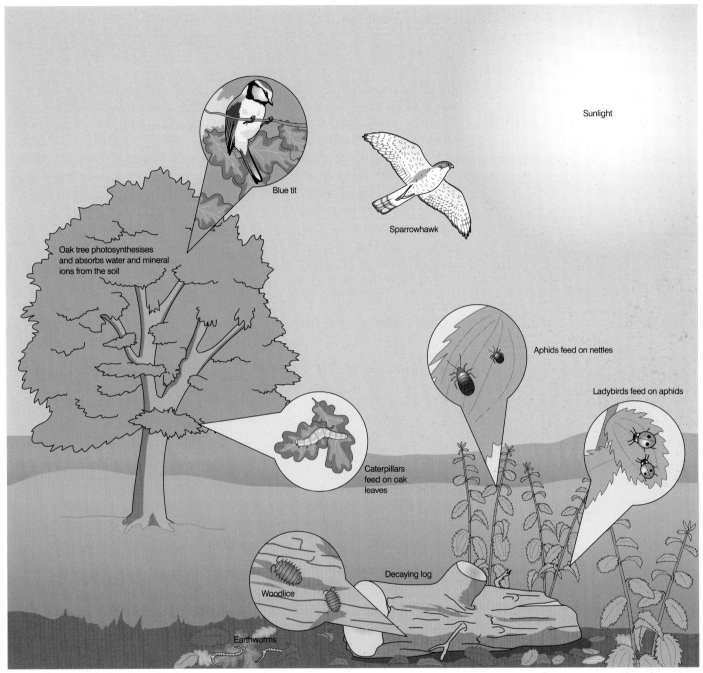

Fig 9.5 Oak woodland ecosystem

Within any ecosystem, there is, unavoidably, competition for the resources available. As a result of this competition, no two species can have exactly the same requirements. Each must occupy a different **niche**.

9.3.4 Ecological niche

This is the ecological role of a species within its community. It is, however, more than the area inhabited by a particular species and the food it eats. It also takes into account the time and area of activity of the organism, e.g. is it nocturnal or diurnal? Does it have a large or small territory? Some species may appear very similar, but their nesting habits or other aspects of behaviour will be different, or they may show different levels of tolerance to, e.g. a pollutant or a shortage of oxygen or nitrates. Any differences in niche, however small, limit competition between species. No two species occupy exactly the same niche.

Food chains and food webs

The organisms found in an oak woodland, or any other **ecosystem**, rely on a source of energy to carry out all their activities. The ultimate source of this energy is sunlight, converted to chemical energy by plants (producers) and then passed as food from one animal (consumer) to another.

9.4.1 Producers

Producers are photosynthetic organisms that manufacture organic substances using light energy, water and carbon dioxide:

$$6CO_2 \quad + \quad 6H_2O \quad \xrightarrow[\text{chlorophyll}]{\text{light energy}} \quad C_6H_{12}O_6 \quad + \quad 6O_2$$

The rate at which they produce this organic food is referred to as their **productivity**:

- **Gross primary productivity** is the total production of organic food in a given area and in a given time. It depends on the types of plant growing, their density and the climate.
- **Net primary productivity** (NPP) is the rate of production of organic food after allowing for that used in respiration by the plant – in other words, the production of material that might be eaten by consumers (section 9.5.1). Table 9.2 shows the mean NPP for a number of different ecosystems.

Table 9.2 *Net primary production in different ecosystems*

Ecosystem	Mean NPP / kJ m^{-2} yr^{-1}
Deserts	260
Oceans	4700
Temperate grasslands	15 000
Intensive agriculture	30 000
Tropical rainforest	40 000

9.4.2 Consumers

Animals can only gain energy and nutrients by eating other organisms. Those that eat plants are **herbivores**, those eating animals are **carnivores**, and those eating both are **omnivores**. Herbivores are also referred to as **primary consumers**, and they are eaten by **secondary consumers**. Tertiary and quaternary consumers are usually predators, but they may also be scavengers or **parasites**

9.4.3 Decomposers

Decomposers are mainly bacteria and fungi. They feed on the dead remains of plants and animals to obtain the energy they require, at the same time releasing valuable minerals into the soil.

9.4.4 Food chains

The term food chain describes a feeding relationship in which plants are eaten by herbivores, which are in turn eaten by carnivores. Each stage in this chain is referred to as a **trophic level**. The first trophic level is represented by producers, the second by herbivores, and all subsequent ones by carnivores. The shortest food chains usually have three levels:

grass → sheep → human

and the longest usually have no more than four or five:

nettle	→	aphid	→	ladybird	→	tit	→	sparrowhawk
(producer)		(primary consumer)		(secondary consumer)		(tertiary consumer)		(quaternary consumer)

Ladybird feeding on aphids

The arrows on food chain diagrams represent the **direction of energy flow**. Many herbivores feed on dead plant material and are part of a **decomposer chain**:

dead leaves $\rightarrow$ earthworms $\rightarrow$ shrews $\rightarrow$ badgers

9.4.5 Food webs

In reality, most animals do not rely upon a single food source, and within a single habitat many food chains will be linked together to form a **food web**. For example, on the edge of an oak woodland, the food chain shown in section 9.4.4 may be combined with others to form the web shown in figure 9.6.

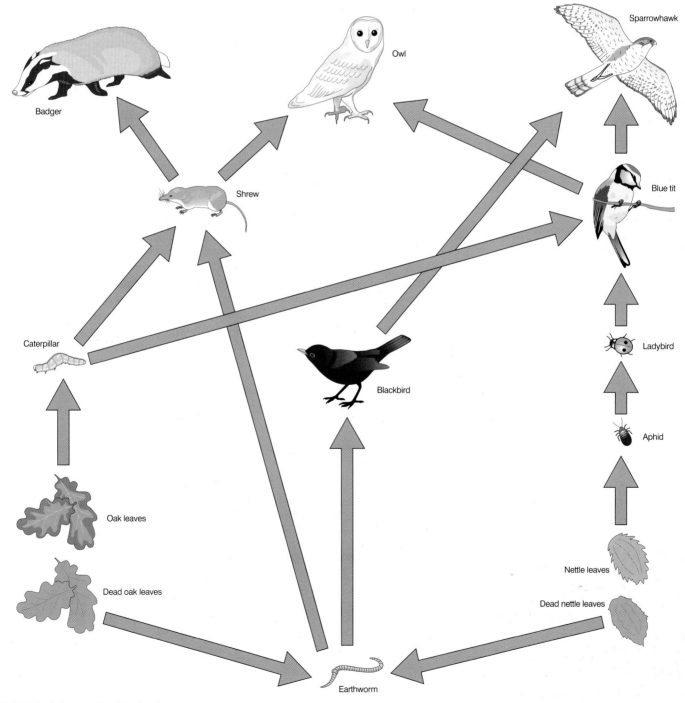

Fig 9.6 *Part of a woodland food web*

Energy transfer in ecosystems

The ultimate source of energy in **ecosystems** is the sun. However, very little of the energy available in sunlight is converted by plants to chemical energy. Similarly, there is inefficient transfer of energy at each step in the food chain.

9.5.1 Energy losses in food chains

Even under the most favourable conditions, **plants** convert only about 5% of the energy available to them into carbohydrates. Losses occur in a number of ways:

- over 90% of solar energy is reflected back into space by clouds and dust, or absorbed by the atmosphere and re-radiated
- not all wavelengths of light can be absorbed and used for photosynthesis
- light may not fall on a chlorophyll molecule
- low carbon dioxide levels may limit the rate of photosynthesis.

Plants then lose 20–50% of their **gross primary productivity** (unit 9.4) in respiration, leaving little to be stored as potential food for herbivores. Even then, only about 10% of the **net primary productivity** of plants is used by **herbivores** for growth. This low percentage results because:

- some of the plant is not eaten
- some parts are eaten, but cannot be digested (lost in faeces)
- some of the energy is lost in excretory materials (e.g. urine)
- some energy losses occur in respiration and heat loss to the environment. These losses are high in mammals and birds due to their high body temperatures.

Carnivores are slightly more efficient, transferring about 20% of the energy available from their prey into their own bodies. It is the relative inefficiency of energy transfer between **trophic levels** that limits the length of food chains to four or five steps. Energy flow along food chains is summarised in figure 9.7. This inefficient energy transfer also limits the number and **biomass** of organisms that can be supported at each trophic level, as well as the amount of energy stored at each stage. It is possible to construct **ecological pyramids** representing the numbers, biomass or stored energy of organisms at different trophic levels in a food chain.

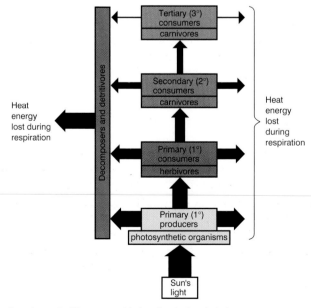

Fig 9.7 *Energy flow through different trophic levels of a food chain*

9.5.2 Pyramids of number

Usually, the numbers of organisms at lower trophic levels are greater than the numbers at higher levels. This can be shown by drawing bars with areas proportional to the numbers present in each trophic level (Fig 9.8a). However, figure 9.8 b and c indicate that there can be a significant drawback to using these pyramids to describe a food chain. For example:

- no account is taken of size – one tree is equated to one aphid and each **parasite** has the same numerical value as its larger host
- the number of individuals can be so great that it is impossible to represent them accurately on the same scale as other species in the food chain. For example, one tree may have millions of greenfly living off it.
- no account is taken of juveniles and other immature forms of a species, whose diet and energy requirements may differ from those of the adult.

9.5.3 Pyramids of biomass

Biomass is the total mass of the plants and/or animals in a particular place. It is normally measured over a fixed period of time. The term is sometimes used to refer to all living organisms on Earth, or a major part of the Earth, such as the oceans. It may also refer to plant or animal material that is exploited as fuel or raw material for industry. A more reliable, quantitative description of a food chain is provided when, instead of counting the organisms at each level, their biomass is measured. The fresh mass is quite easy to assess, but the presence of varying amounts of water makes it unreliable. The use of dry mass measurement overcomes this problem, but because the organisms must be killed it is usually only made on a small sample, and this sample may not be representative. In both pyramids of numbers and pyramids of biomass, only the organisms present at a particular time are shown; seasonal differences are not apparent. This is particularly significant when the biomass of some marine **ecosystems** is measured: over the course of a whole year, the mass of **phytoplankton** (plants) must exceed that of **zooplankton** (animals), but at certain times of the year this is not seen. For example, in early spring around the British Isles, the **standing crop** of zooplankton is greater than that of phytoplankton, as shown in figure 9.9.

9.5.4 Pyramids of energy

Collecting the data for pyramids of energy (Fig 9.10) can be difficult and complex, but the result is a true representation of the energy flow through a food web, with no anomalies. Data are collected in a given area (e.g. 1 square metre) for a set period of time, usually a year. The results are much more reliable than those for biomass, because two organisms of the same dry mass may store different amounts of energy. For example, 1 gram of fat stores twice as much energy as 1 gram of carbohydrate. The energy flow in these pyramids is usually shown as kJ m^{-2} yr^{-1}.

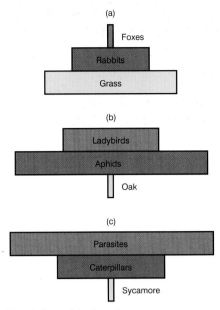

Fig 9.8 *Pyramids of numbers*

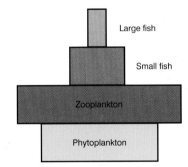

Fig 9.9 *Pyramid of biomass for a marine ecosystem*

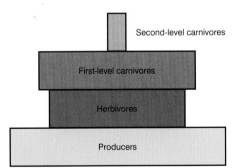

Fig 9.10 *Pyramid of energy, based on oak trees*

SUMMARY TEST 9.5

The sun is the ultimate source of energy for all living organisms on Earth. At best, only about **(1)** % of this solar energy is converted to carbohydrate by plants. Of this amount, around **(2)** % is used by plants during respiration. Of the remainder, known as **(3)**, only around 10% is consumed by **(4)**. Carnivores convert around 20% of the energy in their prey, for their own use. This inefficiency in transferring energy from one **(5)** to the next along a food chain is the reason that each food chain is short and the numbers at each level usually reduce, producing a pyramid of numbers. It is more reliable, however, to use the total mass of organisms, rather than numbers, to give a pyramid of **(6)**. More reliable still is a pyramid of **(7)**.

The flow of energy through living systems is **linear** (unit 9.5), but the flow of matter is **cyclical**. There is a finite (limited) amount of carbon and nitrogen available to living organisms. If they are not to run out, they must be used over and over again. Most nutrient cycles have two components:

- **geochemical** – including rocks and deposits in oceans and the atmosphere
- **biological** – including **producers**, **consumers** and **decomposers** that, in some way, help to convert one form of the mineral into another.

9.6.1 Water cycle

All living organisms require water. Figure 9.11 illustrates the main features of the water cycle. Evaporation, respiration and **transpiration** send water vapour into the atmosphere. It condenses to form clouds, and water returns to the Earth as precipitation (rain, hail and snow). Of all the water on Earth, 97% is in the oceans and only 3% is fresh. Most of this fresh water is not available to living organisms, because it is frozen in glaciers and ice sheets. One threat of **global warming** is that this ice would melt, leading to widespread flooding as sea levels rise (unit 10.4). Most precipitation falling onto the ground seeps through the soil and back to the oceans via the underground water table. Only a small percentage is returned via rivers.

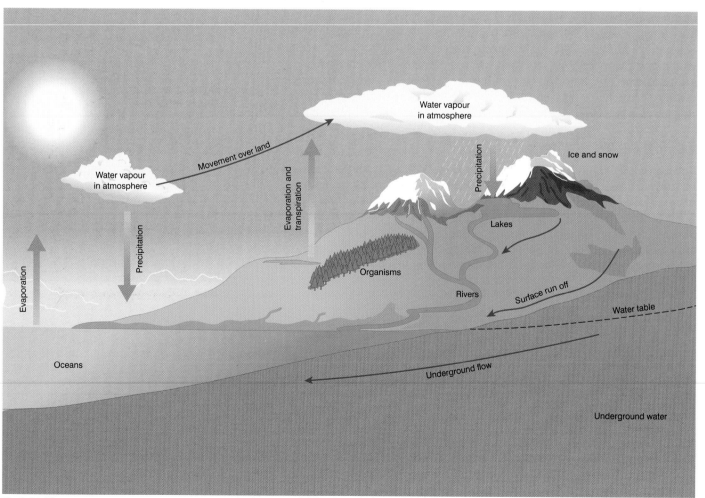

Fig 9.11 The water cycle

9.6.2 Carbon cycle

Despite containing less than 0.04% **carbon dioxide**, the atmosphere is the main source of carbon for terrestrial organisms. It can only be used directly by green plants, which fix it by photosynthesis into carbohydrates and other organic molecules. These are then passed along food chains to other organisms. There are also photosynthetic organisms, phytoplankton, in the oceans. They use dissolved carbon dioxide in the form of hydrogen carbonate ions. The store of carbon in the oceans is 50 times greater than that in the atmosphere; it helps to keep atmospheric levels more or less constant, as 'excess' carbon dioxide dissolves in the water. When atmospheric levels are low, the reverse occurs. The organisms that actively remove carbon dioxide from the environment act as **carbon sinks**. Carbon dioxide is returned to the atmosphere through respiration. Carbon compounds in dead organisms may be decomposed by **saprobiontic** microorganisms, which, in turn, respire and release carbon dioxide. Not all dead organisms decay. The bones and shells of many aquatic animals may sink to the bottom of the ocean, forming, over millions of years, carbonate-rich rocks such as chalk and limestone. This carbon eventually returns to the atmosphere as the sedimentary rocks are weathered. Other undecayed dead organisms may fossilise and produce carbon-rich fuels like coal, oil and gas. It is the increased combustion of these fossil fuels by humans which is one of the factors that is threatening to upset the delicate balance of the carbon cycle and to contribute to global warming (unit 10.4).

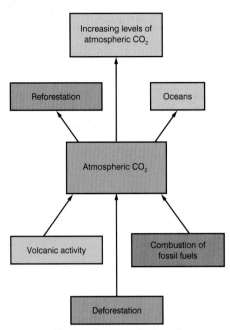

Fig 9.12 *Changes in atmospheric carbon dioxide levels*

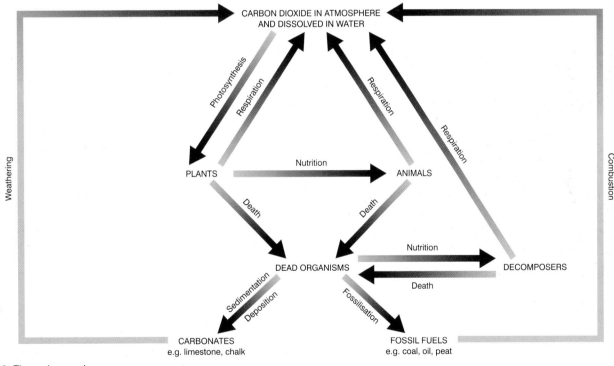

Fig 9.13 *The carbon cycle*

SUMMARY TEST 9.6

Water is vital to all living organisms and yet only **(1)** % of that present on Earth is fresh water and most of this is in the form of **(2)**. Water is made into vapour in three ways, by **(3)**, **(4)** and **(5)**. This vapour condenses to form **(6)** and is later returned to earth as rain or snow.

Carbon exists as carbonate minerals such as **(7)** which make up the **(8)** component of this nutrient cycle. It is also present in living organisms that comprise the **(9)** component, and as carbon dioxide that makes up **(10)**% of the atmosphere.

Nitrogen cycle

All living organisms require a source of nitrogen from which to manufacture proteins, nucleic acids and other nitrogen-containing compounds. Although 78% of the atmosphere is nitrogen, there are very few organisms that can use nitrogen gas directly. Plants take up most of the nitrogen they need in the form of nitrate ions (NO_3^-), from the soil. These ions are absorbed, using **active transport**, by the root hairs (section 7.7.1). Animals obtain nitrogen-containing compounds by eating and digesting plants.

Nitrate ions are very soluble, and easily leach through the soil, beyond the reach of plant roots. One way of rebuilding the nitrate levels is to add fertilisers, but it is also achieved through the natural recycling of nitrogen-containing compounds. When plants and animals die, the process of decomposition begins a series of steps by which microorganisms replenish the nitrate levels in the soil.

Within the nitrogen cycle (Fig 9.14), four main stages can be recognised; they all involve microorganisms:
- **decomposition**
- **nitrification**
- **nitrogen fixation**
- **denitrification**.

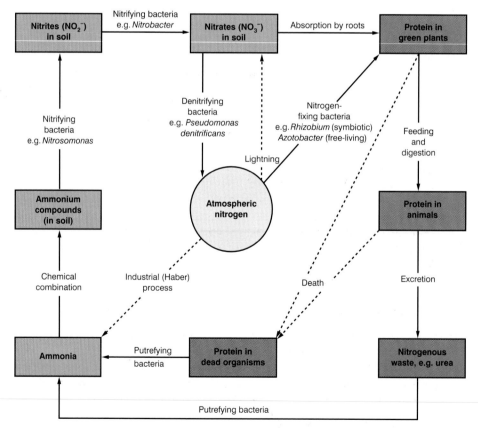

Fig 9.14 *The nitrogen cycle*

9.7.1 Decomposition

Decomposers are mainly bacteria and fungi. They feed on the dead remains of **producers** and **consumers**, as well as on their organic wastes. As they break down these materials, they release **ammonium ions (NH_4^+)** into the soil.

9.7.2 Nitrification

Plants use light energy to produce organic compounds. They are **photoautotrophs** (unit 9.1). Some bacteria are **chemoautotrophs**, using chemical reactions as a source of energy. The conversion of ammonium ions to nitrates involves oxidation reactions, which release energy for the nitrifying bacteria involved. This conversion occurs in two stages:

- **oxidation of ammonium ions to nitrites (NO_2^-)** by nitrifying bacteria such as *Nitrosomonas*, which live freely in well-aerated soil:

$$2NH_3 \quad + \quad 3O_2 \quad \rightarrow \quad 2NO_2^- \quad + \quad 2H^+ \quad + \quad 2H_2O$$

ammonia   oxygen   nitrite ions   hydrogen ions   water

- **oxidation of nitrites to nitrates (NO_3^-)** by other free-living nitrifying bacteria such as *Nitrobacter* and *Nitrococcus*:

$$2NO_2^- \quad + \quad O_2 \quad \rightarrow \quad 2NO_3^-$$

nitrite ions   oxygen   nitrate ions

The oxygen requirements of nitrifying bacteria mean that it is important for farmers to keep soil structure light and well aerated by ploughing. Good drainage also prevents the air-spaces from being filled with water.

9.7.3 Nitrogen fixation

This is the process by which nitrogen gas can be converted into nitrogen-containing compounds. There are three ways of doing this, all of them requiring energy:

- **lightning** allows nitrogen and oxygen to combine, producing oxides of nitrogen. These are washed into the soil by rain, and absorbed by plant roots in the form of nitrates.
- **industrial processes** such as the **Haber process** use high temperatures and pressures to combine nitrogen and hydrogen to produce ammonia. Much of this is added to the soil as nitrogen-containing fertilisers (unit 10.6).
- **fixation by microorganisms** is carried out by many bacteria living freely in the soil, and by some that live in root nodules on the roots of leguminous plants (section 9.1.4):
  - **free-living nitrogen fixers** include the bacteria, *Azotobacter* and *Clostridium*. They reduce gaseous nitrogen to ammonia, which they then use to manufacture amino acids. Nitrogen-rich compounds are released from them when they die and decay.
  - **mutualistic nitrogen fixers** include *Rhizobium*, living in nodules on the roots of plants such as peas and beans. *Rhizobium* obtains carbohydrates from the plant and the plant acquires amino acids from the bacterium. More details of this relationship are given in section 9.1.4.

9.7.4 Denitrification

When soils become waterlogged, and therefore short of oxygen, a different type of microbial flora flourishes. Fewer nitrifying and free nitrogen-fixing bacteria are found, and there is an increase in **anaerobic** denitrifying bacteria. These include *Pseudomonas denitrificans* and *Thiobacillus denitrificans*, both of which convert soil nitrates into gaseous nitrogen. The stages of reduction are as follows:

$$NO_3^- \quad \rightarrow \quad NO_2^- \quad \rightarrow \quad N_2O \quad \rightarrow \quad N_2$$

nitrate   nitrite   dinitrogen oxide   nitrogen

This reduces the availability of nitrogen-containing compounds for plants. As with any nutrient cycle, the delicate balance can be easily upset by human activities. The effects of the activities of humans on the nitrogen cycle are considered in unit 10.2 (deforestation and desertification) and unit 10.5 (water pollution).

Legume with root nodules

SUMMARY TEST 9.7

Nitrogen gas makes up **(1)** % of the atmosphere. A few organisms can convert this gas into compounds useful to other living organisms, in a process known as **(2)**. These organisms can be free-living such as **(3)** or live in a **(4)** relationship with other organisms. *Rhizobium* for example lives in **(5)** on the roots of plants such as **(6)**. Most plants absorb their nitrogen by absorbing **(7)** from the soil through their root hairs. Animals obtain this nitrate when they eat the plants and then convert it into **(8)** in their bodies. On death, **(9)** break down these organisms, releasing **(10)** which can then be oxidised to form **(11)** by nitrifying bacteria such as **(12)**. Further oxidation by other nitrifying bacteria like **(13)** will form nitrate ions. These nitrate ions may be converted back to atmospheric nitrogen by the activities of **(14)** bacteria.

1 The diagram shows how the premolar and molar teeth in the upper and lower jaws of a sheep (ruminant herbivore) fit together.

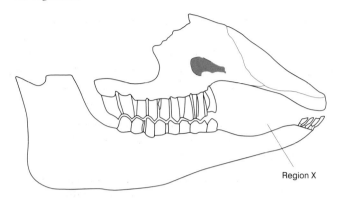

Region X

a Describe how the jaw action of herbivores such as sheep uses the interlocking surfaces of these teeth to chew plant material. *(2 marks)*

b Suggest the function of the region labelled X. *(2 marks)*

c Explain the importance of the rumen in digestion.
(2 marks)
(Total 6 marks)
Edexel 6103/03 June 2001, B B(H) AS/A, No.1

2 Study the passage and data below and then answer the questions that follow.

Grazing in the Serengeti grassland

The Serengeti is a huge area of tropical grassland in Tanzania. Herds of grazing mammals, such as wildebeest, gazelle and zebra, roam freely. Every year, these herds migrate across the Serengeti, in search of fresh grassland. The grazing mammals affect the primary productivity of the grassland. Long-term research has found that the rate of primary production is linked to both the rainfall and the numbers of grazing animals.

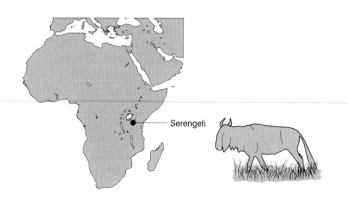

Map of Africa showing the location of the Serengeti

A wildebeest, one of the many types of grazing mammal found in the Serengeti

Serengeti

Grazing can increase the growth rate of many grass species. This is called compensatory growth. The grazing mammals remove the upper parts of the grass leaves and this increases the amount of light reaching the rest of the plant. The smaller leaf area reduces transpiration and this decreases the uptake of water by the roots.

Researches investigated compensatory growth of grasses in a region of the Serengeti. They fenced off several areas to prevent mammals from grazing the grass. During the annual migration, thousands of wildebeest moved into the study region where they grazed intensively for 4 days and then moved on.

The researchers recorded the changes in the fresh biomass of the grasses in the grazed and ungrazed areas over the next 32 days (table). A further investigation studied the effect of grazing intensity on primary productivity (graph).

Table showing the fresh biomass over a 32-day period on grazed and ungrazed grassland. Day 1 is the first day after the wildebeest moved on.

Day	Fresh biomass / g m^{-2}	
	Grazed	Ungrazed
1	50	430
8	55	420
16	100	380
24	120	350
32	200	300

Graph to show the effect of grazing intensity on the primary productivity of Serengeti grassland.

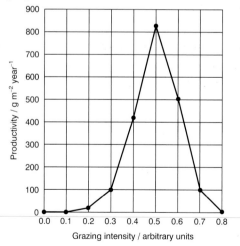

a Define the terms **biomass** and **productivity**. *(2 marks)*

b (i) Using the data in the table, calculate the mean rate of change of fresh biomass per day, for both the grazed and ungrazed grassland, between days 1 and 32. Show your working. *(4 marks)*

(ii) Suggest why the fresh biomass of grass in the ungrazed area decreased. *(3 marks)*

c Study the information provided in the graph.
   (i) What is the optimum grazing intensity for Serengeti grassland? *(1 mark)*
   (ii) Suggest reasons for the very low levels of primary productivity at the highest grazing intensity (0.8 arbitrary units). *(3 marks)*

d In the Serengeti, lions are the top consumers. Suggest and explain how a fall in the number of lions could affect the Serengeti food chain and grassland productivity. *(3 marks)*

e The grazing mammals have an important role to play in the recycling of nutrients on the grassland. Their dung is rich in organic nitrogen-containing compounds. Describe how these compounds are converted to nutrients. *(4 marks)*

(Total 20 marks)

Edexcel 6103/03 June 2001, B (BH) AS/A, No.3

3 a Explain the meaning of the terms *producer* and *trophic level*. *(4 marks)*

The table shows the estimated energy content for four trophic levels of a grassland ecosystem.

	energy content / kJ m^{-2}
producers	5600
herbivores	125
omnivores	15
carnivores	10

b (i) Calculate the percentage energy decrease between the producers and herbivores. (Show your working and give your answers to the nearest whole number.) *(2 marks)*

   (ii) Explain why the energy content of the herbivores is less than that of the producers. *(4 marks)*

c Suggest two factors which might reduce the productivity of the producers. *(2 marks)*

d Suggest:
   (i) why it can be a problem placing omnivores into a trophic level. *(1 mark)*
   (ii) a reason for the difference in energy content between the omnivores and the carnivores in the ecosystem. *(1 mark)*

(Total 14 marks)

OCR 2801 Jan 2001, B(BF), No.3

4 Organisms, both plants and animals, are able to exist together in the same environment. They interact, both with members of their own species and with other species.

a State the word or phrase which best describes each of the following.

   (i) The place where an organism lives. *(1 mark)*
   (ii) A number of **different** species interacting in a particular place. *(1 mark)*
   (iii) All members of the **same** species in a particular place. *(1 mark)*
   (iv) The ecological role of an organism. *(1 mark)*
   (v) The first trophic level in **all** food chains. *(1 mark)*
   (vi) A natural unit of living and non-living parts, interacting to produce a stable system. *(1 mark)*

Some farmers spread waste from sewage works on their fields. This waste contains nitrogen compounds such as protein, urea and ammonia.

b (i) Describe how the nitrogen compounds present in the sewage waste are converted naturally into a form which can be taken up by the plants in the fields. *(4 marks)*
   (ii) Suggest a possible **disadvantage** of spreading this waste on the fields. *(1 mark)*

(Total 11 marks)

OCR 2801 June 2001, B (BF), No.2

5 The table below refers to processes and bacteria in the nitrogen cycle.

Complete the table by filling in the most appropriate word or words in the empty boxes.

Name of process	Chemical change	Name of ONE genus of bacterium responsible for the process
	Ammonia to nitrate	
	Nitrogen to ammonia	*Rhizobium*
Denitrification		

(Total 5 marks)

Edexcel 6049 June 2001, B(H) AS/A, No.1

Human influences on the environment

Energy resources

EDEXCEL

EDEXCEL (Human)

Since the industrial revolution, both the size of the human population and its demand for energy resources have risen dramatically. Energy-rich fuels are used for producing electricity, for heating, and to maintain industrial processes. Some fuel sources, like coal and oil, are basically non-renewable. Others are renewable as long as the rate of use is equal to, or less than, the rate of production.

10.1.1 Fossil fuels

Fossil fuels such as coal and oil are continuously being formed. However, the process takes millions of years and so they are classed as a **non-renewable** resource. Stores of fossil fuels are rapidly being reduced, and it is estimated that available sources may be used up within 50 years. Their use is by no means evenly spread across the Earth: the fewer than 25% of people who live in the most developed areas use more than 80% of them. As supplies of coal and, especially, oil decrease, the use of renewable energy resources becomes an economic prospect.

10.1.2 Fast growing biomass

Carbohydrates built up by the photosynthesis of green plants are energy stores that may be utilised by humans as fuel. In its simplest form, this use might include the burning of wood on a domestic fire, to generate heat. However, it is also possible to create dense plantations of fast-growing trees like willow, hazel and poplar. Their branches can be harvested every 2–4 years, dried, and burnt to generate electricity. This system of growing fuel specifically for power generation is called **short-rotation coppicing**. It has been estimated that an area of 4000 hectares in Britain would be enough to provide power for 20 000 homes indefinitely.

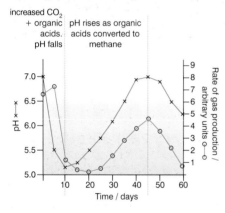

Fig 10.1 *Graph of changes in pH and gas production in a biogas converter*

10.1.3 Biogas

Biogas is a mixture of gases produced by the action of **anaerobic** bacteria on organic waste such as animal manure. It typically contains about
- 65% methane
- 35% carbon dioxide
- traces of ammonia, hydrogen sulphide and water vapour.

There are three stages to the digestion process:
- **Hydrolysis** – **aerobic** bacteria produce enzymes to break down the carbohydrates, proteins and lipids in the waste materials to simple sugars, amino acids, fatty acids and glycerol.
- **Acetogenesis** – as the oxygen is used up, acetogenic bacteria (e.g. *Acetobacterium*) convert the sugars and other substrates to short-chain organic acids, mainly ethanoic acid, with some carbon dioxide and hydrogen.
- **Methanogenesis** – methanogenic bacteria (e.g. *Methanobacterium*) need anaerobic conditions to convert organic acids to methane. Figure 10.1 shows the changes in pH and gas production that occur during biogas formation. For the digestion to produce methane, anaerobic conditions and a stable temperature of around 35°C are essential (Fig 10.2). The industrial process therefore takes place within an enclosed tank called a **digester**. Although there are many varied designs, they share the following features:

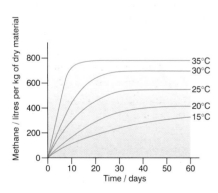

Fig 10.2 *Graph of methane production at different temperatures*

- **strong** enough to hold a large volume of material and withstand the build-up of pressure inside
- **gas-tight**, to maintain anaerobic conditions and to hold methane
- **suitable inlets and outlets**, for loading the material, recovering the gas and removing the residue
- **insulation** to maintain constant temperature. The digester is often buried in the ground.

Biogas is produced in sewage disposal units in the UK as well as on a small scale in rural areas of developing countries such as China, India and Nepal (Fig 10.3). It can also be used as an energy-efficient means of getting rid of animal wastes from intensive farming methods. The methane in biogas burns with a clear flame, producing carbon dioxide and water, but no other hazardous air pollutants.

Open fermentation vessels, in which sucrose is converted to ethanol

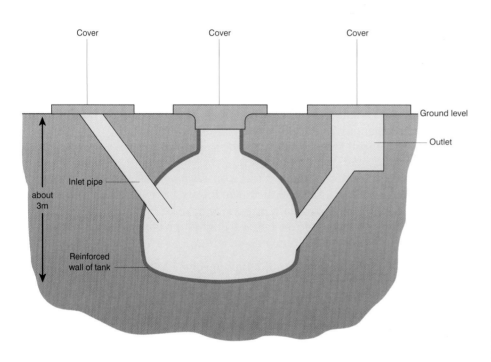

Fig 10.3 A small-scale biogas plant as used in parts of rural China

10.1.4 Gasohol

Gasohol is made up of 80–90% unleaded petrol and 10–20% ethanol. The programme to produce ethanol from **biomass** was begun in Brazil over 25 years ago, but was almost abandoned as too expensive. The threat of climate change linked to the burning of coal and oil has renewed interest in ethanol-based fuels. Ethanol can be produced from starch-rich plants, such as maize and potatoes, sugar-rich ones like sugar cane, or from cellulose waste, such as paper. When sugar cane is used, the following processes are involved:

- growing and harvesting the sugar cane
- extracting the sucrose to sell
- fermenting the remaining molasses using yeast, *Saccharomyces cerevisiae*, to produce dilute ethanol
- distilling the dilute ethanol to give 95% ethanol.

The fibrous waste, bagasse, produced after the extraction of sucrose, can be burnt to fuel the distillation process. Using ethanol as a fuel has environmental benefits:

- carbon monoxide emissions are reduced by up to 30%
- hydrocarbon, sulphur dioxide and particulate emissions are reduced.

SUMMARY TEST 10.1

Fossil fuels such as coal and oil are classed as (**1**), as there is effectively a finite amount available for use. Other fuels can be replaced as they are used. Fast-growing biomass, for example, can be used as a fuel by growing plantations of rapidly growing trees like (**2**) then harvesting them after a few years and burning them to generate electricity. Such a system is known as (**3**). Biogas is an energy source made in an enclosed tank called a (**4**) as result of the action of (**5**) on organic waste. Its production involves the hydrolysis of complex materials into simple ones and then their conversion into organic acids by the process of (**6**). The final stage is the production of (**7**), which makes up around (**8**) % of biogas, and carbon dioxide that forms the bulk of the remainder, at an optimum temperature of (**9**). Gasohol is made by adding ethanol to (**10**). The ethanol can be produced from sugar molasses by the process of (**11**), using the microorganism called (**12**).

Deforestation and desertification

Deforestation and desertification are two environmental problems that can be directly linked to the increase in the size of human populations and our demand for resources such as space and food.

10.2.1 Deforestation

Forests can be harvested as a renewable resource as long as the amount removed is less than the rate of production. **Deforestation** is the term used to describe the permanent clearing of forestland to convert it to non-forest uses. It has been estimated that over 7.5 million hectares is being destroyed every year. Although some loss is accounted for by accidental fires, most is the result of human actions. Trees are removed to provide:

- **fire and shelters** – used by nomadic early humans, this had little effect on the environment
- **grazing** – large areas were cleared for domesticated animals such as sheep, cattle and llamas. Large **cattle ranches** run by multinational companies have been a particular problem in Central and South America
- **agricultural land** – land clearance for the cultivation of crops is a world-wide problem
- **hardwood timber** – used for furniture, construction, paper and fuel. (Many developed countries now protect their own forests, but exploit the forest resources of parts of Africa, South America and Asia)
- **mining areas** – for example the removal of titanium ore in Madagascar
- **building land** for increased human populations.

Air pollution by oxides of nitrogen and sulphur (unit 10.3) has killed trees and caused some loss of forests. Some of the effects of deforestation are as follows:
- **Changes to the oxygen and carbon dioxide balance**. Felled trees no longer remove carbon dioxide from the air, and their burning releases carbon dioxide. This can accelerate **global warming** (unit 10.3).
- **Changes to the water cycle** (unit 9.6). Water is no longer taken up by tree roots, but runs off the surface, causing soil erosion and the sedimentation of water courses. There is less **transpiration**, less rainfall, and prolonged periods of drought.

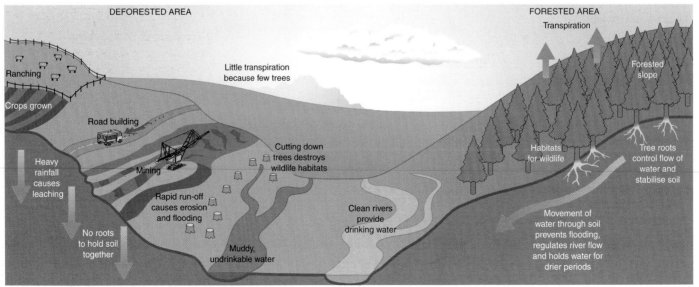

Fig 10.4 *Comparison of forested and deforested areas*

- **Reduced biodiversity**. The loss of habitat and food supplies leads to the extinction of many species and therefore reduces variety. The lost plants and animals may have possessed unique chemicals and genes with potential benefits to mankind.

The most serious problems of deforestation occur in some of the poorest and least developed countries. It may be possible to slow down some of the **habitat** destruction by using some of the following measures:

- **Reducing dependence on income from wood**. This may include finding alternative sources of income from the forest – e.g. nuts, latex and medical compounds – and easing developing countries' debts to developed nations.
- **Forest management**. Controlled timber removal and replanting, so that the long-term yield is not affected.
- **Encouraging tourism**. This could provide income in conserved rainforests.
- **Decreasing pollution**.

Figure 10.4 summarises some of the differences between forested and deforested areas.

10.2.2 Desertification

Desertification is the process whereby the biological productivity of land is so reduced that it leads to desert-like conditions in semi-arid areas. Some of the causes of desertification include:

- **Climate changes**. Extended periods of drought make plant regeneration difficult, and bare soils are then exposed to erosion. Further loss of ground vegetation allows wind erosion to remove the soil.
- **Increased human populations**. This leads to intensive cultivation, overgrazing, lumbering, house building and the cutting of vegetation for fuel. These all leave the land bare and exposed to wind erosion.
- **Deforestation** (section 10.2.1). Trees act as windbreaks, and their roots help bind the soil. Without them, wind erosion increases. Rainwater is no longer absorbed by roots, and transpiration is reduced; the consequent reduction in rainfall leads to a change in the local climate.
- **Poor land management**. Unsuitable irrigation schemes in hot climates lead to a build-up of salt **(salination)** in the soil as the water evaporates. The salts reduce soil productivity, some plants die, and the land is once again exposed to wind erosion.

The part of the world suffering most from desertification is the Sahel, semi-arid land on the edge of the Sahara desert. Over-grazing and poor cultivation techniques also produced the 'Dust Bowl' in the American mid-west. As desertification spreads, it leads to:

- reduced plant variety and deterioration of **ecosystems**
- deterioration of soil structure and fertility
- regular severe droughts
- reduction of land for growing crops and therefore increased chances of a severe famine.

Some of the solutions to desertification include:

- **management of water resources** – irrigation with salt-free water, water wells and pumps, and building irrigation canals
- **planting schemes** to stabilise land and provide shelter belts. **Afforestation** of suitable areas
- **soil conservation schemes**, using contour ploughing and terracing, and trapping water run-off with stone lines
- **reducing overgrazing**
- **GM plants** able to grow in saline soil.

Grazing in a desertified area

Effect of irrigation from the Nile

SUMMARY TEST 10.2

Deforestation is the permanent clearance of forest to convert the land to other purposes. The loss of the trees means that less **(1)** gas is absorbed as a result of photosynthesis, and this in turn contributes to **(2)**. Another effect is the loss of habitats for organisms, and this leads to reduced **(3)**. As less water is taken up from the ground, once the trees have gone the water runs off the surface, causing **(4)** and also the **(5)** of rivers and streams. This in turn may lead to desertification, as water is not held in the soil.

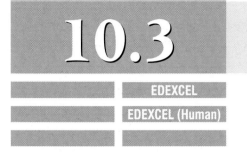

10.3 Acid rain

Air pollution from a coking plant

Acid rain is not a new problem. The damage caused to vegetation by sulphuric acid in rain was noted by Britain's first air pollution inspector in 1851.

10.3.1 Causes of acid rain

All rain is slightly acidic – around pH 5.65 – due to the presence of some weak acids:

$$H_2O \quad + \quad CO_2 \quad \rightarrow \quad H_2CO_3 \quad \rightarrow \quad H^+ \quad + \quad HCO_3^-$$

water   carbon dioxide   carbonic acid   hydrogen ions (lower pH)   hydrogen carbonate ions

The 'more acidic' rain has been blamed on pollutants such as sulphur dioxide (SO_2) and a mixture of nitrogen oxides (NO and NO_2), known as NO_x. Most of these emissions result from the burning of fossil fuels, which contain lots of sulphur and nitrogen. Oxides of sulphur and nitrogen react with water molecules in clouds to produce sulphuric and nitric acids:

- some sulphur dioxide reacts with water droplets to form sulphurous acid:

$$SO_2 \quad + \quad H_2O \quad \rightleftharpoons \quad H_2SO_3$$

- the sulphurous acid dissociates to produce hydrogen ions (giving rain its low pH) and bisulphite ions:

$$H_2SO_3 \quad \rightleftharpoons \quad H^+ \quad + \quad HSO_3^-$$

- the hydrogen sulphite ion is converted to hydrogen sulphate ions by oxidants in the atmosphere, such as ozone (O_3):

$$HSO_3^- \quad + \quad O_3 \quad \rightleftharpoons \quad HSO_4^- \quad + \quad O_2$$

- dissociation leads to the production of more hydrogen ions and a further lowering of the pH:

$$HSO_4^- \quad \rightleftharpoons \quad H^+ \quad + \quad SO_4^-$$

The nitrogen oxides react with water in a similar way.

Acid rain production mainly occurs in industrial regions such as northern Europe and North America. These areas are heavy users of fossil fuels to produce electricity, and this is the main source of sulphur dioxide. The main source of nitrogen oxides is fuel combustion in motor vehicles (table 10.1). A problem with sulphur dioxide and nitrogen oxide emissions is that the wind carries them from their origin, to fall as acid rain on countries far away. Thus pollutants from Britain and Germany affect Scandinavia and those from USA factories affect Canada. It has been estimated that pollution from the northern USA is responsible for 90% of the nitrogen and 63% of the sulphur in the rain falling on Canada.

10.3.2 Effects of acid rain on soils

Most soils in Europe are acidic, but the acids (largely carbonic and organic acids) are immobile and do not readily transfer their acidity elsewhere. Acid rain, however, contains strong negative ions such as sulphate (SO_4^{2-}), which does mobilise hydrogen ions and can transfer them to run-off waters. Nitrate ions would have a similar effect, but most of them are absorbed by plant roots. Many UK soils can neutralise the effects of acid rain. However, where thin soils, lacking the **cations** calcium and magnesium, overlie resistant rock, the acidity is not neutralised and enters water courses. High levels of sulphate ions also displace

Table 10.1 *Sources of acidifying gases*

Source	Percentage contribution	
	Nitrogen oxides	Sulphur dioxide
Motor vehicles	45	1
Power stations	37	71
Industry	12	19
Domestic	3	5
Other sources	3	4

aluminium ions (Al^{3+}) from the soil, leaching them into streams and lakes.

10.3.3 Effects of acid rain on aquatic ecosystems

Acid rain affects fresh-water organisms in a number of ways:
- **Gaseous exchange in fish** – High concentrations of aluminium ions interfere with ion transfer in the gills of fish, so that the gills become covered with thick mucus that limits the transfer of oxygen from water to blood. It also upsets the mechanism which maintains the correct salt balance in the fish's body.
- **Shell and exoskeleton formation** – Acidic streams lack calcium and magnesium. This makes it difficult for molluscs, like the freshwater limpet (*Ancylus*), and crustacea, like the freshwater shrimp (*Gammarus*), to survive. Some insect **nymphs** are less sensitive to acidic water. This variation enables organisms to be used as **indicator species**, giving information about the extent of acidification (Fig 10.5).
- **Fish egg-wall formation** – A water pH of 4.5 or less reduces the effect of an enzyme that softens the egg wall. This makes it difficult for young fish to hatch, and so fish populations decline.
- **Bird egg-shell formation** – Birds feeding in acidic waters not only find their food sources reduced, but they also lay eggs with thinner shells. Populations of birds like dippers therefore decline.

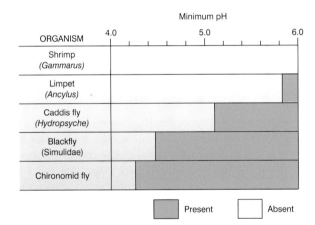

Fig 10.5 *Tolerance of various invertebrates to acidification*

10.3.4 Effects of acid rain on forests

In general, conifers are more affected by acid rain than **deciduous** trees. Acid pollutants have both direct and indirect effects on trees:
- **Atmospheric SO_2 and NO_x** break down the cuticle, causing damage to the mesophyll layers and reducing photosynthesis.
- **Uptake of aluminium ions,** which are toxic to trees, causes **defoliation**.
- **Reduction in calcium and magnesium ion uptake** results in poor middle lamella and chlorophyll formation.

- **Reduction in decomposition of organic material** occurs, because microorganisms are inhibited by heavy metals.

10.3.5 Effects of acid rain on humans

Acid rain can affect humans, because we eat food, drink water and breathe air that have come into contact with acid deposition. Toxic metals such as aluminium **leach** into untreated drinking water, and a link has been found between acid pollutants and asthma. Acid rain also affects humans by weathering buildings and lowering the productivity of fisheries and forestry.

Photochemical smog in Rio de Janeiro

10.3.6 Prevention and cure

Attempts to cure the effects of acid rain have proved both difficult and expensive. Tipping large amounts of limestone into lakes and giving supplementary minerals to forest soils are, at best, short-term solutions. The fitting of catalytic converters to internal combustion engines could reduce the emission of nitrogen oxides by 30%, but this is offset by the increase in the number and use of motor vehicles. Reacting sulphur oxides with water to form a commercial source of sulphuric acid, without it escaping to the atmosphere, has reduced Britain's industrial SO_2 emissions by 40%. Clean air legislation in Britain began in the 1950s, after a series of smogs (smoke and fog with a low pH) that killed thousands. International concern about acid rain as a global problem has been discussed at a series of conferences since one held at Stockholm in 1972. Within the European Union (EU), directives setting maximum permitted concentrations for SO_2 emissions were made in the 1970s. There have also been controls over vehicle emissions, and in the late 1990s EU objectives for air quality control were laid down. Although there are financial penalties to enforce some of these measures, education is likely to bring the most success.

Greenhouse effect

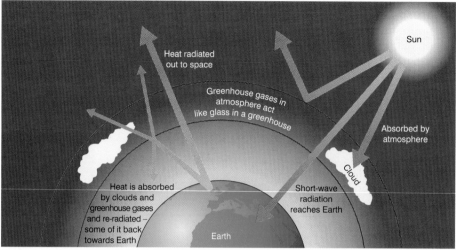

Fig 10.6 *The greenhouse effect*

A natural process, known as the 'greenhouse effect', is occurring all the time and keeps average global temperatures at around 15°C. Without it, the average temperature at the surface of the Earth would be about −18°C. Some solar radiation is reflected back into space, some is absorbed by the atmosphere and, fortunately, some reaches the Earth. Some of the radiation reaching the Earth is reflected back as heat, and is lost into space. However, some is radiated back to Earth by clouds and the surrounding 'greenhouse gases', keeping the Earth warm. This process is illustrated in figure 10.6.

10.4.1 Greenhouse gases

The most important greenhouse gas is carbon dioxide, partly because there is so much of it, and partly because it remains in the atmosphere for so much longer than other greenhouse gases (100 years, compared with 10 years for methane). It has been estimated that 50–70% of global warming is due to carbon dioxide in the atmosphere. It is mainly as a result of human activities that the concentration of carbon dioxide is increasing, enhancing the greenhouse effect and causing environmental concerns (section 10.4.2). Other natural greenhouse gases include water vapour, methane, nitrous oxide and ozone. The concentration of all of these can be affected by humans. In addition, the use of synthetic compounds, like CFCs (chlorofluorocarbons), has been banned, because they were estimated to have contributed up to 14% of the greenhouse effect in the 1980s. A comparison of the relative efficiencies of various greenhouse gases is shown in table 10.2.

Table 10.2 *The relative efficiencies of various greenhouse gases*

Gas	Relative efficiency as a greenhouse gas
Carbon dioxide	1
Methane	30
Nitrous oxide	150
Ozone	2000
CFCs	20 000

10.4.2 Global warming

Analyses of trapped air bubbles deep in the Greenland ice caps show a correlation between high atmospheric carbon dioxide levels and periods of warmth (see figure 10.8 opposite). What we cannot be sure of is which is the cause and which the effect. Human activities such as the burning of fossil fuels (section 9.6.2) and the deforestation of land (section 10.2.1) are generating much more carbon dioxide than natural causes like volcanoes, respiration and natural fires. Results from the Mauna Loa recording station in Hawaii, shown in figure 10.7, reveal an upward trend in atmospheric carbon dioxide. Overall, its concentration has risen from 270 parts per million (ppm) before the industrial revolution to 355 ppm today. Carbon

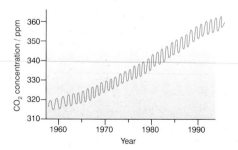

Fig 10.7 *Trends in atmospheric carbon dioxide concentration since 1958 (recorded at Mauna Loa, Hawaii)*

dioxide emissions are greatest in the world's most industrialised countries, although these do not have large populations. The industrialisation of heavily populated countries like China and India could result in much higher emissions. Table 10.3 relates emission levels to population size in 1996.

10.4.3 Consequences of global warming

It has been estimated that a doubling of present-day atmospheric carbon dioxide levels would result in
- a mean annual temperature increase of 4–5°C
- a mean annual rainfall increase of 10–50%.

Different regions of the world would not be affected evenly; in some, temperatures and rainfall could fall. Even if humans continue to release greenhouse gases at the present rate, it is by no means certain what the effects would be. Carbon dioxide levels could even reduce naturally, as phytoplankton and other plants increase their rates of photosynthesis. However, if present trends continue, the following predictions have been made for 2050:
- **Melting of polar ice caps**, causing the extinction of some animals, such as polar bears, and causing sea levels to rise.
- **A rise in sea level** due to the thermal expansion of oceans, which would flood low lying land, including much of Bangladesh. It would also flood many major cities, and fertile land such as the Nile delta. Salt water would extend up rivers, and make cultivation difficult.
- **Melting permafrost** would cause landslides and damage oil-lines.
- **Melting alpine snow** would destroy the skiing industry.
- **Desertification** would spread. The Sahara could extend as far as southern Spain.
- **Greater rainfall and intense storms** would occur in some areas, due to the disturbance of climate patterns.
- **Tropical diseases** would spread towards the poles.

However, it should be remembered that there could also be benefits to some parts of the world. The increased rainfall would fill reservoirs, the warmer temperatures would allow crops to be grown where it is presently too cold, and it might be possible to harvest twice a year instead of once.

10.4.4 Control of global warming

Although we are by no means certain of the effects of increasing carbon dioxide emissions by humans, it would seem sensible to attempt some sort of control to reduce them. A series of global conferences has been held to set targets for the emission of greenhouse gases by different countries. For example the Montreal Protocol (1987) and London Conference (1990) decided to phase out the use of CFCs. The original Kyoto agreement (1998) was to cut the developed world's emissions by 5.5% by 2010 from a 1990 baseline. This will be difficult, because emissions have risen dramatically since 1990, especially in the USA. In 2001, President Bush refused to sign the agreement, but talks will continue in the future.

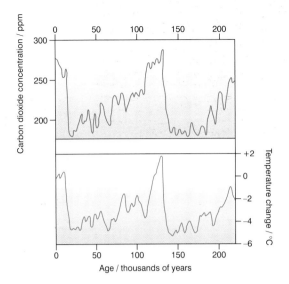

Fig 10.8 *Comparison of long-term changes in atmospheric carbon dioxide (from analysis of gas bubbles in Greenland ice) and possible temperature changes (from oxygen isotope studies)*

Table 10.3 *Carbon dioxide emissions in relation to population size in various regions of the world, in 1996*

Region	CO₂ emissions as % of world total	Population as % of world total
USA	25	4.7
Europe	19.6	9
China	13.6	21.5
India	3.6	16.3
UK	2.5	1.02

SUMMARY TEST 10.4

Carbon dioxide is formed during the process of **(1)** by organisms and is largely absorbed by plants during the process of **(2)**. Carbon dioxide from the burning of **(3)** is, however, adding to the amount present in the atmosphere, which has increased by around **(4)** % since the industrial revolution. As carbon dioxide absorbs energy from the sun and prevents it reflecting back into space, it is known as a greenhouse gas and leads to a rise in **(5)** at the Earth's surface. This in turn, it is feared, will cause a rise in sea levels as a result of the **(6)** of the oceans and the melting of **(7)**. Other natural greenhouse gases include water vapour, nitrogen oxides, methane and **(8)**. While methane is **(9)** times more efficient as a greenhouse gas, its overall effect is reduced because it remains in the atmosphere **(10)** times less than carbon dioxide. One synthetic group of greenhouse gases is **(11)**. To try to control the emission of greenhouse gases, an agreement named after the Japanese city called **(12)**, where world leaders met, was made in 1998. Unfortunately, one country, **(13)**, refused to ratify the agreement, making it unlikely that its targets can be achieved.

EDEXCEL

EDEXCEL (Human)

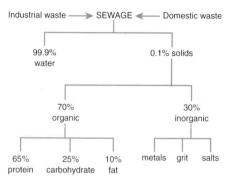

Industrial waste → SEWAGE ← Domestic waste

99.9% water

0.1% solids

70% organic

30% inorganic

65% protein — 25% carbohydrate — 10% fat

metals — grit — salts

Fig 10.9 *Composition of sewage*

Discharge of sewage onto a beach.

Water is essential for life, and needs to be kept as free of pollutants as possible. Pollutants are substances likely to cause harm to living organisms, including humans, or which reduce the amenity value of the environment.

10.5.1 Water pollutants

As the size of the human population has increased, more pressure has been put on the environment. There is a danger of sewage and industrial effluents reaching water courses. Run-off picks up fertilisers, used to increase crop yields, as well as metal ions mobilised by acid rain (section 10.3.2).

- **Sewage.** Untreated sewage – and even treated sewage, to a lesser extent – affects water quality. Figure 10.9 shows the composition of sewage. Bacterial oxidation of sewage releases nitrates (Fig 10.10), which can cause **eutrophication** (section 10.5.2). Sewage also contains phosphates, about half of which come from detergents, which typically contain 5–10% phosphate by weight. These also cause eutrophication. Other components, such as oestrogens (section 5.11.2) can affect fertility in fish – and even in humans, if they reach drinking water.
- **Fertilisers.** Inorganic fertilisers, rich in nitrates (NO_3^-) and phosphates (PO_4^{3-}) are frequently applied to increase crop yields. Being soluble, these ions readily leach into water courses, where they cause eutrophication. For further details on fertilisers, see unit 10.6.
- **Other pollutants.** These include aluminium ions mobilised by acid rain (section 10.3.2). There is some question of a link between these ions in drinking water and the occurrence of Alzheimer's disease. Pregnant women are at risk when acid rain increases the likelihood of copper, lead or mercury ions reaching drinking water and affecting foetal development.

10.5.2 Eutrophication

Waters with a high concentration of salts are termed **eutrophic**. These salts are usually present because of the sewage or fertiliser contamination referred to in section 10.5.1. However, the term eutrophication is frequently applied to a whole series of changes occurring in a body of water as a result of its nutrient content:

- **nitrate and phosphate concentrations increase**
- **algae and green protists multiply rapidly** due to the extra nutrients, and form an **algal bloom**
- **light penetration is reduced**, and so algae and plants in the deeper regions are unable to photosynthesise, and die
- **algae die** and add to the build-up of organic material
- **aerobic bacteria multiply** as they break down the large quantities of organic material
- **oxygen levels fall** as a result of this microbial activity
- **aerobic organisms die**, adding to the mass of organic material. This encourages the growth of even more aerobic **decomposers**, further lowering oxygen levels in the water.

10.5.3 Biochemical oxygen demand

This is also known as biological oxygen demand or deficit (BOD). Aerobic microorganisms in water need oxygen to break down organic matter and release energy. The higher their demand for oxygen, the more polluted the water. **BOD is defined as the difference in the amount of dissolved oxygen before and**

after 5 days of incubation in the dark at 20°C. Results are given in mg dm^{-3}. Eutrophication therefore increases the BOD and leads to deoxygenation of the water. Figure 10.10 shows the effect of sewage on the oxygen content of water.

10.5.4 Water pollution and biodiversity

Where the level of organic matter is high, the concentrations of **saprobiontic** bacteria increase. Although algae levels near sewage outfalls may be low because of the lack of light, they rise as nitrate and phosphate levels rise. The populations of animal species vary according to the levels of oxygen in the water. The bloodworm, *Tubifex*, is tolerant of low oxygen levels; as the concentration of oxygen rises, populations of midge larvae (*Chironomus*), then the water louse (*Asellus*), and then the freshwater shrimp (*Gammarus*), increase. As conditions improve, the number and variety of organisms increase. **Biodiversity** is low in polluted water, and higher when it is cleaner. Some of the changes in flora and fauna downstream of a sewage outfall are shown in figure 10.11.

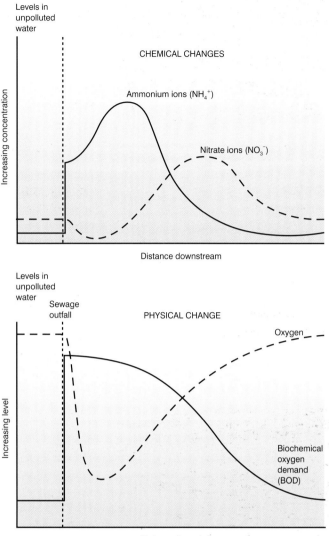

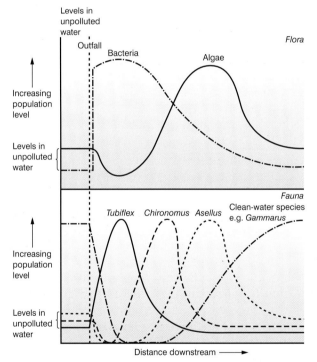

Fig 10.11 *Changes in the fauna and flora of a river, due to sewage effluent*

10.5.5 Prevention and cure

It is difficult and expensive to clean up polluted water. It is far better to prevent pollution in the first place: sewage should be treated before it is released into rivers or the sea; fewer fertilisers should be used; the phosphate content of detergents can be reduced; measures to reduce air pollution will decrease the problems caused by acid rain. The EU defines water quality standards for various things, from clean beaches (awarded Blue Flags), to controls on the levels of nitrates in drinking water, and permitted emissions of sewage and pesticides into rivers.

Fig 10.10 *Chemical and physical changes in a river, due to sewage effluent*

SUMMARY TEST 10.5

Sewage pollutes water because (**1**) microorganisms that thrive on it use the oxygen in the water and so create a (**2**). As a result of the lack of oxygen other organisms die. The microbial breakdown of sewage also leads to the release of (**3**) which, coupled with (**4**) from detergents in the sewage causes (**5**), a process in which additional nutrients lead to an increased population of photosynthetic organisms near the surface, giving rise to (**6**). Plants beneath this dense layer die due to a lack of (**7**) and are then broken down by (**8**), causing a further reduction in oxygen levels. Although the populations of some organisms such as (**9**) increase in the low oxygen levels near sewage outfalls, the overall biodiversity is (**10**) in polluted water than non-polluted ones. Other water pollutants include (**11**) from the run-off from agricultural land, aluminium ions mobilised by (**12**), and (**13**) that affect fertility in fish.

Fertilisers

Crops, like any plants, not only need the raw materials for photosynthesis (carbon dioxide and water), they also need soil nutrients.

10.6.1 Soil nutrients

Mineral salts are absorbed from the soil by plant roots. In the soil solution, mineral salts dissociate into positively charged cations (e.g. K^+, Mg^{2+}) and negatively charged **anions** (e.g. NO_3^-, SO_4^{2-}), which are absorbed independently of one another. Plants require relatively large amounts of seven **macronutrients**, especially nitrogen, phosphorus and potassium. They require smaller amounts of the **micronutrients** (trace elements). These elements are listed in table 10.4.

10.6.2 Depletion of soil minerals

In natural habitats, unmanaged by humans, mineral salts will be recycled. They are taken up from the soil by plant roots, and will be returned to it when the plants die and are decomposed. However, when humans harvest crop plants, the nutrients have no chance to return to the soil, and levels of essential ions will fall. This is particularly noticeable with crops like wheat, which need a lot of nitrate for optimal growth. Similarly, when an area of forest is harvested for timber, the soil left is low in nutrients. When deforestation is followed by fire, the ash returns nutrients to the soil. However, this benefit is short lived if crops are then grown on the cleared land and harvested. Nevertheless, the human population is growing rapidly, and land clearance and crop harvesting must continue in order to supply enough food (Fig 10.12). Soil nutrients must therefore be replenished by the addition of organic or inorganic fertilisers.

10.6.3 Organic fertilisers

Organic fertilisers include:
- farmyard manure
- liquid slurry
- compost
- green manure from ploughing in of young crops.

They **slowly** release anions and cations into the soil as they are broken down by bacteria and fungi. These manures can be added to the surface in a layer thick enough to prevent weed growth – a process called mulching – so preventing additional loss of nitrogen from the soil into the weeds. Table 10.5 shows the advantages and disadvantages of using organic fertilisers.

10.6.4 Inorganic fertilisers

Table 10.4 Essential plant nutrients

Macronutrients	Micronutrients
NITROGEN	Boron
PHOSPHORUS	Chlorine
POTASSIUM	Cobalt
Calcium	Copper
Iron	Manganese
Magnesium	Molybdenum
Sulphur	Zinc

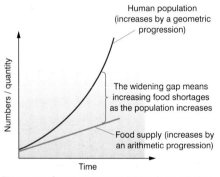

Fig 10.12 Graph showing Malthus's prediction for growth of the human population and its food supply

Table 10.5 Advantages and disadvantages of organic fertilisers

Advantages	Disadvantages
Supply all the necessary nutrients for growth	Not easily obtained
Increase water retention of the soil	Bulky and therefore expensive to transport and apply
Maintain the air content of the soil both directly and by encouraging earthworms	Slow acting and therefore not effective in a single season
Effective over a long period	Difficult to handle
Improve the crumb structure of the soil	Relatively expensive
Improve drainage and so help prevent waterlogging	

Table 10.6 Comparison of content of organic and inorganic fertilisers

Fertiliser	Comparison as % total mass		
	Nitrogen	Phosphorus	Potassium
Organic Cattle manure	0.6	0.1	0.5
Inorganic Compound fertiliser	20.0	12.0	10.0
Ammonium nitrate	34.0	0.0	0.0

Inorganic fertilisers, produced by the agrochemical industry, provide required anions and cations in a concentrated form. Some inorganic fertilisers, such as ammonium nitrate, contain only one of the macronutrients required by plants. Others, known as compound fertilisers, contain all three (nitrogen, phosphorus and potassium). The relative concentrations of macronutrients in some inorganic and organic fertilisers is shown in table 10.6. The advantages of using inorganic fertilisers are shown in table 10.7.

10.6.5 Fertilisers and leaching

The amount of nitrogen fixed by the **Haber process** is over one-third the amount fixed by natural processes. Most of this is used in the manufacture of fertilisers. The use of nitrogenous fertilisers increased dramatically during the 20th century, as shown in figure 10.13. Nitrates are very soluble, and are easily washed through the soil, away from plant roots. This process is called leaching. High levels of nitrates and other ions in water lead to **eutrophication** (section 10.5.2). Further eutrophication occurs because most fertiliser is taken up by plants, and released on decomposition. Decomposition is fastest in the autumn, when the warm, moist soil favours microbial activity. The greatest danger from leaching therefore occurs when

* crop cover, and therefore nitrate uptake, is low
* nitrogen fertilisers are applied in the autumn
* large areas are ploughed (releasing nitrates from organic material in soil)
* rainfall is high.

Slow-release fertilisers that release their nitrogen over a period of time help to overcome the problem of leaching.

10.6.6 Fertilisers and crop yield

Fertilisers are added to increase yield and profitability in agriculture, horticulture and forestry. It is estimated that the use of fertilisers has increased agricultural food production in the UK by around 100% since 1955. This has been achieved by using fertilisers containing the three main macronutrients:

* **Nitrogen** – is a constituent of proteins and is important for leaf growth, thus increasing the rate of photosynthesis.
* **Phosphorus** – is found in **ATP** and nucleic acids. It is very important for cell division and therefore growth.
* **Potassium** – has an important role in both respiration and photosynthesis.

As the amount of nitrogenous fertiliser added to a crop increases, so the yield increases, steeply at first, but then more gradually. There comes a point when the yield levels off and the farmer would simply be wasting money by adding more fertiliser. The process is sometimes referred to as the **law of diminishing returns**; it is illustrated in figure 10.14. This may be because something else, other than a lack of nitrate, is now limiting the growth of the crop. Alternatively, high nitrate levels may actually result in damage to the root system.

Table 10.7 *Advantages and disadvantages of inorganic fertilisers*

Advantages	Disadvantages
Relatively light and therefore easy to transport and apply	Do not improve the physical characteristics of the soil
Quick acting	Can be easily removed by leaching
Easy to handle	Need to be regularly applied
Relatively cheap	Run-off can cause pollution of water courses
Easily obtained	

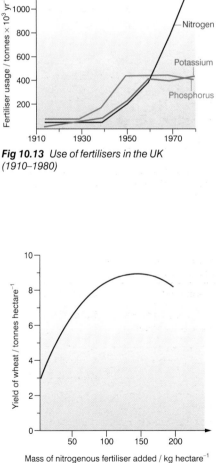

Fig 10.13 *Use of fertilisers in the UK (1910–1980)*

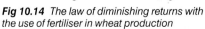

Mass of nitrogenous fertiliser added / kg hectare⁻¹

Fig 10.14 *The law of diminishing returns with the use of fertiliser in wheat production*

SUMMARY TEST 10.6

Organic fertilisers include compost, manure and **(1)**. They add nutrients to the soil and suppress weed growth when added in a thick layer called a **(2)**. Compared with inorganic fertilisers, they have the advantages of increasing **(3)** retention and **(4)** structure of the soil, and encouraging the activities of **(5)** that lead to better aeration and drainage. Inorganic fertilisers are, however, easily obtained, relatively cheap and act **(6)**. The nitrates they contain are, however, very **(7)** and so wash out of the soil by the process called **(8)**. Once in rivers, they can cause **(9)** that leads to rapid growth of algae.

10.7 Controlling pests

Although it is difficult to define what is meant by a 'pest', it is generally taken to be an organism that competes with humans for food or space, or could be a danger to health. As the human population increases, the supply of food becomes more critical. Weeds are pests because they compete with crop plants for water, mineral ions, space and light. Insect pests may damage the leaves of crops, limiting their ability to photosynthesise and thus reducing yield. Alternatively, they may be in direct competition with humans, eating the crop itself. Many crops are now grown in **'monoculture'** and because this enables insect and fungal pests to spread rapidly, various control measures must be taken.

10.7.1 Chemical pesticides

Pesticides are poisonous chemicals which kill pests; they are named after the pests they destroy:

- **herbicides** kill weeds
- **fungicides** kill fungi
- **insecticides** kill insects.

Pesticides are potential pollutants, deliberately produced and dispersed. To limit its environmental impact, a pesticide should be:

- **specific**, so that it is only toxic to the organisms at which it is directed. It should be harmless to humans and other organisms, especially to the natural predators of the pest, to earthworms, and to pollinating insects such as bees
- **biodegradable**, so that, once applied, it will break down into harmless substances in the soil. At the same time, it needs to be chemically stable, so that it has a long shelf-life
- **cost effective**, because development costs are high and new pesticides remain useful only for a limited time. This is because pests can develop genetic resistance making the pesticide useless.

In addition, there should be no **bioaccumulation** of the chemical pesticide. It should not build up, either in specific parts of an organism, or as it passes along food chains. Pesticides work in a number of different ways:

- **Residual pesticides** (e.g. the herbicide, simazine) are sprayed onto soil, or used to treat seeds before they are planted. They kill weed seedlings, fungal spores, and insect eggs and larvae.
- **Contact pesticides** (e.g. the herbicide, paraquat) are absorbed directly through the surface of plants, fungi and insects. They are generally inexpensive but, because their effects are short-lived, they have to be reapplied.
- **Systemic pesticides** (e.g. glyphosate) are often very effective insecticides. They are sprayed onto crops, absorbed, and transported round the plant. They are taken up by sap-sucking insects such as aphids, which are then poisoned. Systemic herbicides are also absorbed by leaves and transported around the plant, thus killing even the underground root system. A variety of pesticides is shown in table 10.8.

Genetically modified crops that are resistant to the effects of certain herbicides can help reduce the amount of herbicide used (section 12.10.1).

10.7.2 DDT

DDT (dichlorodiphenyl-trichloroethane), first synthesised in 1874, was used extensively to control lice and fleas in the Second World War. Later, it was used as a non-selective insecticide. Not only is DDT persistent, it also accumulates along

Weed control. Sugar beet in which weeds on the right have been controlled by chemicals, and those on the left are untreated

Cane toad eating a pygmy possum. Introduced as a means of biological control, the cane toad is now a predator of native species

food chains. If it is sprayed on garden plants to control greenfly, some of the greenfly survive and are eaten by tits, in which it accumulates in the fatty tissues. These tits may then be eaten by a sparrowhawk. If enough of the chemical is ingested, the bird will die. Even relatively low doses result in thin egg shells, which break during incubation. The population of sparrowhawks therefore declines. Although DDT is now largely banned, its use was once so widespread that traces of it now appear all over the globe. Many humans contain more DDT than is permitted, in some countries, in food for human consumption.

10.7.3 Biological control

It is possible to control pests by using organisms that are either predators or parasites of the pest organism. The aim is not to eradicate the pest; indeed, this would be counter-productive. If the pest was reduced to such an extent that there was insufficient food for the predator, the predator would die. The surviving pests would then be able to multiply unchecked. Ideally, the control agent and the pest should exist in balance with one another, at a level at which the pest has no disadvantageous effect. Some examples of biological control are shown in table 10.9.

The advantages of using biological control instead of chemical pesticides are shown in table 10.10. However, sometimes, an introduced predator may itself become a pest, as has been the case with the cane toad in Australia.

10.7.4 Integrated pest management

Although biological control methods can be seen to have some advantages over chemical pesticides, they also have a number of disadvantages:
- **They do not act as quickly**; there is often some time between introducing the control organism and a significant reduction in the pest population.
- **The pest is never eliminated completely**, it is merely maintained at a low level.

Integrated pest management aims to make maximum use of introduced biological control agents (pest and parasites), while at the same time managing the environment to provide suitable habitats, close to the crops, for natural predators. Pesticide use is kept to a minimum, but has an important role if pest populations start to get out of control.

Table 10.8 Some major pesticides

Name of pesticide	Type of pesticide	Additional information
Inorganic pesticides		
Calomel (mercuric chloride)	Fungicide	Used for dusting seeds to control transmission of fungal diseases
Sodium chlorate	Herbicide	Used to clear paths of weeds. Persistent, although not very poisonous
Organic pesticides		
Organo-phosphorus compounds (e.g. malathion and parathion)	Insecticides	Although very toxic, they are not persistent, and therefore not harmful to other animals if used responsibly. May kill useful insects such as bees, however
Organo-chlorine compounds (e.g. DDT, dieldrin)	Insecticides	DDT is fairly persistent, and accumulates in fatty tissue as well as along food chains
Hormones (e.g. 2,4-D)	Herbicides	Selective weedkillers which kill broad-leaved species

Table 10.9 Some examples of biological control

Target (pest)	Harmful effects of pest	Control agent	Method of action
Scale insect	Kills citrus fruit crop	Ladybird	Ladybird uses scale insect as a food source
Codling moth	Ruins orange crop	African wasp	Wasp parasitises moth eggs
Larvae of many butterflies and moths	Consume the foliage of many economically important plants	Bacterium (HD-1 strain of *Bacillus thuringiensis*)	Bacterium parasitises the larvae of moths and butterflies

Table 10.10 Comparison of biological and chemical control of organisms

Biological control	Chemical pesticides
Very specific	Always have some effect on non-target species
Once introduced, the control organism reproduces itself	Must be reapplied at intervals, making them very expensive
Pests do not become resistant	Pests develop genetic resistance, and new pesticides have to be developed

SUMMARY TEST 10.7

A pesticide which kills weeds is called a (**1**). If it is sprayed onto the soil or seeds before planting, as in the case of simazine, it is called a (**2**) pesticide, whereas if it is sprayed directly onto the weed, as in the case of paraquat, it is called a (**3**) pesticide. Those, like glyphosate, that are absorbed by the plant and transported throughout it are called (**4**) pesticides.

Pesticides which kill organisms such as greenfly are called (**5**), and a specific example of such a pesticide is (**6**). The use of another living organism to control pests is called (**7**) control and when used in an environmentally sensitive way along with chemical control, the process is known as (**8**).

1 The graph shows the effect of applying different amounts of nitrogen fertiliser on the yield on wheat.

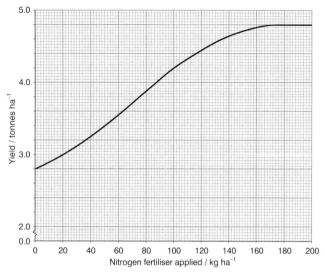

(i) Calculate the percentage increase in yield obtained when the amount of nitrogen fertiliser is increased from 50 to 100 kg ha⁻¹. Show your working.
(2 marks)

(ii) It would **not** be advisable for the farmer to apply more than 160 kg ha⁻¹ of fertiliser to this wheat crop. Use the graph to explain why. *(2 marks)*

(iii) Give **two** advantages of using an inorganic nitrogen fertiliser, rather than an organic fertiliser.
(2 marks)
(Total 6 marks)
AQA Jan 2001, B (A) BYA2, No.8

2 The diagram shows the growth of wheat crops on a British farm.

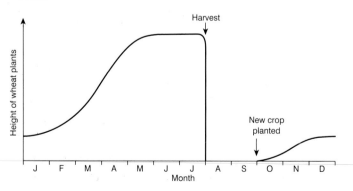

a Explain why it is necessary to add fertiliser if an area is used to grow wheat for a number of years. *(2 marks)*

b In Britain, the rainfall during the autumn is usually high while temperatures are low.

(i) Explain the advantage to a farmer of **not** adding fertiliser in the autumn. *(2 marks)*

(ii) Describe how adding fertiliser during the autumn could damage the environment. *(3 marks)*
(Total 7 marks)
AQA (specimen) 1999, B (A) BYA2, No.2

3 The removal of forest trees to supply wood, or so that land can be used for other purposes, has long been common practice. However, the last 250 years have seen an increase in the rate of deforestation.

One of the results of deforestation may be that there is a net increase of carbon dioxide in the atmosphere. During the period from 1958 to 1980, it is estimated that there was a net increase of 57.3×10^9 million tonnes of carbon in the atmosphere, mainly in the form of carbon dioxide. Much of this increase is thought to be a direct result of the clearance of the forests.

The photograph below was taken from space by satellite, showing an area of Amazonian rainforest partially cleared in 1998.

a State **two** ways in which the wood cleared from forests may be used. *(2 marks)*

b Suggest, giving a reason, **one** possible use for the cleared land shown in the photograph. *(2 marks)*

c Explain how the biodiversity in this area might be affected by deforestation. *(2 marks)*

d Explain why extensive clearance of forests might lead to *'a net increase of carbon dioxide in the atmosphere'* (Paragraph 2). *(3 marks)*

e With reference to a specific method of forestry management, explain what is meant by the sustainable management of forests. *(3 marks)*
(Total 12 marks)
Edexcel 6103/03 June 2001, B B(H) AS/A, No.1

4 Exposure to pollutant gases, such as sulphur dioxide and nitrogen oxides, has a number of effects on the growth of plants and on insect populations.

a A study was carried out to investigate the effect of pollutant gases on plants. Analysis of the sap from plants growing near to a motorway showed that they had a higher amino acid content than those further away.

Suggest an explanation for this difference in amino acid content. *(4 marks)*

b Aphids are small insects that feed on sap from plants, which they obtain by inserting a hollow feeding tube, or stylet, into the phloem. In an investigation to study the growth rate of aphids, a large number of bean plants infected with aphids were divided into two groups. One group was planted very close to a motorway and the other group 300m from the motorway. The mean numbers of aphids per bean plant were then determined at intervals from May 27th until July 15th. The results are shown in the table below.

Date	Mean number of aphids per bean plant	
	Close to motorway	300m from motorway
May 27	5	5
June 3	50	45
June 10	190	100
June 17	280	105
June 25	120	11
July 1	100	10
July 8	10	11
July 15	8	6

  (i) Compare the changes in the aphid numbers on the bean plants near the motorway with those on the plants further away. *(3 marks)*

 (ii) Suggest an explanation for these differences in the numbers of aphids. *(2 marks)*

(Total 9 marks)

Edexcel 6052/01 June 2001, No.2

5 Read through the following passage about pollution of fresh water by raw sewage, then write on the dotted lines the most appropriate word or words to complete the account. Sewage contains mineral ions such as nitrates and ...., and also suspended organic solids. If raw sewage flows into a river, the suspended solids are broken down by ...., resulting in a decrease in the concentration of dissolved oxygen in the water. The volume of oxygen used by a sample of water is known as the ...., which steadily .... as the organic solids start to be broken down. Mineral ions stimulate the growth of algae which can reduce the growth of submerged plants by reducing the amount of .... reaching them. *(Total 5 marks)*

Edexcel Jan 1999, B(H) AS/A, No.2

6 Graph A below shows the global fossil fuel consumption, between 1860 and 1990.

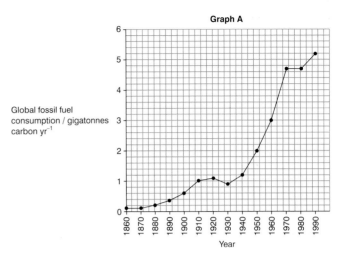

Graph A

Global fossil fuel consumption / gigatonnes carbon yr^{-1}

Graph B shows how the mean global temperature, between 1860 and 1990, varied from that in 1970.

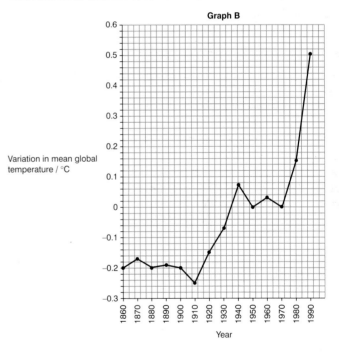

Graph B

Variation in mean global temperature / °C

a Calculate the percentage increase in the use of fossil fuels between 1970 and 1990. Show your working. *(2 marks)*

b State **two** reasons for the increased use of fossil fuels. *(2 marks)*

c To what extent do the graphs support the theory that rising global temperature is due to the increased use of fossil fuels? *(3 marks)*

d Suggest how the use of fast-growing plants, such as willow, as a fuel could help to slow down the rate at which global temperatures are rising. *(3 marks)*

(Total 10 marks)

Edexcel 6049/01 June 2001, B(H), No.6

Xerophytes and hydrophytes

EDEXCEL

AQA.B   OCR

Prickly pear

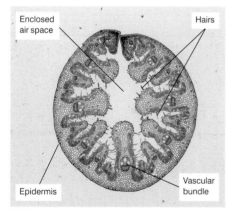

TS leaf of Ammophila, *showing xerophytic features (×3 approx)*

Norway spruce needles

Terrestrial plants all face, to different extents, problems associated with obtaining and retaining water. Most plants live in areas where ample water is available, but some may suffer severe shortages, while others may be partially or completely submerged. Plants may therefore be grouped according to the availability of water in their environment:

- **Mesophytes** are normally able to take up sufficient water to replace what is lost by **transpiration**. Most plants are in this category.
- **Xerophytes** live in areas where their water losses via transpiration may be greater than their water uptake.
- **Hydrophytes** live either wholly or partially submerged in fresh water, and so have no problems of water balance, although obtaining oxygen for respiration can be difficult.

11.1.1 Xerophytes

Xerophytes ('xero' = dry, 'phyte' = plant) are typically thought of as desert plants, showing a whole range of adaptations to cope with hot, dry conditions. However, similar adaptations may also be seen in plants found in sand dunes or other dry, windy habitats in a temperate climate, or in those where water shortage may be the result of freezing conditions, not high rates of evaporation. The structural and physiological modifications shown by all these plants are known as **xeromorphic** features (table 11.1).

Desert cacti such as prickly pears and saguaro have:
- shallow, extensive rooting systems that enable rapid absorption of water over a wide area after rainfall, before it evaporates
- leaves reduced to spines that limit both transpiration and grazing
- thick cuticles that reduce evaporation of water from the surface
- stomata that open at night, not during the day, reducing transpiration
- a specialised 'crassulacean acid metabolism' system, which allows photosynthesis to occur on hot sunny days while the stomata are closed. Carbon dioxide is fixed as malate at night, and released for use in the **Calvin cycle** during the day
- water storage in succulent stems.

Marram grass (*Ammophila*), found on sand dunes which can be both dry and windy, shows the following xeromorphic features:
- rolled leaves that increase humidity and reduce transpiration
- hinge cells that increase rolling in dry conditions
- sunken stomata that increase humidity within the region enclosed by the leaf
- epidermal hairs that slow down air movement and reduce transpiration
- thick cuticle that reduces evaporative losses.

Conifers, such as pine and spruce, are not always found in areas typically thought of as dry, but they are evergreens and, in the winter when ground water is frozen, it is important that they keep water losses to a minimum. Therefore they have:
- leaves reduced to needles
- thick cuticles
- stomata in depressions.

Labels on TS leaf diagram: Enclosed air space, Hairs, Epidermis, Vascular bundle

Table 11.1 *Xerophytic adaptations of plants*

Xeromorphic feature	Mechanism	Examples
Long root system	• absorption of water from deep underground	Cacti
	• absorption of water from near surface after rain	*Acacia*
Reduction in transpiration rate	• reduction in leaf area	Needles of pine and spruce
		Spines of cacti and gorse
	• increasing humidity around stomata	Rolled leaves of marram grass
		Sunken stomata of holly and pine
	• thick cuticle	Evergreens such as pine and holly
Storage of water	• retention of water in stem for use in droughts	Prickly pear and saguaro
Resistance to wilting	• having smaller cells, which makes the proportion of cell wall material greater	Many xerophytes
	• more lignified material in leaf, enabling photosynthesis to continue	*Hakea*

11.1.2 Hydrophytes

Hydrophytes ('hydro' = water, 'phyte' = plant) may be wholly submerged (*Elodea* and water milfoil), have floating leaves (water lilies), or have some leaves submerged and others out of the water (water crowfoot). A common feature of all hydrophytes is the presence of **aerenchyma**. This aeration tissue gives the plant buoyancy, keeping it in a position suitable for photosynthesis. It also aids the transport of gases from aerial or photosynthetic parts to regions in deeper water, where there may be an oxygen shortage. Hydrophytes lack cuticles and so can absorb water over the whole surface. This means that root and vascular systems can be considerably reduced, or even absent. The high density of water supports the plants and makes specialised supporting tissues unnecessary. Submerged leaves lack stomata and most are finely branched and frond-like, which prevents breakage in the water currents. Floating leaves, like those of water lilies, have stomata on the upper, aerial surface, and supporting **sclerenchyma** is present, preventing the leaves from rolling up. They have a thick, upper, layer of palisade cells. To summarise, hydrophytes show the following features:

* air-spaces (aerenchyma)
* no cuticle
* no stomata
* reduced or absent roots
* reduced support and vascular systems
* finely branched submerged leaves.

Water crowfoot (Fig 11.1) shows the significance of these hydrophytic features clearly: its submerged leaves are finely dissected and lack both cuticle and stomata, while its aerial leaves are broad, with cuticle and stomata.

Fig 11.1 *Aerial and submerged leaves of water crowfoot*

*Water lily (*Nymphaea*)*

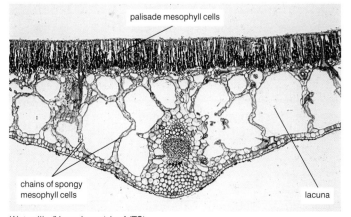

palisade mesophyll cells

chains of spongy mesophyll cells

lacuna

*Water lily (*Nymphaea*) leaf (TS)*

SUMMARY TEST 11.1

Xerophytes are adapted to survive in areas where water loss due to **(1)** is potentially greater than the rate of its uptake. Some achieve this by having thick **(2)** on their leaves to reduce water loss and **(3)** that open at night when evaporation is less. Others, such as **(4)**, store water in succulent stems and yet others like **(5)** reduce their leaves to needles.

Adaptations of crop plants to different climates

About half the human food supply comes from just three crops: wheat, rice and maize. They are important as fodder crops too, and they account for about 75% of the world grain production tonnage, the remaining 25% being made up of sorghum, millet and oats (table 11.2). Cereal grains have a low water content – about 15%, compared with 80% for some root crops. This means that they are less water-demanding than many fruits and vegetables as they ripen. They are also easy to transport and to store, have a high energy value per gram, and a protein content of 7–15% (table 11.3). Cereals, like other plants, are adapted to particular conditions, and there is little geographical overlap between the growing areas of the main crops:

- **Wheat** is frost-tolerant, and is grown in temperate regions with moderate rainfall.
- **Rice** is grown in wet, tropical areas.
- **Maize**, grown mainly in the tropics, requires less water than rice and wheat, but needs high temperatures and bright light.
- **Sorghum**, found in the semi-arid tropics, requires high temperatures but little water, because it tolerates drought well.

11.2.1 Rice (*Oryza sativa*)

Most rice (95%) is grown in less developed countries, most of it in China. It has the following features, which suit it to its wet, tropical environment:

- It grows in regions of high seasonal rainfall and high temperatures (a minimum of 20°C throughout the growing period).
- It thrives in heavy, silty soils.
- Although it has a higher water requirement than other cereal crops, it is not a true **hydrophyte**; it will die if it is totally submerged, and it can grow in non-waterlogged conditions.
- The cells of the embryo are tolerant of the ethanol that builds up as a result of **anaerobic** respiration.
- The stems and roots have many air-spaces between the cells (aerenchyma), which allow oxygen to diffuse from aerial parts.
- If the roots remain short of oxygen, they respire anaerobically and tolerate the build-up of ethanol.
- Artificial selection has led to the development of many varieties, with properties such as disease resistance, salt tolerance and high yields.
- The rice seeds are able to germinate in the anaerobic conditions that prevail in the mud of the paddy fields.

The method of cultivation of rice also has certain advantages:
- The flooding provided by paddy fields reduces many weeds.
- Flooding encourages the growth of cyanobacteria, which are able to fix nitrogen, and so less fertiliser is needed to maintain yields.
- Paddy fields may also be stocked with fish, which eat weeds and insects, as well as providing a second profitable crop.

Ten percent of rice is grown in upland areas of high rainfall without additional irrigation, but higher yields are generated in the lowlands where the fields are flooded to a depth of 5–10cm.

Table 11.2 Harvested area of different crops

Crop	Area harvested in 1999 / millions of hectares
Wheat	215
Rice	155
Maize	139
Sorghum	43

Table 11.3 Food composition of cereals

Crop	Energy / kJ	Protein / g	Lipid / g
Wheat	1420	1210	210
Rice	1296	810	210
Maize	1471	1010	410
Sorghum	1455	1010	510

Rice cultivation in paddy field

11.2.2 Maize (*Zea mays*)

Maize, or corn, originated in Mexico, and is still grown extensively in South and Central America. However, the USA now produces almost half the world tonnage. Maize requires

- a frost-free environment
- high temperatures
- prolonged sunshine.

Without these conditions, the yield will be low, and the crop may only be suitable for animal fodder. Sub-tropical conditions are ideal for maize because, as well as light and warmth, there is sufficient rainfall to provide the moisture needed at germination. Maize has a specialised method of photosynthesis, known as the **C$_4$ pathway**. Photosynthesis takes place in a series of small steps, and most plants, including wheat and rice, use a metabolic pathway known as C$_3$ photosynthesis. In these plants, the first step is the coupling of carbon dioxide with ribulose bisphosphate to form a three-carbon compound (C$_3$). In plants such as maize, carbon dioxide combines with phosphoenol pyruvate (PEP) to form a four-carbon compound (C$_4$). The PEP has a higher affinity for carbon dioxide and is especially efficient at high temperatures. This makes it ideal in warm sub-tropical conditions, and in fields of crops where carbon dioxide levels may be low. There are two further advantages of C$_4$ photosynthesis:

- It allows the plant to accumulate a store of carbon dioxide when it is relatively plentiful, and store it for later use when external supplies are reduced.
- C$_4$ plants avoid photorespiration. Like respiration, this process is the reverse of normal photosynthesis but, unlike respiration, it does not yield **ATP**.

11.2.3 Sorghum (*Sorghum bicolor*)

Sorghum is the most drought-resistant cereal, found in the hot, dry semi-arid tropics such as parts of Africa and central India. It has many **xeromorphic** features to enable it to grow in such difficult conditions:

- a dense root system, giving efficient water absorption
- a thick waxy cuticle, which reduces water loss
- a leaf which rolls in dry conditions, trapping moist air and reducing transpiration
- a low number of stomata, sunk into the leaf surface, keeping the transpiration rate low, but allowing gas exchange.

In addition, sorghum is a C$_4$ plant, like maize, which means it avoids photorespiration and has an increased net rate of photosynthesis. Sorghum not only withstands drought well but, unusually, it can recover quickly when the drought ends, by re-opening its stomata rapidly. Embryonic and mature sorghum plants are able to withstand excessive heat better than other cereals. To avoid denaturation, they are able to synthesise 'heat-shock' proteins rapidly when the temperature rises. These proteins enable sorghum to continue both to respire and to photosynthesise at 35°C.

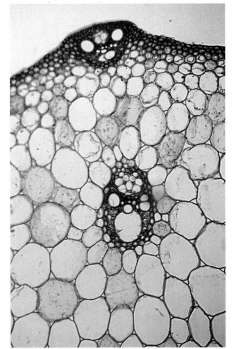
T.S of stem of rice

Maize plants

Sorghum plants

SUMMARY TEST 11.2

Of the main human food crops, **(1)** is the most drought resistant. As it grows in semi-arid areas it has many **(2)** features that allow it to survive in dry areas. These include a thick waxy **(3)** on their leaves which roll up during dry spells to reduce **(4)**. It carries out a special form of **(5)** known as the C$_4$ pathway, in which carbon dioxide combines with **(6)** to give a four-carbon compound rather than with **(7)** to give the more normal three-carbon compound. This process has the advantage of being efficient at low carbon dioxide concentrations, allowing the plant to store carbon dioxide and avoiding **(8)**.

EDEXCEL (Human)

Humans are **endotherms**, able to maintain a more or less constant body temperature within the range 36–37.5°C irrespective of the environmental temperature. Taking the temperature using a thermometer placed in the mouth gives a slightly less reliable estimate of the temperature of **core**, or deep, body structures than taking a rectal temperature.

11.3.1 Variations in normal body temperature

Even without the effects of exercise and variations in environmental conditions, a person's body temperature will fluctuate, often in a cyclical way. The following are examples of daily (diurnal) and monthly variations:

- **Diurnal variation** – core body temperatures reach a peak between mid-day and early afternoon and are at their lowest at night, after midnight (Fig 11.2).
- **Monthly variation** – in women, body temperatures are lowest during **menstruation** and rise at **ovulation**.

11.3.2 Temperature balance

If a person is to maintain a steady body temperature, there must be a balance between the heat gained by their body and the heat lost by it. Humans gain heat by:

- **internal means** – metabolic activities of all body cells, but especially those of the liver and active muscles, generate heat
- **external means** – if the temperature of the environment is higher than the temperature of the skin
- **direct means** – through the consumption of hot food or drinks.

Heat loss mostly occurs across the skin by:

- **conduction** from the body to objects in contact with it
- **convection** to air or water when the skin is warmer than the environment
- **radiation** from the body to the surrounding environment
- **evaporation** of water, usually in the form of sweat.

Some heat is also lost via urine and faeces.

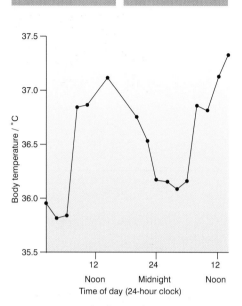

Fig 11.2 Diurnal variation in body temperature

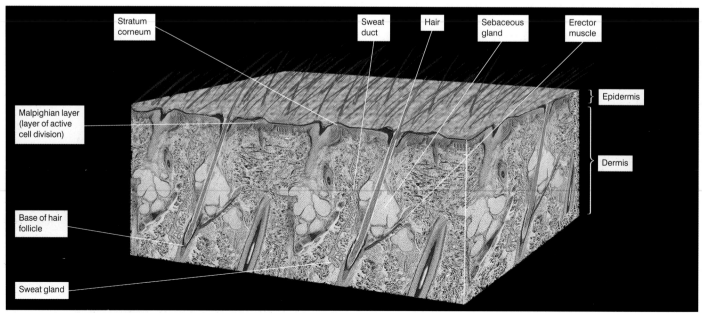

Fig 11.3 Human skin (VS) (×30 approx.)

11.3.3 The structure of the skin

The skin forms the boundary between the external environment and the inside of the body, and so most heat exchange occurs across it. The following structures, shown in figure 11.3, have an important role in thermoregulation in humans:
- **Malpighian layer** – **melanin** is produced here. It determines skin colour, absorbing ultra-violet light and helping to protect the underlying layers.
- **Blood capillaries** – blood flow through these can be controlled, to increase or decrease heat loss.
- **Sweat glands** – cells that absorb fluid from capillaries and secrete it into a tube, from where it passes to the surface via the sweat duct. Sweat consists of mineral salts and urea dissolved in water.
- **Subcutaneous fat** – beneath the dermis, this layer of **adipose tissue** acts as an insulating layer.
- **Temperature receptors** – detect variations in environmental temperatures.

11.3.4 Keeping cool

Humans use a range of adaptations to help them increase heat loss and thus maintain a constant body temperature in warm environments:
- **Vasodilation** – superficial arterioles dilate, bringing blood close to the skin surface. Heat from this blood is conducted through the epidermis, and radiated away from the body (Fig 11.4).
- **Sweating** – humans may produce up to $1 dm^3 hr^{-1}$ from sweat glands, which cover the whole body surface. Evaporation of this sweat requires energy, which is taken from the body in the form of heat. In this way, sweating cools the body.
- **Behavioural mechanisms** – these are of many kinds in humans, but include seeking shade, being less active, wearing fewer clothes, and 'spreading out' to increase the surface area available for heat loss.

11.3.5 Keeping warm

As the temperature of the environment falls, humans make use of a range of mechanisms both to generate heat and to reduce heat loss:
- **Shivering** – skeletal muscles undergo rhythmic, involuntary contractions, which produce metabolic heat **(shivering thermogenesis)**.
- **Increased metabolic rate** – hormones (e.g. thyroxine) are produced which raise the metabolic rate of organs such as the liver, generating heat.
- **Vasoconstriction** – superficial arterioles constrict, reducing blood flow close to the skin surface (Fig 11.5).
- **Insulation** – humans have little hair, but the layer of adipose tissue beneath the skin is effective at reducing heat loss.
- **Behavioural mechanisms** – these include seeking shelter, wearing more clothes, being more active, eating more hot food, and curling up to reduce the surface area exposed to the cold.

11.3.6 The role of the hypothalamus

The hypothalamus, at the base of the brain, monitors the temperature of blood passing through it and also receives nervous impulses, from thermoreceptors in the skin. Heat gain and heat loss centres then trigger the changes required to maintain the core body temperature within the range 36–37.5°C (Fig 11.6). Humans are very sparsely covered with hairs, and so raising and lowering them has little effect on insulation.

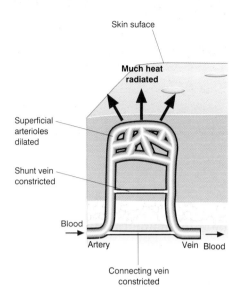

Fig 11.4 *Vasodilation*

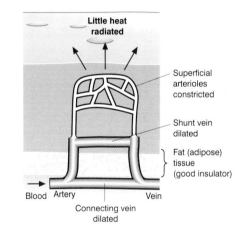

Fig 11.5 *Vasoconstriction*

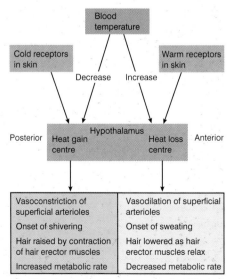

Fig 11.6 *Summary of body temperature control by the hypothalamus*

197

The mean body temperature of a human is 36.7°C. The lower 'normal' temperature is 35.8°C and the upper 'normal' temperature 37.5°C. While the body can tolerate a fall in core temperature of as much as 10°C, a rise in temperature of only 4°C may be lethal (Fig 11.7).

11.4.1 Adaptations to high temperatures

When people from temperate climates visit hot countries, they gradually **acclimatise** to the conditions, as follows:
- **becoming less active and eating less**, which helps to reduce the heat generated, as metabolism slows down
- **drinking more and eating foods with a high water content**, which replaces water lost by sweating and helps prevent dehydration
- **dilating superficial arterioles**, helping to increase heat loss – only if the environmental temperature is lower than normal body temperature
- **increasing the rate of sweating** – an effective way of losing heat, as long as the atmosphere is not humid
- **reducing the salt content of sweat**, preventing excessive loss of ions such as sodium (Na^+) and chloride (Cl^-).

People who normally live in hot climates are **adapted** by having a higher rate of sweating, and their sweat has a lower salt content. They also wear long, loose clothes, which prevent direct exposure to the sun's radiation, but allow the evaporation of sweat.

11.4.2 Heat stress

Humans can compensate for any increase in body temperature up to about 41°C, by such mechanisms as **vasodilation** and increased sweating. However, if these **homeostatic** mechanisms fail, the body temperature continues to rise, causing **hyperthermia**, which may result in death above 43°C. As the body temperature starts to rise, the early symptoms of **heat stress** may include sunburn and prickly heat. The latter is an irritating rash, which results from blocked sweat ducts that prevent sweat reaching the skin surface. As the blood flow to the skin increases, it is diverted away from internal organs, so that the person may be dizzy or suffer from **heat collapse**. Increased rates of sweating can cause severe dehydration if fluids are not replaced, and this results in **heat exhaustion**. A loss of 5–8% of body fluid causes fatigue, but moderate dehydration (10% loss) can result in mental and physical impairment. Severe dehydration may result in damage to cells, as intracellular fluid is withdrawn. In addition, any increase in sweating will result in the loss of ions. If adequate salt is not included in the diet, painful **muscle cramps** will be experienced. If measures are not taken to relieve heat stress, **heat stroke** may occur. This extremely serious condition results from a failure of the sweating mechanism and, without treatment, the body temperature continues to rise, resulting in coma and then death.

11.4.3 Adaptations to low temperature

The human body has no clearly defined mechanisms for acclimatising to cold in the way that it has for hot climates, but it can tolerate a considerable fall in body temperature. To keep warm in extreme cold, humans rely on the increased production of metabolic heat, together with measures to conserve as much heat as possible. The methods used include:

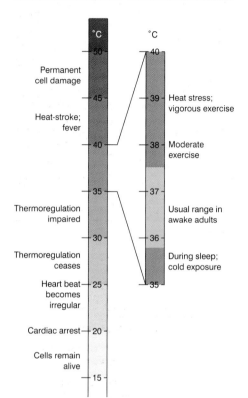

Fig 11.7 *The normal range of body temperature under a variety of conditions*

Table 11.4 *Effect of increased wind speed on the temperature of exposed skin*

Wind speed / km h⁻¹	Actual temperature / °C		
	5	−18	−40
<8	5	−18	−40
24	−6	−36	−66
48	−11	−45	−79
64	−13	−48	−83

- **eating more food** especially lipids
- **increased production of hormones such as adrenaline**, which raise the metabolic rate
- **shivering**, which generates heat
- **increasing the subcutaneous fat deposits**, which conserve heat
- **wearing layers of warm clothing** to insulate the body
- **suitable housing** to provide a warm environment
- **vasoconstriction**, in which reduction of blood flow to the extremities is alternated with periods of vasodilation (the '**hunting reaction**'), to prevent numbness and tissue damage.

People such as the Inuit, living permanently in cold conditions, have a high metabolic rate as a result of their high protein diet, and thick layers of subcutaneous fat; they also have greater blood flow to their hands and feet than visitors do. The risk of cold stress occurring is increased in windy conditions (**wind chill**), as the warm layer of air around the body is blown away. In the wet, the problem is made worse by an increase in evaporative cooling. This is shown in table 11.4: it can be seen that, if properly clothed, a person would be in little danger at $-18\,°C$ with a wind speed less than $8\,km\,h^{-1}$ (skin temperature $= -18\,°C$), but they would risk cold injury if the wind speed was $24\,km\,h^{-1}$ (skin temperature $= -36\,°C$) or more.

11.4.4 Cold stress

As core temperature falls, the body may start to show signs of cold stress, referred to as **cold injury** if actual damage to the tissues occurs. The damage normally results from reduced blood flow to the extremities, so the cells lack nutrients, and normal metabolism stops. In **frostbite**, the tissues may actually freeze; the ice crystals then damage cell membranes and the cells dehydrate. If the effect extends to deep tissues such as muscle and bone, it may cause permanent damage, and gangrene may result in the loss of fingers or toes. **Trench foot** may result from prolonged exposure to cold in wet conditions. Skin becomes blackened, and the injury to muscles and nerves may lead to gangrene. When the core temperature drops to $35\,°C$ or below, **hypothermia** develops. The three groups most at risk are:

- **babies**, who are unable to shiver or to use behavioural methods to warm themselves
- **old people**, who may have limited mobility or a lower metabolic rate, and who may have poor housing or a poor diet
- **hill walkers or climbers**, who may suffer from exposure in cold, wet and windy conditions.

The symptoms of hypothermia become more severe as the core temperature falls, as shown in table 11.5. The symptoms are mainly the result of

- **reduced heart rate**, so supplies of blood to the muscles and brain may be inadequate
- **slower respiration rate**.

Inuit in the Arctic

Table 11.5 *The effects of hypothermia*

Temperature / °C	Symptoms
35	Lack of coordination Muscle weakness
34	Vision disturbed Mental confusion
32–30	Loss of consciousness
28–25	Death

SUMMARY TEST 11.4

Normal human body temperature is **(1)**, although a temperature of up to **(2)** can be controlled using **(3)** mechanisms such as vasodilation and sweating. Vasodilation, however, diverts blood from the internal organs and may lead to dizziness and **(4)**. Sweating can lead to dehydration resulting in **(5)**. Also, the loss of salts in sweat can lead to painful **(6)**. If core body temperature falls, individuals may show the symptoms of **(7)** as a consequence of reduced blood flow to the extremities of the body. If tissues actually freeze as a result, then **(8)** may occur.

Table 11.6 *Changes in atmospheric pressure with altitude*

Altitude / m	Atmospheric pressure / kPa	Partial pressure of oxygen / kPa
0	100	21
3500	65.5	13.8
8500	31.9	6.7

Table 11.7 *Fall in temperature with altitude, assuming temperature at sea level to be 25°C*

Altitude / m	Temperature / °C
0 (sea level)	25
3500	2
8500	−54

In 1953, Hillary and Tenzing conquered Mount Everest, at 8848m the world's highest mountain. Every year, thousands of people climb and trek in the mountains, pushing their bodies to the limits of human endurance.

11.5.1 Environmental conditions in high mountains

- **Low atmospheric pressure** – although the proportion of oxygen remains the same anywhere in the atmosphere (21%), the air pressure falls as we ascend from sea level. As a result, the air at the top of Everest has less than one-third of the oxygen available at sea level (see table 11.6). Such low oxygen pressures make breathing, and the loading of haemoglobin with oxygen, difficult.
- **Low temperature** – environmental temperature falls about 1°C for every 150m, and temperatures below freezing are common (table 11.7).
- **High winds** – air currents replace warm air near the body with cold air, which is warmed by convection from the skin, cooling the body as it loses energy. The wind-chill factor (section 11.4.3) increases the risk of hypothermia and frostbite.
- **Low humidity** – this increases the rate of evaporation. The evaporation of water uses energy, resulting in heat loss from the skin when sweat evaporates, and from the respiratory tract where air is warmed and moistened. Mountaineers may suffer from cracked lips and a persistent cough.
- **Increased solar radiation** – there is less air to filter the solar radiation, and 90% of the radiation is reflected back from snow, so heat gain by the body during the day can be considerable. Mountaineers wear goggles to protect the cornea from the ultra-violet radiation and prevent snow-blindness.

11.5.2 Physiological effects of high altitude

Although sudden exposure to the low oxygen levels at high altitude could lead to death within 10 minutes, if mountaineers ascend slowly, their bodies **acclimatise**. Low oxygen levels in the blood and tissues are referred to as **hypoxia**, and the body needs time to respond to altitude by trying to maximise the amount of oxygen delivered to the tissues:

- **Increase in cardiac output** – this is a first response to altitude, and distributes oxygenated blood to the tissues more rapidly. After a few days, it returns to about the same as that at sea level.
- **Hyperventilation** – increasing the rate and depth of breathing delivers more oxygen to the lungs for absorption. Until acclimatisation is complete, this causes a conflict, because hyperventilation also removes larger than normal volumes of carbon dioxide. The result of this is reduced ventilation (section 6.6.3) and so, overall, the pattern of breathing can be irregular.
- **Increase in pulmonary diffusing capacity** – this is a measure of the rate of exchange between the alveoli and the pulmonary capillaries. It may be increased by having an enlarged lung volume, and by increasing the rate of flow through the capillaries. These changes can be seen in native highlanders, but not in visitors.
- **Increase in red blood cells and haemoglobin concentration** – after a few days at high altitude, water is absorbed from the circulation, concentrating the red blood cells and thickening the blood. Then, after 1 or 2 weeks, the kidneys increase production of the hormone **erythropoietin**, which stimulates the formation and release of more red blood cells from the bone marrow.

Effects of frostbite

11.5.3 Native highlander adaptations

There is little evidence that people who live all their lives at high altitude are genetically adapted to do so. Although they may have a greater lung volume, and consequently be more barrel-chested, than their lowland counterparts, in other respects they are merely better acclimatised. Both natives and visitors will show some hyperventilation, and have a higher red blood cell count and greater concentration of haemoglobin. For these last two reasons, athletes find it an advantage to train at high altitude. The features are lost in both natives and visitors soon after they move to sea level.

11.5.4 High altitude stress

More than half the people trekking to 5000m will experience the altitude-related illness called **acute mountain sickness**. Its many symptoms include:
- headaches and dizziness
- nausea and vomiting
- insomnia
- general lethargy
- dry, irritating cough
- breathlessness.

If ignored, these symptoms may worsen, leading to the following:
- **Mental impairment** – at moderate altitude, many people are shown to have slower reactions than at sea level. As they climb higher, they tend to lack concentration, and they will find it difficult to calculate and make judgements. Still higher, and they may suffer hallucinations, and have a dangerous sense of well-being.
- **Redistribution of body fluids** – severe hypoxia leads to an increased production of ADH (antidiuretic hormone), which reduces urine output and causes more water to remain in the blood. In cold weather at altitude, because the blood flow to the extremities has been reduced, the extra fluid accumulates outside blood vessels, causing swelling or **oedema**. This may be apparent in swollen feet and legs as well as in the face, especially around the eyes. More seriously, fluid may accumulate in the lungs **(pulmonary oedema)**, causing breathlessness and frothing at the mouth. An accumulation of fluid in the brain **(cerebral oedema)** causes it to swell and push against the cranium. Severe headaches may be followed by unconsciousness, and even death, if a return to low altitude does not occur soon enough.

Table 11.8 Scale for acclimatisation

Effect	Minutes	Days	Weeks
Increased cardiac output	████████		
Hyperventilation		████████	
Increased blood concentration			████
Increased red blood cell production			████

SUMMARY TEST 11.5

The amount of oxygen at high altitude is **(1)** % of the total atmosphere. Its partial pressure is lower at high altitude, being only about **(2)** % at 85 000m compared with its partial pressure at sea level. Other environmental differences found at high altitude include lower **(3)** and **(4)** but increased **(5)** and **(6)**. Mountaineers therefore need to slowly **(7)** to the conditions if they are to avoid a drop in blood oxygen levels known as **(8)**. To maintain a high oxygen level, the rate of **(9)** and **(10)** both increase initially. After a few weeks the kidneys produce the hormone **(11)**, which results in an increase in **(12)**. A much longer-term response is an increase in **(13)**. Individuals who do not allow time to adapt to high altitude often suffer **(14)**, the symptoms of which include **(15)** and increased production of **(16)** hormone that causes accumulation of water outside blood vessels – a condition called **(17)**.

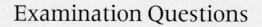

1 a Sorghum and maize are plants which are adapted to grow in tropical areas. Describe and explain the adaptations which enable them to grow in these conditions.

   (i)   Sorghum

   (ii) Maize   *(7 marks)*

b $$\text{The transpiration ratio of a plant} = \frac{\text{mass of water taken up by the plant}}{\text{mass of dry matter produced}}$$

The table shows the transpiration ratios of sorghum and wheat. Both were grown in fertile soil.

Crop	Transpiration ratio
Sorghum	324
Wheat	510

In tropical areas sorghum may be grown in preference to wheat. Use the information in the table to explain **one** reason for this.   *(2 marks)*

(Total 9 marks)

AQA Jan 2001, B (A) BYA2, No.8 a/b

2 a Many varieties of rice are planted in soil which is flooded with water.

   (i)   Suggest why plants of most other species would die if they were planted in these conditions.   *(3 marks)*

   (ii) Ethanol is poisonous. Rice is extremely tolerant and can survive with amounts of ethanol in its tissues which would kill other plants. Explain how this tolerance is an adaptation to growing in swampy conditions.   *(2 marks)*

b The diagram shows a section through the stem of a rice plant.

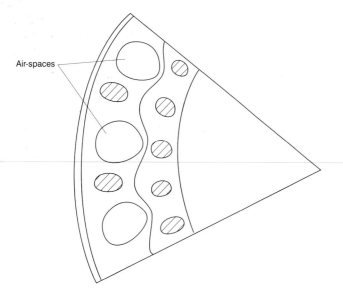

Air-spaces

Give **two** ways in which the air-spaces in the stem help the rice plant to survive in swampy conditions.

(2 marks)

(Total 7 marks)

AQA (specimen) 1999, B (A) BYA2, No.6

3 The photomicrographs show a transverse section through a leaf of *Ammophila*, which is a xerophyte. The large photomicrograph shows details of the tissues inside the box.

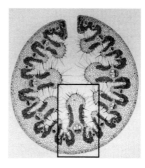

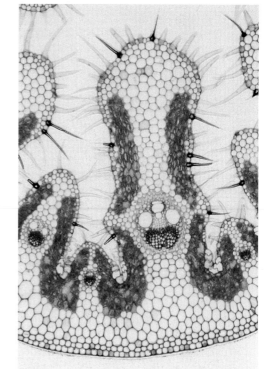

Describe **three** ways in which this leaf is adapted to reduce water loss.   *(Total 6 marks)*

Edexcel 6102/02 June 2001, B AS/A No.5

4 The diagram shows a sweat gland as seen in a vertical section through the skin. The figures show the concentrations of sodium ions in sweat in mmol per dm³, measured in different parts of the sweat duct.

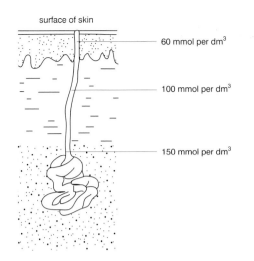

surface of skin

60 mmol per dm³

100 mmol per dm³

150 mmol per dm³

a Describe the changes in the sodium ion concentration as sweat moves from the sweat gland towards the surface of the skin. *(2 marks)*

b Suggest an explanation for these changes. *(1 mark)*

c Explain how an increase in sweating results in a cooling effect in humans. *(2 marks)*

(Total 5 marks)

Edexcel 6112/01 June 2001, B(H) AS/A, No.6

5 Give an account of the ways in which humans acclimatise to life at high altitudes. *(Total 10 marks)*

Edexcel 6112/01 June 2001, B(H) AS/A, No.9

6 The photograph below shows a native woman from a high altitude region (3200 m) in western China.

Describe and explain *three* ways in which native people, such as this woman, are adapted or acclimatised to living at high altitudes. *(Total 6 marks)*

Edexcel Jan 1999, B(H), No.5

7 An investigation was carried out into the effect of exposure to low environmental temperatures on core body temperatures. Two groups of males were studied: Europeans from a temperate climate and aborigines from a climate with hot days and cold nights. Both groups were exposed to an air temperature of 5°C during a night of eight hours. The results are shown in the graph below.

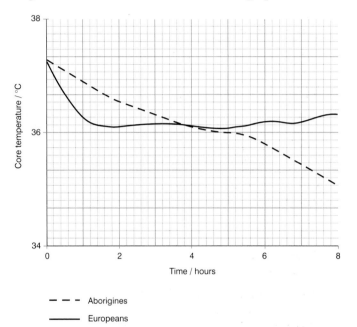

– – – Aborigines

——— Europeans

a Explain the meaning of the term *core body temperature*. *(2 marks)*

b Suggest *one* reason for using male individuals only in this investigation. *(1 mark)*

c (i) Describe the differences between the shapes of the two curves. *(3 marks)*

(ii) Suggest how these differences may be related to the climate in which the aborigines normally live. *(2 marks)*

d Between 4 and 5 hours exposure the mean core body temperatures remain constant. Explain how this is brought about. *(2 marks)*

e Describe the effect on the body if the core temperature drops below 28°C. *(3 marks)*

(Total 13 marks)

Edexcel Jan 1999, B(H), No.6

12.1

Mutations

AQA.B

Deoxyribonucleic acid (unit 3.2) is the hereditary material of cells; it is made up of a sequence of bases which, in groups of three, code for individual amino acids – the **triplet code** (section 3.4.2). Specific lengths of DNA, known as **genes**, are the functional units of inheritance.

12.1.1 Genes and alleles

A gene is a section of DNA that contains coded information which determines the nature and development of organisms. Each gene may have two, or occasionally more, alternative forms. Each of these different forms of a gene is called an **allele**. In pea plants, for example, there is a gene which determines the texture of the coat on the pea seeds. This gene exists in two forms – the allele for a smooth coat, and the allele for a wrinkled coat. Similarly, in cats, the gene for hair length exists as an allele for long hair and an allele for short hair. Alleles are situated in the same relative position on **homologous chromosomes** (section 5.4.2). Each position is called a **locus**. New alleles arise from changes to existing alleles as a result of **mutation**. These mutations, which occur randomly, are caused by a change to the sequence of bases in DNA and are known as **gene**, or point, mutations. There are a number of different types of gene mutation, of which the most common are addition, deletion and substitution.

12.1.2 Addition

A gene mutation by addition arises when an extra nucleotide is inserted into the normal sequence of bases in a DNA molecule. The bases are 'read' in threes (triplet code) when it comes to translating the code into a sequence of amino acids (unit 3.6). The addition of an extra base can completely alter this sequence of amino acids and, consequently, the polypeptide which is manufactured (Fig 12.1).

12.1.3 Deletion

A gene mutation by deletion arises when a nucleotide is lost from the normal DNA sequence. Again, this can completely alter the amino acid sequence which is produced after transcription and translation (Fig 12.1).

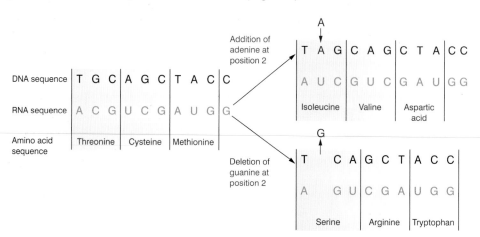

Fig 12.1 *Mutation by addition and deletion, and the effect on the amino acid sequence in the final polypeptide*

12.1.4 Substitution

A gene mutation due to substitution arises when a nucleotide in the DNA sequence is replaced by one with a different base. Take, for example, the triplet of DNA bases, adenine-cytosine-adenine (ACA) which codes for the amino acid, cysteine. A change to one base can cause three possible mutative changes:

- **A silent (synonymous) mutation**, in which the base change still results in the same amino acid being coded for: e.g., if the last base changes from adenine to guanine, then ACA becomes ACG. As both these triplets code for the amino acid, cysteine, there is no change to the polypeptide produced.
- **A mis-sense mutation** occurs when the base change results in a different amino acid being coded for: e.g., if the second base, cytosine, is substituted by thymine, then ACA becomes ATA, which is the DNA code for the amino acid, tyrosine. The polypeptide will therefore be different from the one intended. Whether or not this has an effect on the protein that is finally synthesised, depends on whether this amino acid has a role in formation of the various bonds that determine the protein's three-dimensional shape (section 1.6.3). If it does, then the protein may not function properly – e.g. if it is an enzyme, it may no longer be able to fit to its substrate molecules.
- **A nonsense mutation** occurs if the substitution results in the formation of one of the three stop codons which mark the end of a polypeptide chain (section 3.4.4). In our example, this would arise if the last base, adenine, is substituted by thymine, in which case ACA becomes ACT, which is transcribed into the mRNA stop codon, UGA. This would result in the production of the polypeptide being stopped early. This would almost certainly result in a significant alteration to the final protein, with consequent loss of function.

These examples of substitution mutations are summarised in table 12.1.

Table 12.1 *Examples of various types of substitution mutations*

	Normal triplet of DNA bases	Silent mutation	Mis-sense mutation	Nonsense mutation
Sequence of bases in DNA	ACA	ACG	ATA	ACT
Sequence of bases in mRNA	UGU	UGC	UAU	UGA
Amino acid sequence in polypeptide	Cysteine	Cysteine	Tyrosine	Stop code

12.1.5 Consequences of gene mutations

A gene mutation usually results in the wrong sequence of amino acids in the intended polypeptide. Sometimes, the change will be minor and not affect the structure of the final protein. Often, however, the alteration results in a dramatic change to the organism and its metabolism. If the protein is an enzyme, the change may alter its tertiary structure, making it impossible for its substrate to fit into it. As enzymes usually operate in a sequence, the loss of function by one may completely block a whole chain of metabolic reactions, thereby harming the organism. One example is the disease known as **sickle-cell anaemia**. Here, a mis-sense substitution mutation results in the amino acid valine being substituted for glutamate in the β-polypeptide chains of haemoglobin (section 1.7.2). The resulting haemoglobin gives the red blood cells a bent or sickle shape. These cells are less efficient at carrying oxygen, and may block small capillaries, resulting in tiredness and lethargy. At the same time, however, this cell shape gives some immunity to malaria.

12.1.6 Causes of mutations

Despite being random occurrences, mutations occur with a set frequency. This natural mutation rate varies from species to species, but is typically around one or two mutations per 100 000 genes per generation. Other agents increase this underlying rate. Such agents are called **mutagens**, and include the following:

- **High-energy radiation** such as α-particles, β-particles and neutrons can disrupt the DNA molecule.
- **Chemicals** such as formaldehyde, nitrous acid and mustard gas may alter DNA structure or interfere with **transcription**. Hydroxylamine, for example, causes cytosine to pair with adenine rather than with guanine.

SUMMARY TEST 12.1

Genes may exist in two or more different forms called **(1)**. Gene mutations arise through a change to the sequence of **(2)** bases that make up the gene. They take a number of forms. If the original base sequence was GGCTAGATC, then the types of mutation in the following cases would be: GGCTACGATC = **(3)**, GCCTAGATC = **(4)**, GGCTAGTTC = **(5)**, GGCTAGAC = **(6)**, GGCTAGATCC = **(7)**. The type of mutation known as a substitution takes three forms, depending on the consequences of the change. If the mutation results in a different amino acid being coded for, it is called a **(8)** mutation; if, however, the same amino acid is coded for it is a **(9)** mutation. If one of the three stop codons is formed, it is known as a **(10)** mutation.

12.2

Gene technology

AQA.A

AQA.A (Human)

AQA.B

Gene technology is a general term covering those processes in which genes are manipulated, altered or transferred from organism to organism. Also known as **genetic engineering**, it has enabled the characteristics of an organism to be changed, either by directly altering its genes or by transferring into it genes from another organism of the same, or a different, species. This process of recombining genes to make new combinations is known as **recombinant DNA technology**. The resulting organism is known as a 'genetically modified organism' (GMO).

12.2.1 Recombinant DNA technology

Certain disorders and diseases of organisms are the result of individuals being unable to produce important substances because they have defective portions of DNA (genes). Two such substances in humans are:

- **Insulin** – the protein hormone which stimulates the conversion of glucose to glycogen, and so helps to maintain a constant level of sugar in the blood, and without which the debilitating, and potentially lethal, disorder known as diabetes occurs.
- **Human factor VIII (anti-haemophiliac globulin – AHG)** – a protein essential to the clotting of blood, without which the potentially lethal condition known as haemophilia develops.

In both cases, it is possible to correct the deficiency using DNA technology. There are a number of stages to the process:

- **identification** of the required gene for the desired protein (section 12.2.2)
- **isolation** of this gene from the rest of the DNA (section 12.2.3)
- **multiplication** of the gene by the **polymerase chain reaction** (section 12.3.2)
- **insertion** of the gene into a **vector** in order to transfer it to a host cell (section 12.3.3)
- **introduction** of the gene into suitable host cells (section 12.3.4)
- **identification** of host cells which have successfully taken up the gene (section 12.4.2)
- **growth** of the population of host cells (section 12.5.2)
- **production** of the desired protein by the host cells (section 12.5.2)
- **separation** of the protein from the host cells (section 12.5.4)
- **purification** of the protein for clinical use (section 12.5.4)

12.2.2 Identification of the gene

Before a gene can be transplanted, it must be identified. Given that the required gene may consist of a sequence of a few hundred bases amongst the many million in human DNA, this is no small feat! Three methods are used:

- **Using the protein coded for by the gene**. If the amino acid sequence of the protein coded for by the required gene is known, it is possible to work out the DNA sequence which codes for these amino acids. A machine is then used to arrange nucleotides in the desired sequence, thus making artificial DNA.
- **Using reverse transcriptase**. To make a protein, a cell has first to produce **messenger RNA** from the relevant gene (section 3.5.1). A cell which readily produces the protein is selected: e.g., for insulin, islets of Langerhans from the pancreas are used; for thyroxine, the cells of the thyroid gland are appropriate. These cells will have large quantities of the relevant mRNA, which can be extracted. An enzyme called **reverse transcriptase** is then used to make DNA from RNA, known as **complementary DNA (cDNA)**. This single strand is

Table 12.2 *Advantages of using genetically engineered hormones rather than extracting them from organisms*

- Animals do not need to be slaughtered, e.g. calves do not need to be killed to extract insulin from their pancreases
- The hormones produced are more effective because they are exact copies of the human form rather than animal forms. There is therefore no immune response which might otherwise reject the hormone
- There is no risk of transferring infections such as hepatitis or HIV, as there is when hormones are extracted from human blood

made up of the nucleotides that are complementary to the mRNA. To make the other strand of DNA, the enzyme DNA polymerase is used. This double strand is the required gene.

- **Using DNA probes (gene probes)**. If the base sequence of part of the DNA of a gene is known, this portion can be attached to a radioactive or fluorescent marker. Under particular conditions, this marked section of DNA (the probe) will attach to its complementary sequence of bases in DNA that has been extracted from a relevant cell. The position of the probe is apparent from the position of the radioactivity or fluorescence – this must then be the position of the relevant gene.

12.2.3 Isolation of the gene

Having identified the required gene, the next task is to separate it from the remainder of the DNA. The first stage is to extract DNA from the cell as follows:

- Detergents and enzymes break down cell walls and membranes.
- Cell debris is removed by either filtering or centrifugation.
- The resulting mixture of DNA and proteins is treated with proteases, which break down the protein, leaving pure DNA.

The next stage is to cut up the DNA into small pieces. This is done using enzymes called **restriction endonucleases,** which cut DNA into fragments. They are produced naturally by cells, for protection against viruses. Viruses inject either DNA or RNA into cells, suppressing the host cell's DNA and replacing it by their own. Cells producing restriction endonucleases break down the viral DNA before it can affect the host. Many types exist, and each is named after the microorganism from which it is extracted, using roman numerals to identify different types from the same microorganism. For example, the *Hin*d III nuclease is the third nuclease from *Haemophilus influenzae*. Each type cuts a DNA double strand at a specific sequence of bases. Sometimes this cut occurs between two opposite base pairs. This leaves two straight edges known as **blunt ends**. For example, *Hpa* I nuclease cuts at the sequence GTTAAC (Fig 12.2). Other restriction endonucleases cut DNA in a staggered fashion. This leaves an uneven cut in which each strand of the DNA has exposed, unpaired bases. These are known as **sticky ends**. For example, the *Hind* III nuclease recognises a six-base pair AAGCTT, as shown in figure 12.2.

SUMMARY TEST 12.2

An organism made as the result of recombinant DNA technology is called a **(1)**. The first stage of the process is the identification of the required gene. One method of doing this is to make DNA from RNA using an enzyme called **(2)**. Another method involves a radioactive or fluorescent marker called **(3)**. Having identified the gene, the next stage is to **(4)** it, by breaking down the cell membrane, removing the cell debris by either **(5)** or **(6)**, and then treating the mixture with **(7)** to break down any protein. The DNA is then cut up using **(8)**, leaving fragments with two straight edges, called **(9)** ends, or ones with uneven edges, called **(10)** ends.

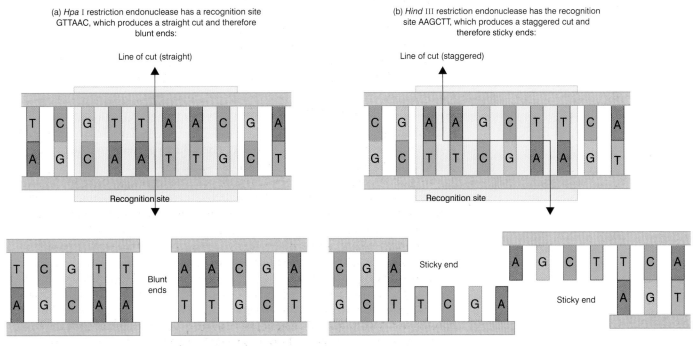

(a) *Hpa* I restriction endonuclease has a recognition site GTTAAC, which produces a straight cut and therefore blunt ends:

(b) *Hind* III restriction endonuclease has the recognition site AAGCTT, which produces a staggered cut and therefore sticky ends:

Fig 12.2 *Action of restriction endonucleases*

Transferring genes

AQA.A

AQA.A (Human)

AQA.B   OCR

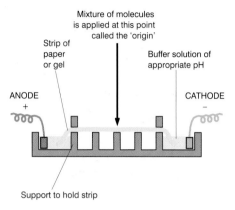

Fig 12.3 *Apparatus for carrying out electrophoresis*

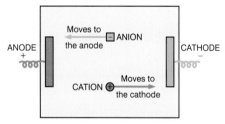

Fig 12.4 *Movement of ions during electrophoresis*

Once the gene for a specific protein such as insulin or human factor VIII has been identified and isolated, the next stage in the production of the protein is to transfer it to the host cell. This transfer is carried out in four stages:
- sorting the DNA fragments
- multiplication of the desired fragments of DNA
- insertion of the DNA into a vector
- introduction of the DNA into host cells.

12.3.1 Sorting the DNA fragments

Once the DNA is cut into lengths using restriction endonucleases, the pieces must be sorted, in preparation for analysis. This is achieved by **gel electrophoresis**. The DNA fragments are placed onto an agar gel and voltage is applied across it (Fig 12.3). The fragments are all naturally negatively charged, as a result of the negatively charged phosphate groups they contain. All fragments therefore move towards the anode (the positively charged electrode) (Fig 12.4). The larger the fragments, the more slowly they move, because of the resistance of the gel. In a given time, therefore, the smaller pieces move nearer to the anode than the larger ones. In this way, pieces of equal length group together in the same position and can be separated from the gel.

12.3.2 Multiplication of the gene – the polymerase chain reaction

Once the relevant fragments have been identified, they need to be multiplied. The **polymerase chain reaction (PCR)** is a method of copying fragments of DNA. The process is automated, making it both rapid and efficient, and requires:
- **the DNA fragment** to be copied
- **DNA polymerase** – an enzyme capable of joining together up to 70 000 nucleotides in a matter of minutes
- **primers** – short pieces of DNA which have a set of bases complementary to those at the end of the DNA fragment to be copied
- **thermocycler** – a computer-controlled machine which varies temperatures precisely over a period of time
- **nucleotides** – from bacteria found in hot springs, which are therefore tolerant to heat (thermostable) and do not denature during the high temperatures of the process.

The polymerase chain reaction is carried out in a series of stages:
- The DNA fragments, primers and DNA polymerase are placed in a vessel in the thermocycler at 73°C.
- The temperature is increased to 95°C, causing the two strands of the DNA fragments to separate.
- The mixture is cooled to 56°C, causing the primers to join (anneal) to their complementary bases at the end of the DNA fragment.
- The temperature is increased to 73°C again, which prompts the DNA polymerase to add complementary nucleotides along each of the separated DNA strands.
- Once the two DNA strands are completed, the process is repeated, to give four strands, and so on until millions of copies have been made.

12.3.3 Insertion of DNA into a vector

With the relevant DNA isolated and copied many times, the next task is to join it into a carrying unit, known as a **vector**. This vector is used to transport the DNA into the host cell. There are three types of vector – plasmids, phages and cosmids.

Phages are viruses that infect bacteria, while cosmids are hybrids of phages and plasmids. Most commonly used is the **plasmid**, which is a circular length of DNA found in bacteria. Plasmids almost always contain genes for antibiotic resistance, and it is at one of these antibiotic resistance sites that restriction endonucleases are used to break the plasmid loop. The same restriction endonuclease is used as the one that cut out our DNA fragment. This ensures that the 'sticky ends' of the opened-up plasmid are complementary to the 'sticky ends' of the DNA fragment. When the DNA fragments are mixed with the opened plasmids removed from bacterial cells, they may become incorporated into the plasmid. Where they are, the join is made permanent using the enzyme, **DNA ligase**. These events are summarised in figure 12.5.

12.3.4 Introduction of DNA into host cells

With the DNA incorporated into at least some of the plasmids, these must then be re-introduced into bacterial cells. This process is called **transformation**, and involves the plasmids and bacterial cells being mixed in a medium containing calcium chloride. The calcium chloride makes the bacterial cell wall permeable, allowing the plasmids to pass through into the cytoplasm. However, not all the bacterial cells will possess the DNA fragments, and the task now is to identify which bacterial cells do contain the gene (DNA fragment). This identification can be carried out using
• DNA probes (section 12.2.2)
• **Gene markers** (unit 12.4).

SUMMARY TEST 12.3

The transfer of a gene from a cell manufacturing it into a host cell is an essential aspect of gene technology. It begins with the cutting of the desired DNA into lengths using enzymes called **(1)**, and then sorting them by **(2)** using **(3)**. The fragments produced are then copied by a method known as the **(4)**. This method uses a computer-controlled machine called a **(5)** that raises the temperature from an initial **(6)** to 95 °C, causing the two strands of the DNA fragment to **(7)**. A drop in temperature to 56 °C causes short pieces of DNA called **(8)** to join to the original DNA fragments.

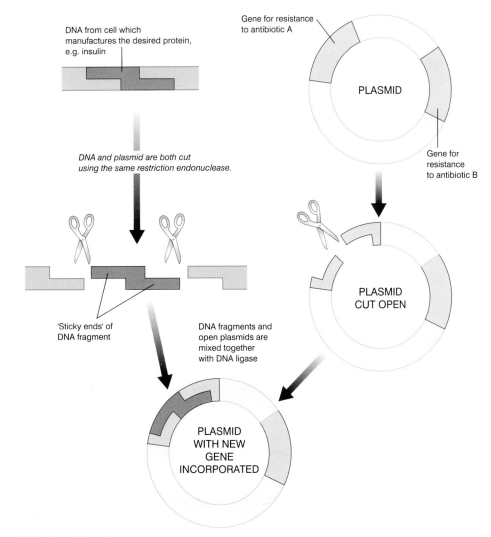

Fig 12.5 *Inserting a gene into a plasmid vector*

Gene markers and replica plating

Gene markers are used to identify which bacterial cells have taken up a desired gene (unit 12.3). One example is the use of **plasmids** that possess genes giving antibiotic resistance.

12.4.1 Antibiotic resistance

The antibiotic penicillin was first discovered by Alexander Fleming in 1929; antibiotics have been commercially produced on a large scale ever since. They destroy bacterial cells and so have been highly effective in controlling diseases caused by bacteria. Bacteria have, over the years, evolved mechanisms for resisting the effects of antibiotics, typically by producing an enzyme that breaks down the antibiotic before it can destroy the bacterium. The genes for the production of these enzymes are found in plasmids – small circular pieces of DNA within bacterial cells, which are separate from the main bacterial DNA. Some plasmids carry genes for resistance to more than one antibiotic. An example is the R-plasmid, which is a vector (section 12.3.3) for genes for ampicillin resistance and for tetracycline resistance. This makes it especially valuable in **gene technology**.

12.4.2 Detection of genetically modified bacteria

There are two reasons why a gene may not always be transferred into a host cell in the process described in unit 12.3:
- some plasmids will have simply annealed (closed up) without incorporating the DNA fragment
- only a few bacterial cells (as few as 1%) take up the plasmids when the two are mixed together.

One way of detecting which cells have incorporated the new gene operates as follows:
- Plasmid vectors normally carry a gene for antibiotic resistance.
- The antibiotic that the plasmid in question gives resistance to is identified.
- The bacterial cells are treated with the same antibiotic.
- The cells that have taken up the plasmid vector are resistant to the antibiotic, and therefore survive.
- The cells that have not taken up the plasmid are not resistant, and die.

This method is effective in showing which bacterial cells have taken up the plasmid and which have not. However, some cells will have taken up plasmids that have closed up without incorporating the new gene. These now need to be eliminated, by a process called **replica plating**.

12.4.3 Replica plating

Some plasmid vectors such as the R-plasmid have two genes, each giving resistance to a different antibiotic. When used in genetic modification, the plasmid is opened by a **restriction endonuclease** at the site of one of these genes (Fig 12.5, section 12.3.3). If the plasmid closes up without taking up the new gene, this antibiotic resistance gene will work normally, and the bacterium will be antibiotic resistant. If, however, the new gene has been incorporated into the middle of the antibiotic-resistant gene, it will be made ineffective, and the bacterium will be killed by the appropriate antibiotic. The problem is that treatment with the antibiotic will destroy the very cells that are required. Replica plating overcomes this problem, as follows:

- Those bacterial cells which survived treatment with the first antibiotic are known to have taken up the plasmid.
- These cells are cultured by spreading them very thinly on nutrient agar plates.
- Each separate cell on the plate will grow into a genetically identical colony.
- A tiny sample of each colony is transferred onto a second (replica) plate in **exactly the same position** as the original colonies.
- This replica plate contains the second antibiotic – the one for which the resistance gene will have been made useless if the new gene has been taken up.
- The colonies killed by the antibiotic must be the ones which have taken up the new gene.
- The colonies in exactly the same position on the original plate are the relevant ones, and are therefore carried forward to the next stage – large-scale culturing.

The process of detecting genetically modified bacteria is summarised in figure 12.6.

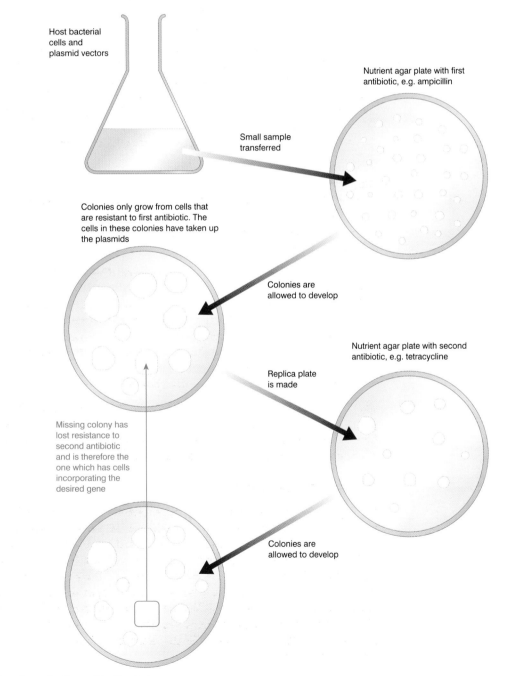

Fig 12.6 *Detection of genetically modified bacteria*

The processes described in units 12.2, 12.3 and 12.4 result in the isolation of a tiny quantity of bacterial cells which contain a gene, often human, that is able to produce a desired protein. The final task is the large-scale production and purification of this protein.

12.5.1 Gene cloning

The first stage in the manufacture of a protein is the production of many identical copies of the required gene **(gene cloning)**. This occurs in two ways:
* Multiplication of the plasmid within the bacterial cell. In optimum conditions, up to 200 copies of the plasmid can be made within a single bacterial cell.
* Multiplication of the bacterial cells by growing them in an industrial fermenter.

12.5.2 Industrial fermenters

A typical fermenter is a large stainless steel vessel, with a capacity of around 500 000 dm³ and around which is a jacket of circulating water to control its temperature. A series of flat blades is rotated, to ensure that the contents of the fermenter are constantly mixed, in order to:
* keep the bacterial cells from settling out at the bottom
* ensure that nutrients are kept in contact with the bacterial cells.

If oxygen is needed, it can be forced in through tiny holes in a plate at the bottom of the tank. A series of ports allows materials to be added or removed at suitable stages of the process. Probes are used to monitor constantly the conditions such as temperature and pH, which can then be adjusted to bring them back to the optimum if necessary. The construction of a stirred-tank fermenter is shown in figure 12.7. There are two main methods of cultivation using a fermenter:
* **Batch cultivation** involves loading the fermenter with all the necessary materials, allowing the bacteria to grow and produce their products. The process is then stopped at a specific stage, and the product removed, before the fermenter is cleaned and sterilised in readiness for the next batch.
* **Continuous cultivation** involves a constant stream of nutrients being added and the products being continuously removed, over a period of many weeks. Although the process is quicker than batch cultivation, it is not suitable for certain products, e.g. antibiotics.

12.5.3 Problems with large-scale culturing

There are a number of difficulties associated with large-scale production of materials by genetically modified bacteria in fermenters:
* The genetically modified bacteria are often weakened mutant strains, with rather slow growth. This can be overcome by incorporating in the bacteria a gene that switches on the gene for the product, only in the presence of a specific chemical. The bacteria can then be grown until their numbers build up, before the addition of the same chemical. Production therefore takes place only when numbers are large.
* The protein being produced often inhibits growth of the bacteria. The solution is as above, keeping the gene for the protein inactive until bacterial numbers have built up.
* Growth of the bacteria generates heat, which can kill them. Although use of cooling water in the jacket around the fermenter can help, the use of genetically engineered bacteria which are tolerant of high temperatures is a cheaper, more effective, solution.

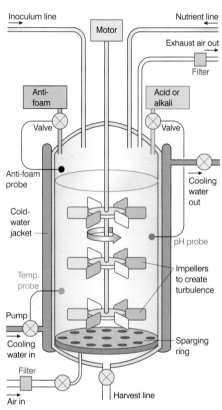

Fig 12.7 A stirred-tank fermenter

Fermenter used for protein production by cloned genetically modified bacteria

12.5.4 Downstream processing

Downstream processing involves the separation of the product, and then its purification. If the desired product is contained within the bacterial cells, they must first be disrupted to release their contents. Separation of a protein such as insulin from the genetically modified bacteria that produced it can be achieved by:

- **settlement** – the bacteria settle out by gravity – assisted by chemicals known as flocculating agents, if necessary
- **centrifugation** – bacteria are removed by spinning the fermenter contents at high speed, causing the bacteria to separate out
- **ultrafiltration** – the contents are forced through filters with a pore size of less than 0.5µm, trapping the bacteria, but allowing the liquid with the protein to pass through.

Purification involves separating the protein from other chemicals, such as metabolites and enzymes. This is achieved using precipitation, **chromatography** or solvent extraction techniques.

12.5.5 Useful substances produced by genetic engineering

The three main substances produced using genetically modified bacteria grown in fermenters are antibiotics, hormones and enzymes.

- **Antibiotics** are produced by bacteria as **secondary** metabolites, i.e. they are made when growth is slowing. This makes batch cultivation more appropriate. A single batch of penicillin production will take around 6 days. Other antibiotics produced in this way are listed in table 12.3.
- **Hormones** – insulin is needed daily by more than 2 million diabetics, in order for them to lead normal lives. Previously, insulin extracted from cows or pigs was used, but this could produce side effects such as immune responses to it. With genetic engineering, bacterial cells have the human insulin gene incorporated into them and so the insulin produced has no adverse effects on the patient; this method also avoids killing animals. Insulin is produced in a modified form, which needs to be changed before it can be injected. Other hormones produced in this way include human growth hormone, cortisone and the sex hormones – testosterone and oestradiol.
- **Enzymes** – many enzymes used in the food industry are manufactured by genetically modified bacteria. Some are listed in table 12.4.

SUMMARY TEST 12.5

Industrial fermenters are usually made of (**1**) and have a capacity of (**2**) dm³. Cultivation is carried out by two methods, (**3**) and (**4**). Fermentation can be used to make antibiotics such as penicillin which is made by the fungus (**5**), and enzymes such (**6**) made by the *Candida spp*. The separation and purification of the products made by fermentation is called (**7**) and is achieved by forcing the fermenter contents through filters, a process called (**8**), or by settling out the bacteria by gravity, called (**9**) or by centrifugation.

Table 12.3 Antibiotics and the organisms that produce them

Antibiotic	Producer organism	Type of organism
Penicillin	*Penicillium notatum*	Fungus
Griseofulvin	*Penicillium griseofulvum*	Fungus
Streptomycin	*Streptomyces griseus*	Fungus
Chloramphenicol	*Streptomyces venezuelae*	Fungus
Tetracycline	*Streptomyces aureofaciens*	Fungus
Colistin	*Bacillus colistinus*	Bacterium
Polymyxin B	*Bacillus polymyxa*	Bacterium

Table 12.4 Enzymes and the organisms that produce them

Enzyme	Examples of micro-organisms involved	Application
α-Amylases	*Aspergillus oryzae*	Breakdown of starch in beer production. Preparation of glucose syrup. Thickening of canned sauces
Lipase	*Candida* spp.	Flavour development in cheese
Pectinase	*Aspergillus* spp.	Clearing of cider, wines and fruit juices
Protease	*Bacillus subtilis*	Meat tenderisers
Pullulanase	*Klebsiella aerogenes*	Soft ice cream manufacture
Sucrase	*Saccharomyces* spp.	Confectionery production

Gene therapy

Many human diseases such as haemophilia, sickle-cell anaemia and thalassaemia are the result of a gene not expressing itself properly, or being missing altogether. There are around 4000 such diseases, affecting 1–2% of the population. The ability to transfer genes (unit 12.3) opens the way for many of these diseases to be cured, by replacing defective genes with ones which have been cloned from healthy genes. One such disease that can be treated in this way is **cystic fibrosis**.

12.6.1 Cystic fibrosis

With one person in 200 in Britain suffering from cystic fibrosis and one in 22 people being carriers, it is the most common genetic disease amongst the white European and North American populations. It is caused by a mutant recessive gene on chromosome 7, in which three DNA bases, adenine-adenine-adenine are missing. It is therefore an example of a deletion mutation (section 12.1.3). The normal gene, called the **cystic fibrosis transmembrane-conductance regulator (CFTR) gene**, normally produces a protein of some 1480 amino acids. The deletion results in a single amino acid, number 508 which is phenylalanine, being left out of the protein. This, however, is enough to make the protein unable to perform its role of transporting chloride ions across epithelial membranes. The protein transports chloride ions out of epithelial cells, and water naturally follows, by the process of **osmosis**. In this way, epithelial membranes are kept moist. In a patient with cystic fibrosis, the defective gene means that the protein is either not made or doesn't function normally. The epithelial membranes are therefore dry, and the mucus they produce remains viscous and sticky. The symptoms this causes include:

- **mucus congestion in the lungs**, leading to a much higher risk of infection as the mucus, which traps disease-causing organisms, cannot be removed
- **breathing difficulties** and less efficient gaseous exchange
- **blocked pancreatic ducts**, which prevents pancreatic enzymes reaching the duodenum, and leads to the formation of fibrous cysts (hence the name of the disease)
- **blocked sperm ducts in males**, leading to possible infertility.

12.6.2 Testing for cystic fibrosis

It is possible to discover whether an individual is a carrier for cystic fibrosis in the following way:

- using a blood sample, or cells from a mouthwash, a DNA sample is collected
- DNA polymerase is used to make copies of a DNA fragment of the relevant part of the two CFTR genes
- these copies are run on an **electrophoresis** gel
- if either of the CFTR genes has the three-base deletion, then some of the fragments will be smaller, and these will move more quickly across the gel (section 12.3.1).

12.6.3 Treatment of cystic fibrosis using gene therapy

Gene therapy may be used to treat cystic fibrosis in two ways:

- In **gene replacement**, the defective **allele** is replaced with a healthy one.
- In **gene supplementation**, one or more copies of the healthy allele are added alongside the existing ones. As the added alleles are dominant, the effects of the defective recessive alleles are over-ridden.

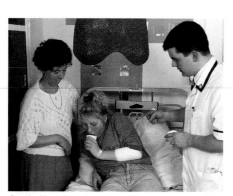

Cystic fibrosis patient coughing up mucus, after treatment to loosen it

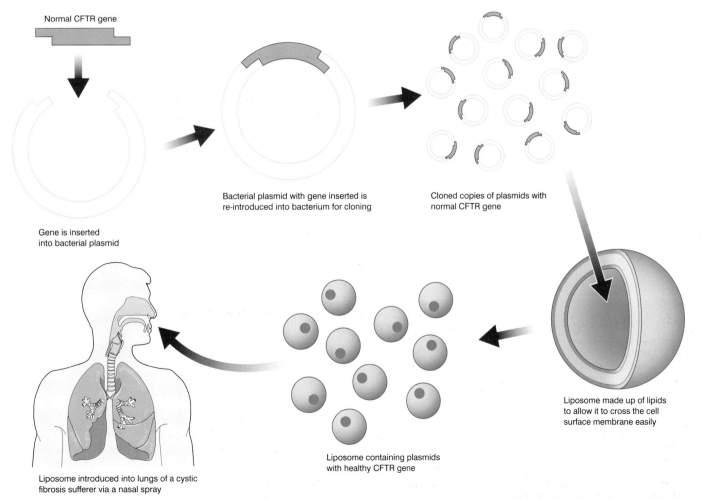

There are two approaches to either type of gene therapy:

- **Germ-line gene therapy**, in which the defective allele is replaced or supplemented in the fertilised egg. This ensures that all cells of the organism will develop normally, as will all those of any offspring they bear. This is therefore a much more permanent solution, affecting future generations. However, the moral and ethical issues of manipulating such a long-term genetic change mean that the process is currently prohibited.
- **Somatic-cell gene therapy** targets just the affected tissues, e.g. lungs, and its effects are therefore not evident in sperm or eggs, and so not passed on to future generations. As the cells of these tissues are continually dying and being replaced, the treatment needs to be repeated periodically. The long-term aim is therefore to target undifferentiated **stem cells**, which give rise to mature tissues. The treatment would then be effective for the life-span of the individual.

There are two main methods for introducing healthy CFTR genes into lung epithelial cells:

- **Using a harmless virus (called an adenovirus) as a vector**, to transfer the healthy gene into the host cells (sections 12.3.3 and 12.3.4). The viruses are treated to make them harmless, the gene is inserted, and then the viruses are introduced into the nostrils of the patients.
- **Wrapping the gene in lipid molecules**, to enable it to pass through the membranes of lung epithelial cells. Again, these packages, called **liposomes**, are sprayed into the nostrils of the patient, and are drawn down into the lungs during breathing.

These events are summarised in figure 12.8.

Normal CFTR gene

Gene is inserted into bacterial plasmid

Bacterial plasmid with gene inserted is re-introduced into bacterium for cloning

Cloned copies of plasmids with normal CFTR gene

Liposome made up of lipids to allow it to cross the cell surface membrane easily

Liposome containing plasmids with healthy CFTR gene

Liposome introduced into lungs of a cystic fibrosis sufferer via a nasal spray

Fig 12.8 *Summary of treatment of cystic fibrosis by gene therapy*

12.7 Genetically modified organisms

AQA.B

Plants and animals have been used for many products since the earliest time of civilisation. Crop plants and domestic animals have been genetically manipulated, by selective breeding, for thousands of years. This has produced crops with higher yields, meat and milk with a lower fat content, and more docile animals. It has only been with the re-discovery of Mendel's genetic work in 1900, and with modern gene technology, that humans can now achieve, in weeks, genetic changes which once took hundreds of years.

12.7.1 Genetic modification

The genetic make-up of organisms can now be altered in several ways:
- by increasing the activity of genes
- by decreasing the activity of genes
- by transferring genes in various ways, e.g.
  - between individuals of the same species
  - between organisms of different species.

These modifications can be made for a variety of purposes, including:
- increasing the yield of plant or animal crops
- improving the nutrient content of foods
- introducing resistance to disease and pests
- making crop plants tolerant to herbicides
- developing tolerance to environmental conditions, e.g. extreme temperatures, drought
- making vaccines
- producing medicines for treating disease (table 12.4).

12.7.2 Examples of genetically modified plants

- **Genetically modified tomatoes** have been developed using anti-sense technology: e.g., a gene is inserted which has a sequence the reverse of that of the gene causing softening. The mRNA of this anti-sense gene matches the mRNA of the softening gene. The two therefore combine, making the mRNA of the softening gene useless. This allows the tomatoes to develop flavour without the problems associated with harvesting, transporting and storing soft fruit.
- **Herbicide-resistant crops** have a gene introduced that makes them resistant to a specific herbicide. The herbicide can then be sprayed on the crops and, while competing weeds die, the crop is unaffected.
- **Disease-resistant crops** have genes introduced that give resistance to specific disease. Genetically modified rice, for example, is unaffected when infected by the (usually highly infectious) rice stripe virus.
- **Pest-resistant crops** – maize can have a gene added which allows the plant to make a toxin such as the natural toxin pyrethrum. This kills insects that try to eat the maize, but is harmless to other animals.
- **Plants that produce plastics** – a possibility currently being explored. It is hoped to genetically engineer plants that have the metabolic pathways to make the raw material for plastic production.

12.7.3 Transgenic animals

A transgenic animal is one in which genetic material from another organism has been artificially introduced. For example, genes from an organism which has natural resistance to a disease may be introduced into a totally different organism.

Table 12.6 *Medicinal proteins made using genetically modified organisms*

Medicinal protein	Genetically modified organism	Medical function
Insulin	Bacteria	Treating diabetes
Growth hormone	Bacteria	Treating pituitary dwarfism
Hepatitis B vaccine	Yeast	Immunisation against hepatitis B
Plasminogen	Mammalian cells grown in culture	Dissolving thromboses (blood clots)
Erythropoietin	Mammalian cells grown in culture	Stimulating red blood cell production to treat anaemia
Alpha-1-antitrypsin (AAT)	Transgenic sheep	Treating emphysema and cystic fibrosis

This second organism is then made resistant to that disease. Another example is in the production of rare and expensive proteins for use in human medicine. In this second case, domesticated milk-producing animals such as sheep are used, and the gene for the protein is inserted alongside the regulator of the lactoglobulin gene, which encodes proteins in sheep's milk. In this way, the desired protein is produced in the milk of the sheep. The gene can be inserted into the fertilised egg of a sheep, so that all the female offspring of that individual will be capable of producing the protein in their milk. One example of a protein made in this way is **alpha-l-antitrypsin (AAT)**. This protein is essential in making the lungs elastic, so that they can stretch during inhalation. Individuals unable to produce this gene develop emphysema. AAT cannot be produced commercially, so the gene has been introduced into sheep. The process is as follows:

- Ewes are given fertility drugs to produce additional eggs.
- The eggs are removed, and fertilised by sperm.
- The healthy gene for AAT from a human is added alongside the milk regulator gene.
- These genetically engineered eggs are implanted in receptive ewes.
- The resulting lambs are screened for the presence of the AAT gene.
- Those lambs with the AAT gene are cross bred, to give a flock in which the ewes produce milk rich in AAT.
- The AAT is extracted from the milk and purified.
- AAT is administered to humans unable to manufacture their own and to sufferers of emphysema and cystic fibrosis (section 12.6.1).

The same technique is used to produce the human blood clotting factor, **factor IX**, which is used to treat a form of haemophilia.

Transgenic ram with the human gene for AAT production. The gene is inherited by its offspring and AAT is secreted in the milk of the female offspring

✗ 12.7.4 Cloning

A **clone** is a group of genetically identical offspring produced by asexual reproduction. It is possible to produce a whole organism from a single cell and therefore if a single cell is first allowed to divide mitotically, many identical individuals can be produced. This is known as **cloning**. An example of cloning in animals was the production of Dolly the sheep in 1997, the principles of which are illustrated in figure 12.9.

A form of cloning in plants is **vegetative propagation**. Vegetative propagation involves the separation of part of a parent plant that then develops into a new individual. Any part of the plant may serve this purpose. Examples of crops produced in this way are given in table 12.7. No plant is likely to survive for long in a changing environment if it relies solely on vegetative propagation and therefore plants exhibiting it continue to produce flowers, to provide genetic variety through the sexual process.

Table 12.7 Crop plants produced by vegetative propagation

Crop plant	Organ of vegetative propagation	Description of organ
Onion	Bulb	Swollen, fleshy leaf bases with buds that give rise to new plants
Potato	Stem tuber	Swollen tips of an underground stem that give rise to new plants
Strawberry	Runner	Thin lateral stems on the surface of the soil that produce new plants where they touch the ground
Blackberry	Stolon	Long vertical stem that bends over as it grows and develops into a new plant where it reaches the soil

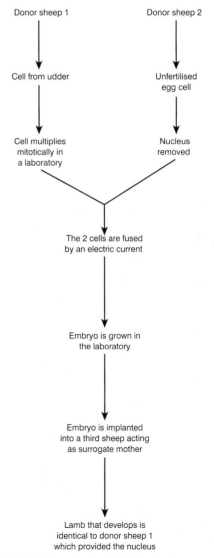

Fig 12.9 Example of cloning in animals

Genetic fingerprinting

Genetic fingerprinting, or **DNA profiling** as it is often known, was developed by Professor Alec Jeffreys of Leicester University in 1984. The technique is based on the fact that 95% of human DNA does not code for any characteristic. These non-coding DNA bases are known as **introns** (section 3.5.2); they contain repeating sequences of DNA called **core sequences**. Every individual has a unique pattern of number and length of core sequences. Although different in all individuals except identical twins, these core sequences are more similar the more closely related are any two individuals. The making of a genetic fingerprint has five main stages – **extraction**, **digestion**, **separation**, **hybridisation** – and, finally, **development**.

12.8.1 Extraction

Even the most tiny sample of animal tissue, e.g. a drop of blood or a hair root, is enough to give a genetic fingerprint. Whatever the sample, the first stage is to extract the DNA it contains by soaking it in a mixture of chloroform and water-saturated phenol. This precipitates out the proteins, leaving the DNA in the layer of water. If the amount of DNA is small, it can be increased using the **polymerase chain reaction** (section 12.3.2).

12.8.2 Digestion

The DNA is now cut into fragments, using restriction endonucleases (section 12.2.3). The endonucleases are chosen for their ability to cut close to, but not within, groups of core sequences known as mini-satellites.

12.8.3 Separation

The different-sized fragments now have to be separated, by gel electrophoresis (section 12.3.1). The fragments are separated by size under the influence of an electrical voltage. The gel is then immersed in alkali, to separate the double strands into single ones. The single strands are then transferred onto a nylon membrane by a technique called **Southern blotting**, which involves:
- laying a thin sheet of nylon over the gel
- covering it with several sheets of absorbent paper
- drawing up the liquid containing the DNA by capillarity
- transferring the DNA fragments to the nylon membrane in precisely the same relative positions they had on the gel
- fixing the fragments to the membrane, using ultra-violet light.

12.8.4 Hybridisation

Radioactive **DNA probes** are now used to bind onto the core sequences. The probes have base sequences which are complementary to the core sequences, and bind to them under specific conditions such as temperature and pH. The process is carried out with different probes, each of which binds with a different core sequence.

12.8.5 Development

Finally, an X-ray film is put over the nylon membrane. The film is exposed (fogged) by radiation from the radioactive probes. As these points correspond to the position of the DNA fragments as separated during electrophoresis, a series of bars is revealed. The pattern of the bands (Fig 12.10) is unique for every individual except identical twins.

Fig 12.10 DNA fingerprint of a child and each parent. Note that each band on the child's fingerprint corresponds to a band on one or other parent's fingerprint

The complete process of genetic fingerprinting is summarised in figure 12.11.

12.8.6 Interpreting the results

DNA fingerprints from two samples, e.g. from a suspect and from blood found at the scene of a crime, are visually checked. If there appears to be a match, the pattern of bars of each fingerprint is passed through an automated scanning machine, which calculates the length of the DNA fragments from the bands. It achieves this using data obtained by measuring the distance travelled during electrophoresis by known lengths of DNA. Finally, the odds are calculated of someone else having an identical fingerprint. The more the two patterns match, the greater the probability that the two sets of DNA have come from the same person.

12.8.7 Uses of DNA fingerprinting

DNA fingerprinting is used extensively in forensic science, to indicate whether or not an individual is connected with a crime, e.g. from blood or semen samples found at the scene. It can also be used to help resolve questions of paternity. Every individual inherits half their genetic material from their mother and half from their father. Each band on a DNA fingerprint of any individual should, therefore, have a corresponding band in the parents' DNA fingerprint (Fig 12.10). This can be used to establish whether someone is the genetic father of a child. Other uses of genetic fingerprinting include:

- screening for genetic disorders (section 14.1.1)
- tracing animal pedigrees
- establishing the identity of corpses, even those that have been buried for a long period
- checking the identity of an applicant, e.g. for immigration
- identification of family relationships.

SUMMARY TEST 12.8

Genetic fingerprinting, also known as **(1)**, is a successful forensic technique based upon the fact that most human DNA is of a non-coding variety called **(2)** which contains repeating patterns called **(3)** sequences. To make a genetic fingerprint it is first necessary to isolate the DNA from the sample by precipitating out the **(4)** and then increasing the amount of DNA using the **(5)**. The DNA is then cut into fragments using **(6)** enzymes. The fragments are then separated using the process of **(7)** and each is immersed in **(8)** to separate the double strands into single ones. These strands are transferred to a nylon membrane by the technique of **(9)**. Parts of the strands are then combined with radioactive **(10)** which cause exposure of **(11)** to reveal a pattern of bars that is unique to all individuals except **(12)**. DNA fingerprinting can be used in forensic science, to determine the genetic father of a child and in **(13)** for genetic disorders.

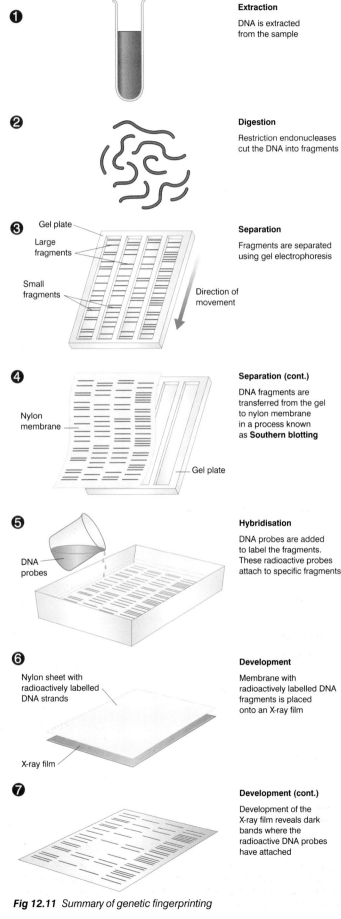

❶ Extraction
DNA is extracted from the sample

❷ Digestion
Restriction endonucleases cut the DNA into fragments

❸ Separation
Gel plate
Large fragments
Small fragments
Direction of movement
Fragments are separated using gel electrophoresis

❹ Separation (cont.)
Nylon membrane
Gel plate
DNA fragments are transferred from the gel to nylon membrane in a process known as **Southern blotting**

❺ Hybridisation
DNA probes
DNA probes are added to label the fragments. These radioactive probes attach to specific fragments

❻ Development
Nylon sheet with radioactively labelled DNA strands
X-ray film
Membrane with radioactively labelled DNA fragments is placed onto an X-ray film

❼ Development (cont.)
Development of the X-ray film reveals dark bands where the radioactive DNA probes have attached

Fig 12.11 *Summary of genetic fingerprinting*

Human genome project

The human genome project is arguably the most exciting piece of scientific research in recent times. The information it provides will help mankind to tackle problems of pollution, disease and ageing, and provide immense benefit to health and the environment. It also presents mankind with social, economic and ethical problems as to how the information revealed might be used.

12.9.1 What is the human genome project?

The human genome project is the largest piece of international collaboration in biology that has ever taken place. The aim is to sequence accurately every one of the 3 billion bases which make up the DNA of a single human being (unit 3.2). The project, which began in 1990, is so vast that it was agreed that specific laboratories in more than 50 countries would each take a different portion of a human chromosome and simultaneously sequence its DNA. Even then, it was expected to take 15 years at least, but, in the event, the first draft of the human DNA base sequence was published on 26th June 2000. Certain details still need to be finalised, but the complete picture should be available in 2003. So vast is the amount of information, that it is never likely to be published on paper. To do so would take nearly 500 000 pages – about 2000 books the size of this one. Instead, the information will be stored electronically and made available over the Internet.

THE HISTORY OF THE HUMAN GENOME PROJECT

1977 - Fred Sanger develops a method for sequencing DNA

1986 - Early discussions on the human genome project take place in the USA

1988 - James Watson (who with Francis Crick discovered the structure of DNA) is appointed head of the human genome project

1990 - The project gets under way

1996 - The first gene map is produced

2000 - The draft of the human genome is published

2003? - The final, highly accurate, version of the human genome is expected

12.9.2 Why sequence the human DNA bases?

As with many aspects of research, one reason for sequencing human DNA is curiosity. However, given the immense costs – between 2 and 4 billion pounds – there clearly has to be more to it than mere curiosity. Some likely benefits include:
- prevention and treatment of human genetic diseases, e.g. cystic fibrosis, muscular dystrophy
- treatment of diseases which are influenced by heredity, e.g. heart disease, cancer, Alzheimer's disease
- development of reliable genetic tests to identify whether an individual is a carrier of a genetic disorder, e.g. thalassaemia (a blood disease), so that individuals can make a decision, based on the risk of their offspring being affected, whether to have children or not
- providing information on human evolution, through the study of the genomes of different racial and ethnic groups.

12.9.3 How are human DNA bases sequenced?

There are about 100 000 genes on human chromosomes. Of the 3 billion bases in human DNA, only around 5% are 'expressed', i.e. have some function. These are called **exons**. The remaining 95% of the DNA bases are not expressed, and are called **introns** (section 3.5.2). These are sometimes referred to as **junk DNA**. The first maps of a chromosome were made in the late 1960s, by use of a stain that attached to the guanine base in DNA. This acted as a marker to identify guanine. With the development of more markers, it is possible to produce a **physical map** of a chromosome. The sequencing of DNA bases was developed by Fred Sanger in 1977 and entails breaking DNA molecules into fragments. The base pair sequences of these fragments are then worked out, using a **DNA sequencing machine**, which can read around 500 bases each time. Computers are used to carry out the mammoth task of fitting together the jigsaw of each of the 500 base pair pieces which make up the 3 billion bases in the complete genome.

12.9.4 The implications of the human genome project

The knowledge gained from the human genome project has moral, ethical, social and cultural implications, and produces a series of dilemmas. Some of these dilemmas are listed below:

- Should insurance companies be allowed to see the results of genetic tests carried out on individuals, or even insist that such tests are undertaken, before agreeing to provide life insurance?
- Should companies who have financed research into a specific gene be allowed to patent that gene and so keep the information to themselves, or are our genes the property of us all?
- Would the money being spent be better used in providing better hospitals or basic necessities, such as clean water, for people in poorer countries who are unlikely to benefit from the advantages of sequencing our DNA?
- Should the information on human DNA be used to allow us to have 'designer babies' by the selection of beneficial genes and elimination of detrimental ones?
- Could the information on an individual's DNA be used to discriminate against them, e.g. by employers, who may reject applicants with known genetic diseases? Should employers be able to require applicants to take a genetic test?
- How can we ensure that DNA information is not used maliciously by unscrupulous individuals for profit or political power?
- How can we ensure that information on racial or ethnic backgrounds revealed by DNA sequencing will not lead to discrimination and prejudice against particular groups?

Research being carried out on the human genome project

The central issues are therefore:
- who should have access to genetic information
- what rules should be observed in using that information.

Knowledge of an individual's genetic make-up is perhaps the ultimate invasion of privacy. Keeping records of individual genotypes creates major problems of confidentiality, which will need to be confronted. The information that the human genome project will provide will add considerably to our knowledge of life. Used in a proper way, this knowledge can lead to major improvements in food production, health and our environment. In so doing, it can relieve suffering. Used inappropriately, it may have the opposite effect. The challenge is to ensure that the information adds to, rather than detracts from, making the world a better place.

SUMMARY TEST 12.9

The Human Genome Project is a massive international effort to sequence all the genes, which number around **(1)**, on the 23 pairs of chromosomes in each human cell. These genes are made up of a total of **(2)** DNA bases of which **(3)** % are not 'expressed' and are known as **(4)**. Those that are 'expressed' are called **(5)**. Determining the sequence of DNA bases entails using markers that attach themselves to each of the organic bases in DNA to produce a **(6)** of the chromosome. The DNA is then broken up into fragments determined using a **(7)**. Computers are then used to piece together these fragment sequences into the completed genome. The knowledge gained can be used to treat genetic diseases such as **(8)**, or to test whether individuals are carriers for other genetic disorders like **(9)**. It can aid the treatment of diseases such as **(10)** in which hereditary factors influence the occurrence. There are however moral, ethical, social and cultural implications that come with this new and immense body of knowledge.

12.10 Evaluation of genetic engineering

AQA.A

AQA.A (Human)

AQA.B

MORALS

are individual or group views about what is right or wrong. Such views refer to almost any subject, such as it is wrong to hunt foxes, work on a Sunday, to swear, to tell lies. Morals vary from country to country and individual to individual, and change over time. Some of the accepted moral values of a hundred years ago in Britain, most people would now disagree with

ETHICS

is a narrower concept than morals. Ethics are a set of standards which are followed by a particular group of individuals and are designed to regulate their behaviour. They determine what is acceptable and legitimate in pursuing the aims of the group

SOCIAL ISSUES

relate to human society and its organisation. They concern the mutual relationships of human beings, their interdependence and their cooperation for the benefit of all

ECONOMIC ISSUES

relate to the production and distribution of wealth and the application of these to a particular group or subject, e.g. a community or a business

Throughout unit 12, we have considered many aspects of genetic engineering, in particular the many medical and economic advantages that manipulating DNA can bring. The technology is not without its problems, however, and its application raises many moral, ethical, social and economic issues. These terms are defined in the adjacent boxes.

12.10.1 Applications of genetic engineering

Genetic engineering undoubtedly brings many benefits to mankind:
- Microorganisms can be modified to produce a range of substances that can be used to treat diseases and disorders: e.g. antibiotics, hormones and enzymes (section 12.5.5).
- Microorganisms can be used to control pollution, e.g. to break up and digest oil slicks or destroy harmful gases released from factories. Care would need to be taken to ensure that such bacteria did not destroy oil in places where it is required, e.g. car engines. To do this, a 'suicide gene' could be incorporated that would cause the bacteria to destroy themselves once all the oil slick had been digested.
- Genetically modified plants (section 12.7.2) can be made to produce a specific substance in a particular organ of the plant. These organs can then be harvested and the desired substance extracted. If a drug is involved, the process is called **plant pharming**. One promising application of this technique is in the production of transgenic plants that manufacture antibodies to a particular disease, or manufacture antigens which, when injected into humans, induce natural antibody production.
- Genetically modified crops can be engineered to have economic and environmental advantages. These include:
  - making plants more tolerant to environmental extremes, e.g. able to survive drought, cold, heat, salt or polluted soils, etc. This permits crops to be grown commercially in places where they are not at present. Each year, an area of land equal to half the United Kingdom becomes unfit for normal crops, because of increases in soil salt concentrations. Growing of genetically modified plants, such as salt-tolerant tomatoes, could bring this land back into productivity
  - providing resistance to herbicides, so that when a crop is sprayed with a particular herbicide it is unaffected, while competing weeds are destroyed
  - providing resistance to pests such as viruses and insects, so that crop yields are improved because the plants are healthy
  - improving storage of plants, such as tomatoes, so they are less likely to rot before they reach the consumer
  - engineering plants to produce chemicals from which plastics can be manufactured, reducing use of our limited oil reserves.
- **Transgenic animals** are able to produce expensive drugs such as **AAT** relatively cheaply (section 12.7.3).
- **Gene therapy** can be used to cure certain genetic disorders, such as cystic fibrosis (unit 12.6).
- **Genetic fingerprinting** can be used in forensic science (unit 12.8).

12.10.2 Implications of genetic engineering

Against the benefits of genetic engineering, must be weighed the disadvantages – both real and potential. Some specific issues are listed below:

- It is impossible to predict with complete accuracy what the ecological consequences will be of releasing genetically engineered organisms into the environment. The delicate balance that exists in any **habitat** may be irretrievably damaged by the introduction of organisms with engineered genes. There is no going back once an organism is released.

- An engineered gene may pass from the organism it was introduced into, to a completely different one. We know, for example, that viruses can transfer genes from one organism to another. What if a virus were to transfer genes for herbicide resistance and vigorous growth from a crop plant to a weed which competed with the crop plant? How would we then be able to control this weed?

- Any manipulation of the DNA of a cell will have consequences for the metabolic pathways within that cell. We cannot be sure until after the event what unforeseen by-products of the change might be produced. Could these lead to metabolic malfunctions, cause cancer, or create a new form of disease?

- Genetically modified bacteria often possess **antibiotic resistance** as marker genes. These bacteria can spread antibiotic resistance to harmful bacteria.

- All genes mutate. What, then, might be the consequences of our engineered gene mutating? Could it turn the organism into a pathogen which we have no means of controlling?

- What will be the long-term consequences of introducing new gene combinations? We cannot be certain of the effects on the future evolution of organisms. Will the artificial selection of 'desired' genes reduce the genetic variety which is so essential to evolution?

- What might be the economic consequences of developing plants and animals to grow in new regions? Developing bananas which grow in Britain could have disastrous consequences for the Caribbean economies which rely heavily on this crop for their income.

- How far can we take gene therapy (unit 12.6)? It may be acceptable to replace a defective gene to cure cystic fibrosis, but is it equally acceptable to introduce genes for intelligence, more muscular bodies, cosmetic improvements, or different racial features?

- Will knowledge of, and ability to change, human genes lead to **eugenics**, whereby selection of genes leads to a means of selecting one race above another?

- What will be the consequences of the ability to manipulate genes getting into the wrong hands? Will unscrupulous individuals, groups or governments use this power to achieve political goals, control opposition or gain ultimate power?

- Is the cost of genetic engineering justified, or would the money be better used fighting hunger and poverty, which are the cause of much human misery. Will sophisticated treatments, with their more high-profile images, be put before the everyday treatment of rheumatoid arthritis or haemorrhoids? Will such treatments only be within the financial reach of the better-off?

- Genetic fingerprinting, with its ability to identify an individual's DNA accurately, is a highly reliable forensic tool. How easy would it be for someone to exchange a DNA sample maliciously, leading to wrongful conviction?

- Is it immoral to tamper with genes at all? Should we let nature take its own course in its own time?

- How do we deal with the issues surrounding the **human genome** project (section 12.9.4)?

It is inevitable that we remain inquisitive about the world in which we exist, and that we will seek to try to improve the conditions in which we live. Genetic research is bound to continue, but the challenge will be to develop the safeguards and ethical guidelines which will allow genetic engineering to be used in a safe and effective manner.

SUMMARY TEST 12.10

Genetic engineering has many benefits. Genetically modified plants can be used to manufacture substances such as drugs – a process known as (**1**). Crops such as tomatoes can be engineered to grow on soils made unusable because of their high concentration of (**2**). Expensive drugs like (**3**) can be made using (**4**) animals, diseases such as cystic fibrosis can be tackled using (**5**), while (**6**) is an extremely useful forensic tool. The benefits are not without their potential drawbacks, however. For example, could the manipulation of genes lead to selective breeding of humans, otherwise known as (**7**).

1 The polymerase chain reaction is a process used to make many copies of a piece of DNA. The process is outlined in the diagram.

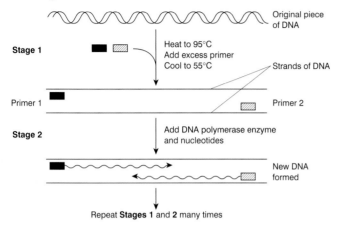

a Describe and explain what happens to the DNA when it is heated to 95°C in **Stage 1**. *(2 marks)*

b The primer is a short sequence of nucleotides that binds to the DNA.

   (i) Explain why the primer only binds to the DNA at a particular position. *(1 mark)*

   (ii) Describe the role of the primer in the polymerase chain reaction. *(1 mark)*

   (ii) Suggest why two **different** primers are needed. *(1 mark)*

c It is important that the original DNA sample is not contaminated with any other material, such as bacteria or human skin. Explain why. *(2 marks)*

d The DNA polymerase used comes from a bacterium which lives in hot springs. Explain the advantage of using DNA polymerase from this organism. *(1 mark)*

(Total 8 marks)

AQA Jan 2001, B (A) BYA2, No.7

2 A new variety of tomato has been produced by genetic engineering. This variety contains a synthetic gene that blocks the action of a natural gene that would make the fruit soften rapidly once ripe. It also contains a marker gene.

The marker gene added by the scientists makes this variety of tomato resistant to the antibiotic, kanamycin. It is possible that this gene could be taken up by disease-producing bacteria in the human gut. In humans, kanamycin is used to treat certain types of gut infections.

Using information from this passage, explain the advantages and disadvantages of putting this new variety of tomato on the market. *(5 marks)*

AQA Jan 2001, B (B) BYB2, No.4

3 About one in 20 000 humans has the condition, albinism. This is caused by the absence of melanin, the dark brown pigment normally present in the human skin, hair, and eyes.

Albinism arises from a gene mutation that causes skin cells to produce a different version of the enzyme tyrosinase. This different version of tyrosinase is faulty.

These events are shown in the diagram.

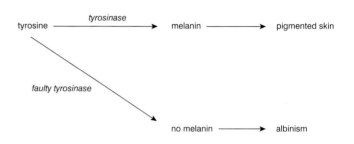

a Explain how a gene mutation may result in a different version of tyrosinase. *(4 marks)*

b The faulty tyrosinase does not produce melanin.

Suggest an explanation for this. *(2 marks)*

(Total 6 marks)

AQA Jan 2001, B (B) BYB2, No.6

4 DNA sequencing is a technique used to find the sequence of nucleotides in a sample of DNA. Enzymes are used to cut the DNA sample into fragments of different lengths.

The ends of these fragments are then labelled using radioactive probes.

Four different probes attach to the end of DNA fragments in which the terminal nucleotide is adenine, cytosine, thymine and guanine respectively. The labelled fragments are then separated.

The diagram shows apparatus used to separate the DNA fragments and the end result of the process.

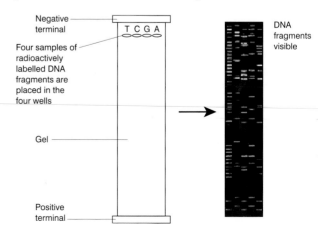

a (i) Use information from the diagram and your knowledge of DNA sequencing techniques to draw a flow chart for the process. The first box has been completed for you.

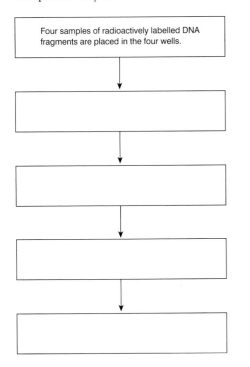

(4 marks)

(ii) Why do the DNA fragments move different distances in the gel? *(1 mark)*

(iii) This technique is being used to find the DNA sequences of human chromosomes. Give **one** advantage to humans of determining this DNA sequence. *(1 mark)*

b (i) Explain how the symptoms of cystic fibrosis are related to faulty CFTR. *(3 marks)*

(ii) Describe how genetic engineering is used to produce alpha-1-antitrypsin which is used to treat cystic fibrosis. *(6 marks)*

(Total 15 marks)

AQA Jan 2001, B (B) BYB2, No.8

5 The figure outlines the way in which the gene for human factor VIII, a protein which is necessary for the clotting of blood, is incorporated into bacterial DNA and inserted into a bacterium.

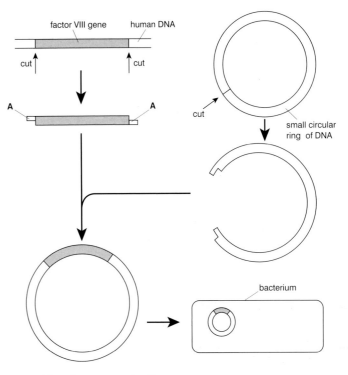

a With reference to the figure, state

(i) the type of enzyme used to cut the factor VIII gene from human DNA; *(1 mark)*

(ii) the name of the small circular ring of DNA; *(1 mark)*

(iii) the term used to describe the regions labelled **A**; *(1 mark)*

(iv) why these regions have been added to the factor VIII gene; *(1 mark)*

(v) the word used to describe the bacterial DNA which now contains the human factor VIII gene. *(1 mark)*

b Suggest why it is considered better to use genetic engineering as a source of human factor VIII rather than material obtained from human blood. *(1 mark)*

c Describe how DNA replicates. (You may use annotated diagrams to assist your explanation if you wish).

(In this question, 1 mark is available for the quality of written communication). *(8 marks)*

(Total 14 marks)

OCR 2801 Jan 2001, B(BF), No.6

13

13.1

Health and disease

Types of disease

OCR

The World Health Organisation, established in 1948, defines health as '**a state of complete physical, mental and social well-being and not merely the absence of disease or infirmity**'. In other words, to be healthy, a person should not only have all body organs working efficiently, but should also feel well. The word **disease** is difficult to define, but it does suggest a malfunction of the body or mind which leads away from good health. Like health, it has physical, mental and social aspects. Some diseases, like malaria, have a single cause but others, like heart disease, are **multifactorial** – having a number of causes. There are many ways of classifying diseases: for example, mental and physical, infectious and non-infectious. Whatever the scheme used, the groups tend to overlap, so that one disease may be put into a number of different categories. The nine main groups in common use in medicine, and examples of each, are given in table 13.1.

Table 13.1 *Main groups of disease and examples of each*

Category	Disease					
	Scurvy	Measles	Lung cancer	Coronary heart disease	Alzheimer's disease	Cystic fibrosis
Physical	✔	✔	✔	✔	✔	✔
Mental					✔	
Social			✔	✔		
Infectious		✔				
Non-infectious	✔		✔	✔	✔	✔
Degenerative				✔	✔	
Inherited						✔
Self-inflicted			✔	✔		
Deficiency	✔					

13.1.1 Physical disease

Almost all diseases have some physical component. In other words, they involve permanent or temporary damage to some part of the body. Examples include leprosy and multiple sclerosis.

13.1.2 Mental diseases

These illnesses involve changes to the mind, and they may not always be associated with a known physical cause. Examples include schizophrenia, depression, claustrophobia (fear of confined spaces), Creutzfeld–Jacob disease (CJD) and Alzheimer's disease. In the last of these, some brain cells that normally secrete the neurotransmitter **acetylcholine** degenerate, and 'plaques' of protein are seen. Such clear changes in brain structure are not seen in schizophrenia or manic depression, although changes in blood flow are apparent, and there are imbalances in the secretion of **neurotransmitters**.

13.1.3 Social diseases

This category includes diseases in which the social environment or a person's behaviour play some part. It can include almost all infectious diseases, because poor sanitation and overcrowding increase the risk of them spreading. Many multifactorial conditions (diseases caused by many factors) such as cardiovascular disease and obesity can also be included. Some occupations carry a direct risk of certain diseases, e.g. glassblowing and silicosis; coal mining and pneumoconiosis.

13.1.4 Infectious diseases

Infectious, or **communicable**, diseases can be spread from person to person or from animals to people. They are caused by disease-causing organisms called **pathogens**. These pathogens include viruses, bacteria, fungi, **protoctists**, worms and insects. They may be transmitted in a variety of ways, such as through water, food, sexual contact or normal social interactions. Examples of infectious diseases include measles and malaria. Carriers are people who have the pathogen in their body, but show no disease symptoms. More details of infectious diseases are given in unit 13.7.

13.1.5 Non-infectious diseases

These are **non-communicable** diseases that are not caused by pathogenic organisms. They include all those in the categories mental, social, degenerative, inherited, self-inflicted and deficiency. Some of them have a single cause, like sickle-cell anaemia (inherited) and others are multifactorial, like strokes (section 13.5.2).

13.1.6 Degenerative diseases

These are most commonly associated with old age, and include problems with poor circulation, reduced mobility and memory loss. They may also occur when the immune system (unit 14.4) starts to attack the body's own cells. Some details of the effects of ageing are given in section 5.15.3 for the skeletal system, in section 5.15.4 for the cardiovascular system, and in section 5.15.5 for the reproductive system. Cancers are considered in units 13.9 and 13.10.

13.1.7 Inherited diseases

Also known as genetic disorders, these diseases are caused by genes, and can be passed from parents to their children. In Britain, the most common genetic disease is cystic fibrosis, caused when sufferers have two copies of the recessive **allele** in their cells (section 12.6.1). Huntington's disease is caused by having just a single copy of the dominant causative allele. Other inherited conditions include sickle-cell anaemia and haemophilia.

13.1.8 Self-inflicted diseases

These are diseases associated with a person's own decisions and behaviour. They cover a range of different conditions, from attempted suicide linked to poor mental health, to the misuse of drugs such as alcohol, nicotine or heroin. Obesity and heart disease may result from a high intake of fatty food, and skin cancer from sunbathing.

13.1.9 Deficiency diseases

These diseases are associated with a lack of food (starvation) or, more often, an unbalanced diet. One or more nutrients, such as particular vitamins or minerals, may be in short supply. If there is a lack of vitamin C, it results in the deficiency disease, scurvy; a lack of iron can lead to anaemia. More details of deficiency diseases are given in sections 14.9.3 and 14.9.4.

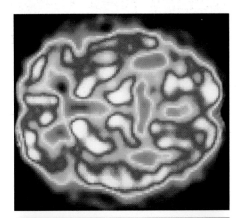
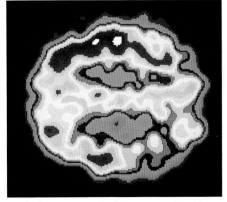

Colour-coded scans of (top) a normal brain and (below) the brain of a patient with Alzheimer's disease, in which the pattern of activity is less symmetrical

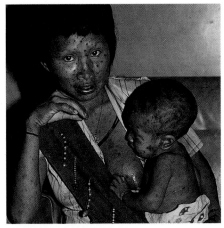

Symptoms of measles

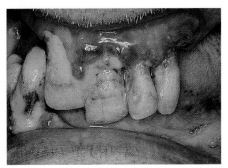

Symptoms of scurvy

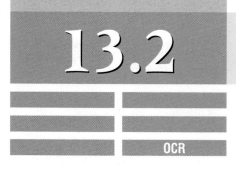

Health statistics

Epidemiology is the study of the spread of disease and its pattern of distribution. These studies are based, not just on the incidence of the disease in individuals, but also on the factors that affect the disease in the population as a whole. Such a detailed study requires the collection and analysis of a large amount of data.

13.2.1 Terminology

In order to refer clearly and unambiguously to health statistical data, a number of terms need to be understood:

- **Incidence** – this is the number of new cases of a disease in a population in a given time, for example a year.
- **Prevalence** – this is the number of people in a population suffering from a particular disease at a given time.
- **Mortality** – the number of people who have died of a particular disease in a given time.
- **Endemic** – infectious diseases are referred to as **endemic** if they are always present in a population. Tuberculosis is endemic in many countries, most people carrying the bacteria even if they do not show symptoms of the disease.
- **Epidemic** – periodically, there are severe and widespread outbreaks of a particular disease, known as an epidemic. Influenza epidemics are common.
- **Pandemic** – this term is used to describe worldwide epidemics, such as human immunodeficiency virus/acquired immunodeficiency syndrome (HIV/AIDS).

13.2.2 Using health statistics

Statistics on the health of populations, rather than of individuals, are collected and expressed in a form that allows fair comparisons to be made, e.g. death rate per thousand population per year. The information gives a valuable insight into the health of a population or a country. Comparisons can be made between different geographical areas, ethnic groups, ages or occupations. This enables governments to set priorities for spending, targeting particular diseases or areas, or setting up screening programmes. The World Health Organisation (WHO) operates similar procedures on a global scale, identifying areas with specific problems and coordinating international action. The collection of accurate data not only enables suitable action to be taken, but also enables the effectiveness of various programmes to be monitored. Such data inform us that no more than six deadly infectious diseases caused half of all premature deaths in 1998 (table 13.2).

13.2.3 Diseases in developing countries

In 1999, the World Health Organisation reported that about one-third of the world's population (1.3 billion people) had an income of less than US $1.0 per day. A similar proportion lacked access to essential drugs, and one in three children were malnourished. Densely populated cities in the developing world had widespread poverty, unsafe water and poor sanitation. Little wonder then that, in the least developed parts of the world, infectious, maternal and perinatal, and deficiency diseases cause more than 50% of deaths. In the poorest countries, it is the young who make up the largest proportion of these deaths. Figure 13.1 shows the main causes of death in low-income countries. It should be noted that some self-inflicted conditions are increasing in developing countries, notably those associated with smoking, considered by the WHO to be reaching epidemic proportions.

Table 13.2 *Millions of deaths worldwide due to various infectious diseases in 1998*

Infectious disease	Deaths / millions
Acute respiratory infections (e.g. pneumonia and influenza)	3.5
HIV/AIDS	2.3
Diarrhoeal diseases (e.g. cholera, typhoid and dysentery)	2.2
Tuberculosis	1.5
Malaria	1.1
Measles	0.9

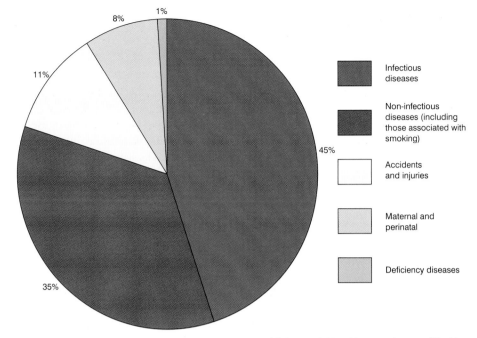

Fig 13.1 *Main causes of death in low-income countries (SE Asia and Africa) in 1998 (source: World Health Organisation Report 1999)*

13.2.4 Diseases in developed countries

In developed countries, both the incidence of infectious disease and the mortality associated with it have been reduced. Living conditions are substantially better than in developing countries, with improved hygiene, sanitation and nutrition. There are successful vaccination programmes, and antibiotics are easily available to cure bacterial infections. As can be seen from figure 13.2, only about 2% of deaths result from infectious diseases. However, relative affluence brings an increase in deaths from cardiovascular diseases, cancer and road accidents. Also, most of the populations live longer than in developing countries, and so the degenerative diseases associated with old age are often seen.

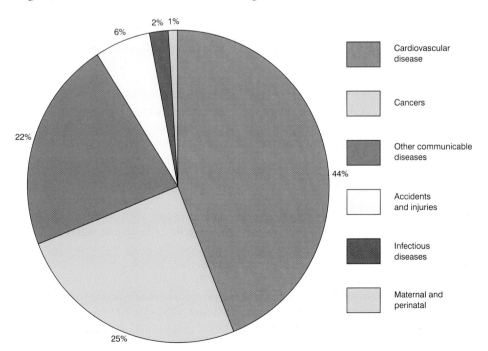

Fig 13.2 *Main causes of death in developed countries in 1998 (source: World Health Organisation Report 1999)*

SUMMARY TEST 13.2

The study of the spread and distribution of disease is called **(1)**. The number of people in a population suffering from a disease at any one time is called its **(2)**, while the number dying from it is called its **(3)**. Where a disease is always present in a population it is said to be **(4)** and an example of such a disease is **(5)**. Diseases such as **(6)** which occur all over the world are said to be **(7)**. The main cause of death in developed countries is **(8)**, while in developing countries it is **(9)**.

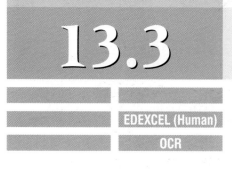

EDEXCEL (Human)

OCR

Table 13.3 *Decline in smoking in Britain*

Year	% of UK population who smoke	
	Men	Women
1972	52	41
1992	29	28

Although the percentage of the British population who smoke has fallen (table 13.3), numbers remain high among the young, and in the less well-off social groups. However, in developing countries, aggressive cigarette marketing has led to a rise in sales. The World Health Organisation considers smoking to be a disease reaching epidemic proportions in the developing world. The links between smoking and cardiovascular disease are discussed in unit 13.5.

13.3.1 Components of tobacco smoke

Tobacco smoke consists of 'mainstream smoke' from the mouth end of the cigarette, and 'sidestream smoke' from the burning tip. About 85% of the smoke released is sidestream smoke. It has a high concentration of toxic substances, and is a severe health hazard for people nearby. Breathing in this smoke is known as **passive smoking**. There are at least 4000 different substances in tobacco smoke, many of them harmful. The most significant are:

- **Carbon monoxide**. This diffuses through the walls of alveoli and into the red blood cells. It combines with haemoglobin to form a stable compound, carboxyhaemoglobin. This prevents haemoglobin from becoming fully oxygenated, reducing blood oxygen levels by 5–10%. This can put a strain on the heart muscle during exercise. It also raises the heart rate and blood pressure of the foetus, slowing growth and possibly causing premature birth.
- **Nicotine** is a drug with complex effects on behaviour. During smoking, its level in the blood rises steeply, and it reaches the brain in less than 30 seconds. It indirectly causes increased adrenaline secretion, leading to raised heart rate and blood pressure, possibly damaging pulmonary capillaries. There is reduced blood flow to the extremities, and increased stickiness of the **platelets** encourages blood clotting.
- **Tar** enters the respiratory tract as minute droplets, and about 70% of it is deposited. It is an irritant, causing inflammation of the mucous membranes lining the airways. Extra mucus is secreted and the cilia are paralysed (the structure of the respiratory system is shown in unit 6.4). Tars also contain **carcinogens**, which greatly increase the chance of developing lung cancer (sections 13.3.4 and 13.10.3).

13.3.2 Chronic bronchitis

Chronic bronchitis is inflammation of the air passages, with symptoms that worsen and persist for many years. It kills about 30 000 people a year in Britain, and about a million others are affected. Tars cause extra mucus secretion; coughing is the only way to remove it, as the cilia lining the air tracts are damaged or destroyed.

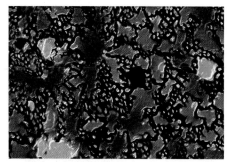

Normal lung tissue

13.3.3 Emphysema

Together with chronic bronchitis, this is often referred to as **chronic obstructive pulmonary disease**. Prolonged infections encourage phagocytes to move from the blood into the respiratory tract. In order to do this, they 'clear a pathway', using the protease, elastase. As this destroys the elastin in the alveoli, they become unable to stretch and recoil, and often burst (see photo and figure 13.3). This condition is emphysema. People with emphysema can exchange very little air with each breath; consequently, their breathing rate increases, blood pressure increases, and the right side of the heart enlarges. The condition is irreversible, and sufferers may have to use an oxygen mask, and may not be able to get out of bed.

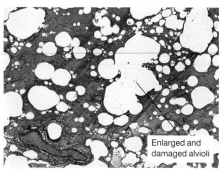

Enlarged and damaged alvioli

Lung tissue damaged by emphysema

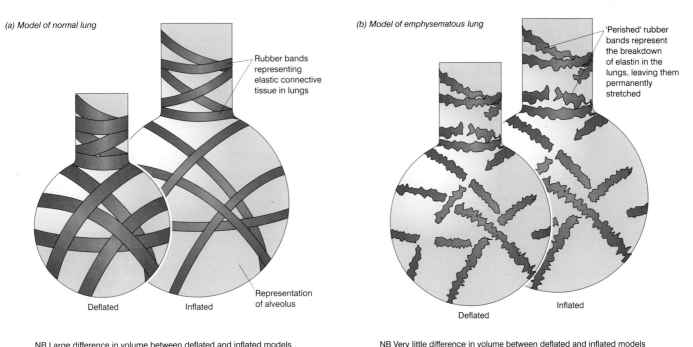

(a) Model of normal lung

Rubber bands representing elastic connective tissue in lungs

Representation of alveolus

Deflated   Inflated

NB Large difference in volume between deflated and inflated models

(b) Model of emphysematous lung

'Perished' rubber bands represent the breakdown of elastin in the lungs, leaving them permanently stretched

Deflated   Inflated

NB Very little difference in volume between deflated and inflated models

Fig 13.3 *'Rubber band' models of normal and emphysematous lungs*

13.3.4 Lung cancer

The tars in tobacco smoke contain a number of carcinogens. They cause mutations in the epithelial cells of the respiratory tract, leading to the formation of a tumour. At first there are no symptoms, and by the time the sufferer becomes aware (by coughing up blood), surgery may be the only possible cure. Further details of cancer are in unit 13.9, and of lung cancer in section 13.10.3.

13.3.5 Evidence for the links between smoking and lung disease

It has been estimated that 25–30% of environmentally associated cancers are the result of smoking. The relative risk rises rapidly in proportion to the number of cigarettes smoked per day (Fig 13.4). Epidemiologists have collected a range of data pointing to a link between smoking and lung diseases, including cancer. These include:

- up to 50% of smokers die of smoking-related diseases
- 98% of people with emphysema are smokers
- 25% of smokers die of lung cancer
- only 0.3% of those who die of lung cancer have never been smokers.

In addition, experiments have demonstrated both the presence of carcinogens in tobacco smoke, and that these substances cause cancer in experimental animals.

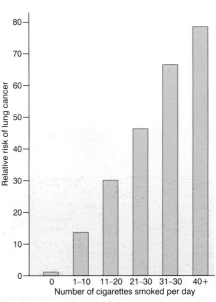

Fig 13.4 *Relative risk of lung cancer according to daily cigarette consumption (Souhami R, Tobias J. Cancer and its Management. 1986)*

SUMMARY TEST 13.3

Of the 400 or more substance in tobacco smoke, the three most important are carbon monoxide, that combines with the blood's main respiratory pigment to form **(1)**, nicotine that increases secretion of the hormone **(2)** and increases the stickiness of **(3)** in the blood and finally **(4)** that contain **(5)** which may cause lung cancer. **(6)** % of smokers die of diseases related to their smoking, which include inflammation of the air passages known as **(7)**, lung cancer and emphysema. Smokers make up **(8)** % of emphysema sufferers, a disease that destroys the **(9)** of the alveoli, making them unable to stretch and recoil. The symptoms include an increase in both **(10)** and **(11)** and enlargement of the **(12)** of the heart.

Any interruption to the heart's beating can have serious, often fatal, consequences. Of the many possible disorders affecting the heart, by far the most common is **coronary heart disease (CHD)**, which affects the pair of blood vessels, the coronary arteries, supplying the heart muscle itself (heart structure is dealt with in unit 8.6). Blood flow through these vessels may be impaired by the build-up of fatty deposits known as **atheromas** (section 13.4.1). If blood flow to the heart muscle is interrupted, it can lead to **myocardial infarction**, in other words, a heart attack (section 13.4.4).

13.4.1 Atherosclerosis

Atherosclerosis is a condition in which the walls of arteries thicken and lose their elasticity. This is due to the build-up of fatty deposits in the artery, known as **atheromas**. These begin as fatty streaks, often after damage to the artery by, for example, high blood pressure. These streaks enlarge to form irregular patches, or **plaques**. Atheromatous plaques are made up of deposits of cholesterol (section 13.4.2), fibrous tissue, blood platelets and dead muscle cells. They bulge into the lumen of the artery, causing it to narrow and blood flow through it to be reduced, as shown in the photo opposite (left). Over many years, calcium may also become deposited in the plaques, causing the artery wall to harden. This condition is known as **arteriosclerosis**, and is particularly associated with ageing. The build-up of atheroma is shown in figure 13.5.

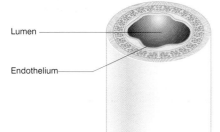

Lumen

Endothelium

13.4.2 Role of cholesterol

Cholesterol is an essential component of plasma membranes and steroid hormones. As such, it is an essential biochemical which must be transported in the blood. It is carried in the plasma as tiny spheres of lipoproteins (lipid and protein). There are two main types:
- **High-density lipoproteins (HDLP)** – remove cholesterol from tissues and transport it to the liver for excretion. They help protect arteries against atherosclerosis.
- **Low-density lipoproteins (LDLP)** – transport cholesterol from the liver to the tissues, including artery walls. They increase the risk of atherosclerosis.

Build-up of atheroma

13.4.3 Thrombosis

If an atheroma breaks through the lining (endothelium) of the blood vessel, it forms a rough surface that interrupts the otherwise smooth flow of blood. This may result in the formation of a blood clot, or **thrombus**, in a mechanism known as **thrombosis**. This thrombus may block the blood vessel, preventing the supply of blood to tissues beyond it. Sometimes, the thrombus is carried from its place of origin to lodge in, and block, another artery. It is then known as an **embolus**. The photo opposite (right) shows a thrombus in a coronary artery.

13.4.4 Myocardial infarction

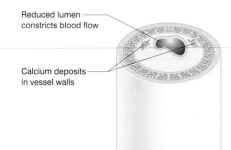

Reduced lumen constricts blood flow

Calcium deposits in vessel walls

More commonly known as a heart attack, the expression 'myocardial infarction' refers to a reduced supply of oxygen to the muscle (myocardium) of the heart. It results from a blockage in the coronary arteries. If this occurs close to the junction of the coronary artery and the aorta, the heart will stop beating. If the blockage is further along the coronary artery, the symptoms will be milder, because a smaller area of muscle will suffer oxygen deprivation. In Britain, about half a million people a year have a heart attack, although fewer than one-third of them die as a result. Almost all show signs of atherosclerosis, and many have coronary thrombosis (clots forming in the coronary arteries).

Fig 13.5 *Build-up of atheroma*

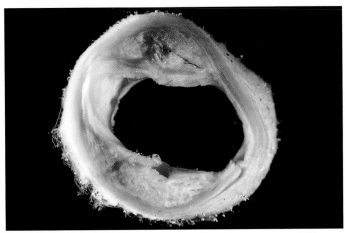

Human aorta with a fatty atheroma partially obstructing the interior (TS)

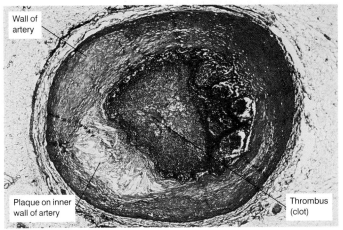

Wall of artery

Plaque on inner wall of artery

Thrombus (clot)

Human coronary artery containing a thrombus

13.4.5 Aneurysm

Atheromas that lead to the formation of clots also weaken artery walls. These weakened points swell to form balloon-like structures called **aneurysms**. A brain aneurysm is known as a cerebrovascular accident (CVA) or a **stroke** (section 13.2).

13.4.6 Risk factors associated with coronary heart disease

Some of the factors which make CHD more likely to develop are unavoidable. These include being male, being older, and having a history of heart disease in the family.

However, there are avoidable risk factors, including:
- **smoking** (unit 13.5)
- **excess cholesterol** and saturated fat in the diet
- **high blood pressure**
- **lack of physical exercise**.

Figure 13.6 illustrates the links between some of the risk factors, and table 13.4 shows how combinations of the three major risk factors – smoking, high cholesterol and high blood pressure – increase the likelihood of heart disease.

***Table 13.4** How the combinations of risk factors for heart disease affect death rate*

Number of risk factors	Death rate for men aged 30–59 years
0	1 in 50
1	1 in 20
2	1 in 11
3	1 in 6

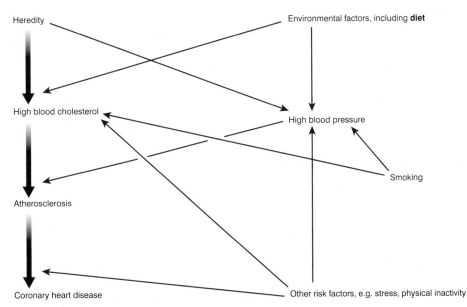

Heredity

Environmental factors, including **diet**

High blood cholesterol

High blood pressure

Smoking

Atherosclerosis

Coronary heart disease

Other risk factors, e.g. stress, physical inactivity

***Fig 13.6** The links between some of the risk factors associated with the development of coronary heart disease*

13.5

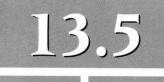

Smoking and cardiovascular disease

AQA.A (Human)

OCR

GOVERNMENT MEASURES TO REDUCE THE INCIDENCE OF HEART DISEASE

- anti-smoking campaigns
- warnings on tobacco products
- health awareness measures
- establishment of clinics to monitor blood pressure and cholesterol levels
- dietary information promoting low cholesterol foods and reduction of salt in the diet
- encouraging exercise and physical fitness
- warning of the dangers of obesity

Cardiovascular diseases are multifactorial degenerative diseases of the heart and circulatory system (unit 13.1). Smoking is just one of the risk factors that increases the chances of developing one of the cardiovascular diseases.

13.5.1 Tobacco smoke and cardiovascular disease

The main components of tobacco smoke that increase the risk of cardiovascular disease are nicotine and carbon monoxide (section 13.3.1). They both increase production of the blood clotting factor, fibrinogen, and at the same time reduce production of enzymes that break down blood clots. They also have the following additional effects:

- **nicotine** – indirectly increases adrenaline secretion, leading to a raised heart rate and blood pressure, with possible damage to the pulmonary capillaries; it also reduces the ability of arteries to dilate, which further increases blood pressure
- **carbon monoxide** – combines with haemoglobin and reduces the ability of the blood to carry oxygen.

Smoking also increases the concentration of cholesterol in the blood – a major risk factor for coronary heart disease (CHD) (unit 13.4).

13.5.2 Strokes

CHD (unit 13.4) is not the only cardiovascular disease. Aneurysms (section 13.4.5) that occur in the brain may burst and lead to a **cerebrovascular accident (CVA)**, otherwise known as a **stroke**. The resulting haemorrhage leads to some areas of the brain suffering a severe oxygen shortage. Some strokes are very minor, but others have crippling effects and there are over 40 000 deaths from CVA in Britain each year.

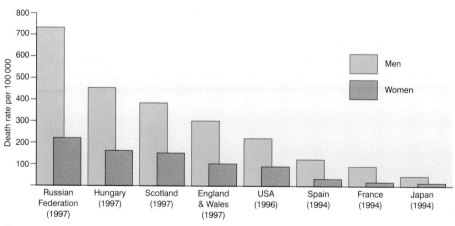

Fig 13.7 Death rates from coronary heart disease in various countries

13.5.3 Global distribution of coronary heart disease

Death rates for CHD vary both within countries and between countries in different parts of the world. It seems that people are not equally at risk of developing the disease. Figure 13.7 shows that the death rate from CHD is highest in northern Europe and lowest in Japan and France. It is also a disease associated more with developed countries than with developing ones. This may be because improved public health in developed countries means that people are not dying young from infectious diseases; they survive long enough to die of degenerative illnesses, like

CHD, in later years. Diseases of the cardiovascular system account for 20% of deaths worldwide, and 50% of deaths in developed countries. Countries like Cuba, Uruguay and Argentina show increasing levels of heart disease as their living standards and health services improve. Within the UK, the incidence of the disease is greatest
- in men
- in Scotland, Northern Ireland and the north of England
- among the poor
- among south Asians.

Some of the major risk factors associated with CHD are given in section 13.4.6.

13.5.4 Prevention of coronary heart disease

Although we can do nothing about our age, sex, or inherited risk of developing CHD, there are many other things that we can do to reduce the chances of developing the disease:
- **do not become overweight** – obesity puts a strain on the heart and increases blood pressure
- **reduce intake of saturated fats and cholesterol** – people with high levels of saturated fats and cholesterol in their diet tend to have high blood cholesterol (above 250 mg cholesterol per 100 cm³ of blood is high)
- **reduce salt intake** – a high salt intake can lead to high blood pressure
- **increase antioxidant and fibre intake** – antioxidants such as vitamin E protect the artery walls against **atherosclerosis** (section 13.4.1)
- **do not smoke** – smoking combines with other risk factors to increase the chance of developing CHD.

Death rates from CHD have fallen in some developed countries in which the governments have actively pursued these policies.

13.5.5 Treatment of coronary heart disease

Prevention, in the long term, is cheaper than attempting to treat CHD. However, treatments are available:
- **Drugs** – can be taken to reduce blood pressure, the risk of blood clots, and blood cholesterol levels.
- **Coronary bypass surgery** – a piece of vein from the leg is grafted onto the aorta and then onto the coronary artery at a point beyond the blockage. Sometimes a number of bypasses are carried out at the same time. This is a widely used operation; more than 24 000 were performed in the UK in 1997.
- **Heart transplant operation** – this is an expensive treatment, and it is difficult to find enough donor hearts. Tissues must be carefully matched to avoid an antigen–antibody response (unit 14.3) and drugs are administered to suppress the immune response (unit 14.4) and prevent rejection.

All treatments for CHD are expensive, and health service money needs to be used wisely. Increasingly, emphasis is being put on
- education to encourage healthy lifestyles
- screening, to target health care at those most at risk of developing the disease.

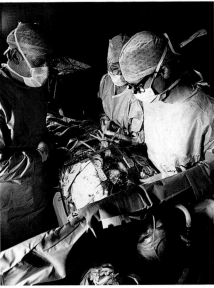
Heart bypass operation

SUMMARY TEST 13.5

Death from coronary heart disease (CHD) accounts for **(1)** % of all deaths worldwide and varies from country to country, being highest in **(2)** and lowest in **(3)**. In general, men are **(4)** times more likely to die from CHD than women, wherever they live. Strokes, also known as **(5)**, are a form of cardiovascular disease that cause **(6)** deaths in Britain each year. To lower the risk of CHD, individuals can change their diet by eating less **(7)**, **(8)** and **(9)** and increasing their intake of **(10)** and **(11)**. Treatment entails drugs, heart transplants and coronary bypass surgery, in which a piece of vein is grafted between the **(11)** and the **(12)**.

'Microorganism' is a general term for a non-multicellular organism which is too small to be seen without the use of a microscope. Microorganisms may be **bacteria**, fungi, protoctists (unicellular algae and protozoa) or **viruses**.

13.6.1 Bacteria

Bacteria are **prokaryotes**; details of their structure are given in unit 4.4. As bacteria reproduce by splitting in two (binary fission), their population can increase rapidly. Figure 13.8 illustrates an S-shaped (sigmoid) growth curve like that of a bacterial population. In it, the following phases can be recognised:

* **the lag phase** – this is a period of slow growth, when the bacterial cells are adjusting to the nutrient solution they are in by synthesising the enzymes they need to make use of it
* **the exponential (logarithmic) phase** – in this phase, the rate of cell division is at its maximum – the bacterial cells doubling every 10–20 minutes (table 13.5)
* **the stationary phase** – as the nutrients are exhausted and waste products build up, the rate at which new cells are produced is equal to the rate at which others die
* **the death phase** – while the total number of cells remains constant, the number of living ones decreases, as an ever-increasing number die from a lack of nutrients or a build-up of toxins.

Table 13.5 *Increase in numbers of bacterial cells over 150 minutes*

Time / minutes	Number of divisions	Number of cells
0	0	1
15	1	2
30	2	4
45	3	8
60	4	16
75	5	32
90	6	64
105	7	128
120	8	256
135	9	512
150	10	1024

13.6.2 Factors affecting growth of bacterial populations

Many factors, including pH and oxygen availability, affect the growth rate of bacterial populations, but the two most important are:

* **temperature** – most cells grow best in the range 20–45°C. This is because growth is governed by enzymes that normally work best within this range (section 2.2.4). Some, however, continue to grow at 0°C and others at 90°C.
* **nutrient availability** – although some bacteria have specific nutrient requirements, in general, the more nutrients there are, the faster the bacterial population will increase. As nutrients run out, growth slows (section 13.6.1). Bacteria require:
  – **carbon** – normally as an organic substance, such as glucose or fatty acids
  – **nitrogen** – as amino acids, ammonia or nitrate ions
  – **other macronutrients** – including oxygen, hydrogen, sulphur and magnesium
  – **micronutrients** – usually metals like copper, manganese and iron
  – **growth factors** – such as amino acids, vitamins, purines and pyrimidines.

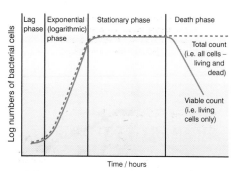

Fig 13.8 *Bacterial growth curve*

13.6.3 Viruses

Viruses lack cellular structure, consisting only of nucleic acid (either DNA or RNA) and protein. They are smaller than bacteria, ranging in size from 20nm to 300nm. They can only multiply inside living cells; outside cells, they are inert particles known as **virions**. Figure 13.9 illustrates the structure of the human immunodeficiency virus (HIV), which is a **retrovirus**. 'Retro' means behind or backwards and refers to the fact that these viruses synthesise DNA from RNA, the reversal of the normal process. HIV contains single-stranded RNA and the enzyme **reverse transcriptase** (section 12.2.2), which it uses to produce double-stranded DNA (known as a **provirus**). This can then be incorporated into the DNA of the host cell. It may remain inactive, or **latent**, for long periods before actively producing new viral particles.

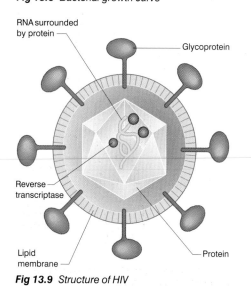

Fig 13.9 *Structure of HIV*

During the latent phase, any division of the host cell results in proviral DNA being duplicated also. In this way, the number of potential retroviruses can increase considerably. Figure 13.10 illustrates the life cycle of a retrovirus such as HIV. HIV and acquired immunodeficiency syndrome (AIDS) are also considered in section 13.7.1.

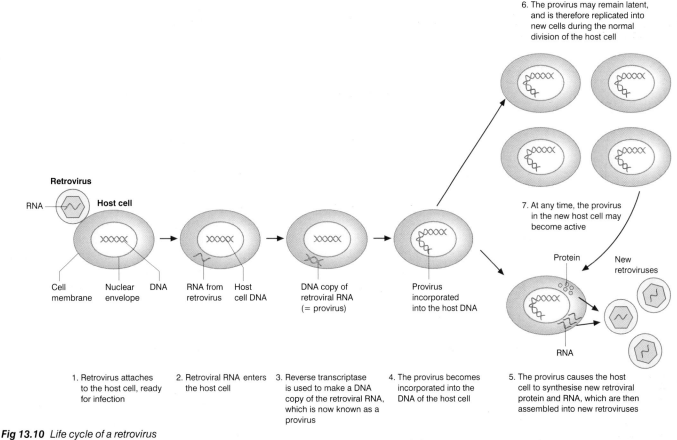

6. The provirus may remain latent, and is therefore replicated into new cells during the normal division of the host cell

7. At any time, the provirus in the new host cell may become active

Protein

New retroviruses

RNA

Retrovirus

RNA

Host cell

Cell membrane

Nuclear envelope

DNA

RNA from retrovirus

Host cell DNA

DNA copy of retroviral RNA (= provirus)

Provirus incorporated into the host DNA

1. Retrovirus attaches to the host cell, ready for infection

2. Retroviral RNA enters the host cell

3. Reverse transcriptase is used to make a DNA copy of the retroviral RNA, which is now known as a provirus

4. The provirus becomes incorporated into the DNA of the host cell

5. The provirus causes the host cell to synthesise new retroviral protein and RNA, which are then assembled into new retroviruses

Fig 13.10 *Life cycle of a retrovirus*

13.6.4 Koch's postulates

It is now widely accepted that infectious diseases are caused by microorganisms. However, it was only during the 19th century that Pasteur and Koch put forward the 'germ theory' of disease. Koch stated four conditions (postulates) that must be met to indicate that a disease is caused by a pathogen (a disease-causing microorganism):

- the causative agent (direct cause of the disease) must be present in all cases of the disease, but absent in healthy individuals
- the agent of the disease can be isolated from the diseased organism and grown in pure culture
- the disease can be reproduced by inoculating some of the pure culture into healthy individuals
- the agent of the disease can be re-isolated from the infected individual.

In order to cause disease, the pathogenic organism must enter the body, e.g. through the skin, or via the nose, anus, mouth or urinogenital tract (Fig 13.11). Once inside, the pathogen produces toxins, which damage the host cells. This damage, and the host's response to it, cause the symptoms of the disease.

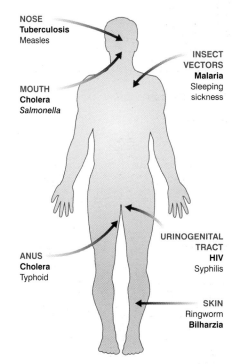

NOSE
Tuberculosis
Measles

INSECT VECTORS
Malaria
Sleeping sickness

MOUTH
Cholera
Salmonella

ANUS
Cholera
Typhoid

URINOGENITAL TRACT
HIV
Syphilis

SKIN
Ringworm
Bilharzia

Fig 13.11 *Routes by which some pathogens enter and leave the body*

Infectious diseases

Infectious diseases are **communicable**, i.e. they can be passed from one person to another. They are caused by living organisms, usually microorganisms.

13.7.1 HIV/AIDS

Some information about the structure and reproduction of the human immunodeficiency virus (HIV) is given in section 13.6.3. After HIV enters the body the following sequence of events occurs:

- HIV infects T-helper lymphocytes.
- HIV makes copy DNA from its RNA using reverse transcriptase.
- HIV DNA becomes part of cell DNA.
- When activated by cell division, HIV DNA manufactures HIV proteins.
- These proteins become incorporated into the plasma membrane of the T-helper cells.
- B-lymphocytes produce antibodies to destroy infected T-helper cells. The person is now HIV positive.
- A balance is established whereby the rate of T-helper cell production matches the rate of their destruction. This can last for 20 years.
- When the rate of destruction exceeds the rate of production, the symptoms of **Acquired Immune Deficiency Syndrome (AIDS)** are experienced.

- **Transmission** – in semen and vaginal fluid during sexual intercourse, infected blood, contaminated syringes, breast milk, across the placenta.
- **Distribution** – worldwide, but especially in south-east Asia and sub-Saharan Africa. The World Health Organisation estimates that by 1998, more than 47 million people had been infected, and 14 million of them had died.
- **Symptoms**:
  - HIV: flu-like symptoms and then none.
  - AIDS: (may be 10 years after initial HIV infection) pneumonia, tuberculosis, cancers, weight loss, diarrhoea, and possibly dementia.
- **Prevention and control** – there is no cure for AIDS and, although drugs can be used to delay the onset of symptoms, education is the best way to prevent its spread. The expenses associated with HIV/AIDS are estimated to have put back the economy of some African countries by 15 years. Public health messages include:
  - avoid unprotected intercourse
  - avoid intravenous drug use or sharing needles
  - screen all blood donations
  - introduce HIV testing.

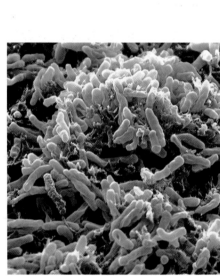

Mycobacterium tuberculosis *cluster (SEM)* (×8500 approx.)

13.7.2 Tuberculosis (TB)

Tuberculosis is a bacterial infection caused by *Mycobacterium tuberculosis* (see photo) and *M. bovis*. The pathogens initially inhabit lung cells, but may spread to the lymph nodes, gut and bone. The bacteria, found in 30% of the world's population, only become active when the person is weakened by, for example, HIV or malnutrition.

- **Transmission** – airborne droplets (coughs and sneezes) from people with the active form. *M. bovis* may be spread from cattle via unpasteurised milk.
- **Distribution** – worldwide, but especially in developing countries, and amongst the poor in the developed world.
- **Symptoms** – dry cough, coughing up blood, chest pain, weight loss.

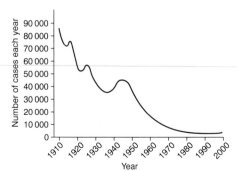

Fig 13.12 *Graph showing number of TB cases for the period 1910 to 2000*

- **Prevention and control** – TB is a notifiable disease, and contacts of infected patients are traced. Lung X-rays (see photo) can indicate infection requiring treatment by antibiotics. Children are vaccinated against the disease (section 14.5.3). However, a crucial weapon in defeating TB is an improvement in social conditions, to reduce malnutrition and over-crowding.

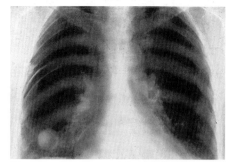

Chest X-ray of a person infected with TB

13.7.3 Cholera

Cholera is caused by the bacterium, *Vibrio cholerae*. The bacteria survive stomach acids if the pH is above 4.5, and pass on to reach the small intestine. Here, they multiply and secrete a toxin, choleragen, which upsets the functions of the epithelia, so that salts and water are removed from the blood, causing severe diarrhoea.

- **Transmission** – ingestion of water or food contaminated by faeces of patients with cholera.
- **Distribution** – Asia, Africa and South America; especially areas of large populations with inadequate sewerage systems.
- **Symptoms** – severe diarrhoea with grey, watery faeces (known as 'rice water'), dehydration and loss of salts leading to cramps.
- **Prevention and control** – proper sanitation and personal hygiene, isolation of patients. Vaccinations are effective for 6 months.

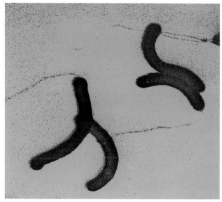

Vibrio cholerae

13.7.4 *Salmonella*

Salmonella are a large group of bacteria causing a variety of diseases, including typhoid, enteric fever and food poisoning (Salmonellosis). When infected food is eaten, the bacteria stick to the surface of epithelial cells in the intestine. The cells then take up the bacteria by **phagocytosis** (section 4.12.3), and the bacteria multiply and produce endotoxins (poisons that are part of the bacterial cell wall and released when the cell is broken down), which damage the cells, producing inflammation and fever.

- **Transmission** – by eating products of infected animals, or through contact with faeces of infected people.
- **Symptoms** of food poisoning are diarrhoea and vomiting, usually with abdominal pain and fever.
- **Prevention and control** – proper sewage disposal, meat inspection, hygiene at home and in the food trade.

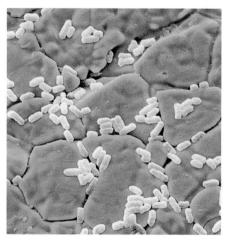

Salmonella spp.

SUMMARY TEST 13.7

AIDS is a communicable disease caused by the **(1)** virus. It can be transmitted during sexual intercourse in **(2)** and **(3)** fluids, through infected blood, in breast milk or across the **(4)** during pregnancy. The virus destroys the **(5)** cells in the blood system, thus lowering an individual's immunity. Another communicable disease, tuberculosis, is caused by *Mycobacterium* spp., which inhabits **(6)** cells. It is transmitted through the **(7)** and also from cattle via **(8)** milk. It can be prevented by vaccination and improvement in social conditions such as reducing **(9)** and **(10)**. Cholera, however, is transmitted via **(11)** or **(12)** contaminated with the faeces of a cholera sufferer, which are grey and watery and called **(13)**. Vaccination against cholera is effective for a period of **(14)**. Food poisoning may be caused by the **(15)** group of bacteria that produce poisons called **(16)**, which give symptoms of **(17)** and **(18)** which include fever and abdominal pain.

A number of parasitic **eukaryotic** organisms (unit 4.4) have to invade living cells in order to reproduce. Examples include the malarial **parasite**, *Plasmodium*, which needs mosquito and human tissues in order to complete its life cycle, and the blood fluke, *Schistosoma*, which invades snails and humans.

13.8.1 *Plasmodium* – the malarial parasite

Malaria is caused by one of four species of the protoctist, *Plasmodium*. This parasite gets through the human skin barrier by first invading the salivary glands of female *Anopheles* mosquitoes; the parasite then gains entry to the human body when the mosquito feeds. An organism that transfers a parasite to its main, or primary, host is called a **vector**. The life cycle of *Plasmodium* is shown in figure 13.13.

13.8.2 Malaria – the disease

Malaria is not just transferred by infected mosquitoes. The infective stages in human blood can also be transmitted by blood transfusion, non-sterile needles, or across the placenta. *Plasmodium* can survive long periods in the blood because it lives within red blood cells and is therefore not detected by the immune system. As *Plasmodium* multiplies in both hosts, there are huge numbers of the parasite, increasing the chances of infection. After being bitten by an infected mosquito, the host has no symptoms for between a week and a year. The symptoms then occur in cycles, coinciding with the release of merozoites. The cycle normally includes:

* headache, tiredness, aching and sometimes vomiting
* shivering and feeling cold for about 2 hours
* sudden onset of fever, with rapid heart beat and breathing rate, lasting 4 hours
* sweating for 2–4 hours as the body temperature returns to normal.

If untreated, the anaemia and acute exhaustion may lead to death, especially in those already weakened by, for example, malnutrition.

Plasmodium *merozoites parasitising a red blood cell (EM) (×7000 approx.)*

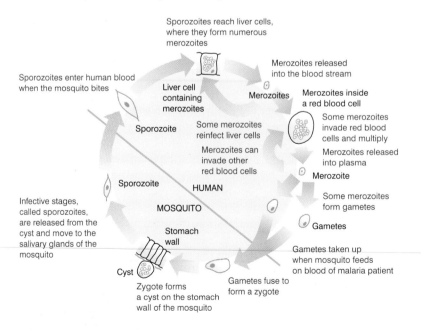

Fig 13.13 *Plasmodium vivax – simplified life cycle*

13.8.3 Adaptations of *Plasmodium* to its way of life

- Use of a **vector** to penetrate the skin of the primary host.
- Rapid multiplication, improving chances of infecting another human or mosquito.
- Very resistant to **antibodies** when inside liver or red blood cells.
- Ability to change surface **antigens** genetically, making development of a vaccine difficult.
- Absence of locomotory structures as it is transported in the blood within its host and between hosts by the mosquito vector.
- Absence of a water regulation mechanism as it is surrounded by blood plasma that is **isotonic** with its cell contents.

13.8.4 Incidence and control of malaria

Found throughout the tropics and subtropics, malaria kills 1.5–2.7 million people each year. There are three main ways to control malaria:

- **Reduce the number of mosquitoes**
  - by draining marshes
  - by stocking ponds with fish to eat the mosquito larvae
  - by spraying with the biological control agent, *Bacillus thuringiensis* (section 10.7.3).
- **Avoid being bitten by mosquitoes**
  - by using insect repellents
  - by wearing covering clothing
  - by sleeping under mosquito nets.
- **Use drugs to prevent infection**
  - quinine and chloroquinine are used to treat infected people

- preventative (**prophylactic**) drugs such as quinine and mefloquine can be used, but the side effects can be unpleasant.

The incidence of sickle-cell anaemia (section 12.1.5) is high in malarial areas, where it offers some resistance to malarial attacks because *Plasmodium* finds it difficult to survive in sickled red blood cells.

13.8.5 *Schistosoma* – the blood fluke

Schistosoma, which causes **schistosomiasis** or **bilharzia**, is one of a small number of parasites to have developed a mechanism for getting through intact human skin. This is a free-swimming larval form, known as a **cercaria**. The life cycle of *Schistosoma* is shown in the annotated diagram (Fig 13.14).

13.8.6 Adaptations of *Schistosoma* to its way of life

- Female always held in groove on male's body, so the different sexes do not need to find each other for sexual reproduction.
- Adult flukes move in large blood vessels, but have suckers for attachment.
- Flukes synthesise chemicals which 'switch off' the host's immune system.
- Flukes feed extracellularly (section 9.1.2) on blood and cells.
- Eggs, produced in large numbers, have spines for penetration of intestinal wall.
- Miracidia and cercariae are free-swimming, increasing chances of finding a new host.

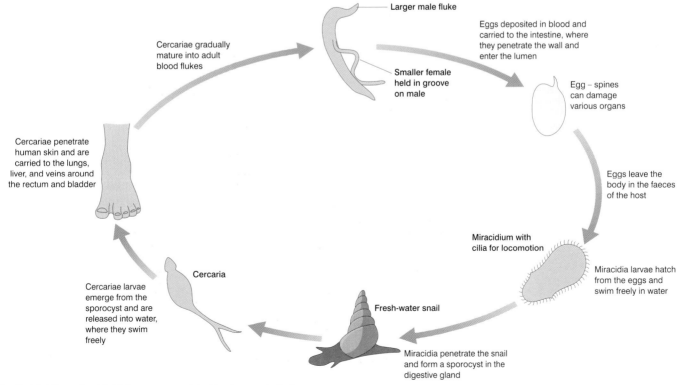

Fig 13.14 *Life cycle of* Schistosoma, *annotated to show main features*

13.9

Biological basis of cancer

AQA.A (Human)

OCR

In normal cells, both growth and division are tightly controlled. Cancers result from mutations which lead to uncontrolled cell division and the eventual formation of groups of abnormal cells, known as **tumours**.

13.9.1 Genetic control of cell division

DNA analysis of tumours has shown that, in general, the cells are derived from a single mutant cell. The initial mutation caused it to divide excessively and, often years later, a further mutation in one of the descendant cells led to structural changes. Subsequent cells will therefore differ from normal, in growth and appearance. The two genes thought to be most concerned with maintaining the normal balance in cell division are **proto-oncogenes** and **tumour suppressor genes**.

13.9.2 Role of proto-oncogenes

Proto-oncogenes stimulate normal cell division, but they sometimes mutate to form **oncogenes** ('oncos' = tumours) that upset cell growth and division. In a normal cell, growth factors attach to a receptor protein on the cell membrane and, via relay proteins in the cytoplasm, 'switch on' the genes necessary for DNA replication (unit 3.3). This process is shown in figure 13.15. When proto-oncogenes mutate into oncogenes, they can affect this mechanism in two possible ways:

* the receptor protein on the cell surface can be permanently activated, so that cell division is switched on in the absence of growth factors
* the oncogene may code for a growth factor that is then produced in excessive amounts, again stimulating excessive cell division.

13.9.3 Role of tumour suppressor genes

Research into hereditary forms of cancer led to the discovery of tumour suppressor genes. These genes have the opposite role in the cell, compared with oncogenes: they inhibit cell division. Some tumour suppressor genes seem to be specific to particular cancers but one, known as **p53**, has been found to be mutated in a wide range of cancers. A normal p53 gene stops the cell cycle, and also destroys cells in which the DNA has become permanently damaged. Tumour cells that no longer have a functional p53 gene are therefore particularly difficult to destroy.

13.9.4 Types of tumour

Most mutated cells die. However, any that survive are capable of making clones of themselves and forming tumours. Not all tumours are harmful **(malignant)**; some are harmless **(benign)**. These tumours are compared in table 13.6.

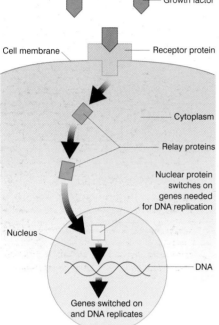
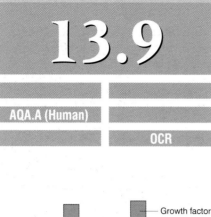

Fig 13.15 *Normal cell receiving signals from growth factors that tell it when to divide*

Growth factor

Cell membrane

Receptor protein

Cytoplasm

Relay proteins

Nuclear protein switches on genes needed for DNA replication

Nucleus

DNA

Genes switched on and DNA replicates

Table 13.6 A comparison of benign and malignant tumours

Benign tumours	Malignant tumours
Remain within the tissue from which they arise	Tend to spread to other regions of the body
Grow very slowly	Grow rapidly
Have cells which are often well differentiated (specialised)	Have cells which are not well differentiated (unspecialised)
Cells tend to stay together, surrounded by a capsule of dense tissue	Cells are not surrounded by a capsule
Not usually life-threatening, but can disrupt functioning of a vital organ	May be life-threatening, as abnormal tumour tissue replaces normal tissue

13.9.5 Primary and secondary tumours

- **Primary tumours** are masses of abnormal tissue in their place of origin. As may be seen in figure 13.16, they can be large and develop their own blood supply. Because they are localised, treatment of these tumours is often very successful.
- **Secondary tumours** are more difficult to treat. As shown in figure 13.16, these arise as cells break away from the primary tumour and migrate via blood or lymphatic vessels. This method of spreading is called **metastasis**. Cells that do not metastasise may spread by growing rapidly, extending the tumour into nearby tissues.

13.9.6 Causes of cancer

Cancer is not a single disease and, likewise, it does not have one single cause. Currently, the following factors are known to play a part:

- **Genetic factors** – more than a dozen forms of cancer are known to be directly inherited. Many more develop in those with genetic predispositions to specific forms of cancer. Oncogenes (section 13.9.2) and tumour suppressing genes (section 13.9.3) are thought to be involved. Cancers with known genetic risk factors include forms of skin cancer and breast cancer.
- **Carcinogens** – these are chemicals that affect genetic activity in some way, causing abnormal cell division. Some carcinogens are **mutagens**, causing changes in the cell's DNA structure (unit 12.1). Carcinogens include:
  - **chemicals** such as the polycyclic hydrocarbons found in soot and cigarette smoke
  - **short-wavelength radiation** such as X-rays, gamma rays and some ultra-violet rays (UVA and UVB – unit 13.10).
- **Age** – some cancers, e.g. leukaemia, are found primarily in young people. Others, such as colon cancer, are found in older adults. Cancers in later life may result partly from the accumulated effects of cell damage.
- **Environment** – exposure to certain types of radiation or chemicals can cause cancer to develop. For example, breathing in asbestos fibres can cause lung cancer, and sunlight can cause skin cancer (unit 13.10).
- **Viruses** – some cancers have now been shown to have a viral origin. This makes sense, because viruses interfere with the genetic make-up of infected cells (section 13.6.3). For example, the papilloma virus has been linked to some cases of cervical cancer.

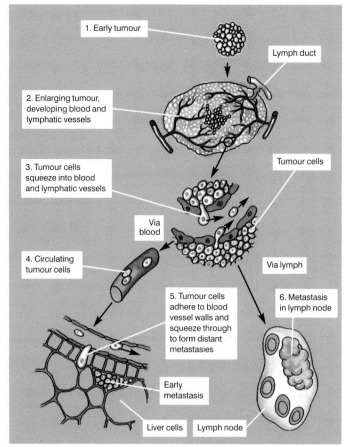

Fig 13.16 *Early primary tumour, and its development and spread as a secondary tumour*

SUMMARY TEST 13.9

Any chemical or form of energy which causes abnormal cell division through altered genetic activity is called a **(1)** and the group of abnormal cells produced is called a **(2)**. These groups of abnormal cells may grow rapidly and be unspecialised and life-threatening, in which case they are said to be **(3)**. Cells may break away and spread to other parts of the body, a process called **(4)**. A susceptibility to some forms of cancer is inherited and involves mutations to either those genes which increase cell division, called **(5)**, or those which inhibit cell division, called **(6)**. Viruses may cause cancer – the papilloma virus for example is thought to cause certain forms of **(7)** cancer. Forms of short-wavelength radiation such as **(8)** or **(9)**, may also cause cancer, while certain environmental factors also contribute, such as **(10)** fibres implicated in causing lung cancer or **(11)** causing skin cancer.

AQA.A (Human)

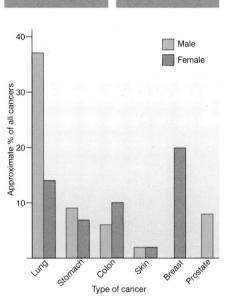

Fig 13.17 *Some different cancers as causes of death in males and females (Mortality Statistics for England and Wales 1985, OPCS)*

As a cause of death, cancer is today second only to cardiovascular disease (unit 13.2). The most common cancers of the developed world are lung, breast, skin, gut and prostate. As can be seen in figure 13.17, the biggest killer of men is lung cancer, while for women it is breast cancer. Some of the factors influencing the development of cancers have been discussed in section 13.9.6. The proportion of cancer deaths as a result of some environmental factors is shown in table 13.7.

13.10.1 Skin cancer and UV light

The sun gives off ultra-violet (UV) radiation. The types are shown in table 13.8 opposite. The effects of UVB include:

- sunburn – reddening due to dilation of blood vessels and damage to cell membranes
- damage to proteins, including those in cell membranes and enzymes
- damage to RNA and DNA, resulting in mutations which may lead to skin cancer.

Body hair offers good protection to UVB, as does the dark skin pigment, **melanin**. Skin cancer is 200 times more frequent in parts of Australia than in India. The long exposure of pale (caucasian) skin to sunlight is the main reason for the high incidence of the disease in Australia. Individuals with dark skin are nevertheless at risk from skin cancer, especially on the paler soles of the feet and palms of the hands. There is some evidence that sunburn or sun exposure in childhood increases the risk of developing skin cancer in later life. Similarly, short periods of intense exposure to sunlight, for example a 2-week beach holiday, increase the risk. The Imperial Cancer Research Fund has recently warned that the UVA emitted by sunbeds may also lead to long-term damage.

13.10.2 Types of skin cancer

There are two main types of skin cancer: **non-melanoma** and **melanoma**.

- **Non-melanoma skin cancer** is a slow-growing malignant tumour (section 13.9.4) within the cells of the skin. It rarely spreads to other tissues. It is the most common cancer in the UK, with approximately 44 000 new cases recorded each year. It is also the most successfully treated cancer. Some of the risk factors for non-melanoma skin cancer include:
  - exposure to ultra-violet light
  - age – it mainly occurs in people older than 50
  - skin colour – more likely to develop in people with fair skin
  - sex – more common in men than women
  - weakened immune system – Kaposi's sarcoma can develop in people with AIDS (section 13.7.1).
- **Melanoma skin cancer** is a malignant tumour that usually starts in moles or in areas of normal-looking skin. If it is not treated, the cancer cells break away from the rapidly growing tumour and spread throughout the blood or lymphatic system, to form secondary cancers (section 13.9.5). Melanoma is the 12th most common cancer in men, and the 11th most common in women in the UK. In 1995, there were 5386 new cases reported. The risk factors are the same as for non-melanoma skin cancer, but in addition include:
  - having a large number of moles (50–100)
  - having other members of the family diagnosed with melanoma.

Table 13.7 *Proportion of cancer deaths attributable to some different environmental factors*

Environmental factors	Approximate %
Tobacco smoke	30.0
Diet	30.0
Occupation	4.0
Atmospheric pollution	1.0
Ultra-violet light	0.5

13.10.3 Lung cancer and smoking

Lung cancer is the second most common cancer diagnosed in the UK (after non-melanoma skin cancer). More men than women are diagnosed with lung cancer, although the number of women with the disease is increasing. In 1995, the total annual number diagnosed was about 40 000. It has been estimated that smoking causes more than 80% of lung cancers; some of the links are discussed in sections 13.3.4 and 13.3.5. Smoking is not just a danger to the smoker. Research in 1997 suggested that a non-smoker who lives with a smoker has a 24% greater risk of developing lung cancer than a person who does not live with a smoker. Other risk factors associated with lung cancer include:

- **radon gas** – a naturally occurring radioactive gas with levels that vary in different parts of the country
- **exposure to certain chemicals** – asbestos, arsenic and nickel chromate for example
- **diet** – low levels of fruit and vegetables
- **past history of lung cancer**.

13.10.4 Ethical and moral issues

The link between lung cancer and smoking was established more than 40 years ago, but there are widely differing responses towards restrictions on smoking. In most Scandinavian countries, smoking-related illnesses are declining, because

- it is socially unacceptable to smoke in public places
- cigarettes are very expensive.

However, in Britain,

- cigarettes are relatively cheaper than they were 25 years ago
- society has been reluctant to accept restrictions.

Arguments have become highly politicised, and focus around:

- health education
- costs of treatment
- loss of tax revenue
- loss of productivity – cancer exceeds all other diseases in terms of years of working life lost
- loss of individual freedom.

With the exception of loss of tax revenue, these arguments are relevant to both smoking (lung cancer) and sunbathing (skin cancer).

Table 13.8 *Types of ultra-violet radiation emitted by the sun*

Type	Wavelength / nm	Notes
UVC	200–280	Absorbed by ozone and other gases in the atmosphere
UVB	280–320	80% absorbed in atmosphere. Harmful to organisms
UVA	320–400	70% reaches the Earth's surface. Formerly thought to be safe, but increasing evidence of skin damage leading to skin cancer

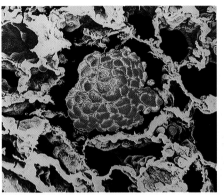

False-colour SEM of bronchial carcinoma (lung tumour) filling an alveolus

SUMMARY TEST 13.10

More men in the developed world die of (**1**) cancer than any other form, whereas for women (**2**) cancer is the biggest cause of cancer deaths. Smoking has been linked to around (**3**) % of lung cancer deaths, although other factors, such as the naturally occurring gas, (**4**), and exposure to certain chemicals such as (**5**) and (**6**), are also implicated. In both sexes, skin cancer is on the increase. The two main forms of skin cancer are (**7**) and (**8**) and both are linked to exposure to sunlight, which comprises three forms of (**9**) light – types A, B and C. Of these, type B causes the most harm, despite (**10**) % being absorbed by the Earth's atmosphere and much of the rest being absorbed by (**11**) in the skin.

Examination Questions

1 *Escherichia coli* bacteria live in the intestines of humans. The graph shows a growth curve for a population of these bacteria. They were grown in a nutrient broth at a temperature of 20°C.

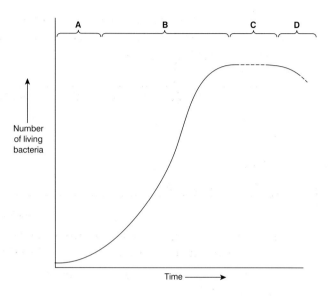

a Name phases **A** and **B** in the growth curve. *(1 mark)*

b Sketch on the graph the curve you would expect if the broth had been kept at a temperature of 35°C. *(2 marks)*

c Give **two** reasons why the numbers of living bacteria decrease in phase **D**. *(2 marks)*

(Total 5 marks)

AQA June 2001, HB (A) BYA3, No.1

2 Coronary heart disease is a major cause of death in the western world.

a The diagram shows an external view of a human heart with a blood clot in one of the main coronary arteries.

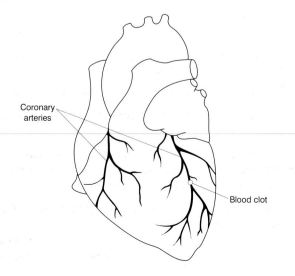

(i) Shade, on the diagram, the area of heart muscle which is likely to receive a reduced supply of blood because of the blood clot. *(1 mark)*

(ii) Explain why a blood clot in a coronary artery is likely to result in a heart attack. *(3 marks)*

b Three important risk factors associated with coronary heart disease are cigarette smoking, high blood pressure and a high plasma cholesterol level. Explain how each of the three factors increases the risk of heart disease. *(6 marks)*

c The graph gives information about the effects of cigarette smoking, plasma cholesterol concentrations and high blood pressure on the incidence of heart disease in American men.

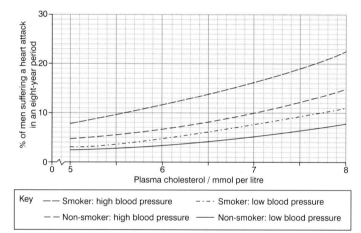

(i) A non-smoker with low blood pressure has a plasma cholesterol concentration of 5 mmol per litre. Over a period of time this concentration increases to 8 mmol per litre. By how many times has his risk of heart disease increased? Show your working. *(2 marks)*

(ii) Two non-smoking men with low blood pressure both have plasma cholesterol concentrations of 5 mmol per litre. One of them starts to smoke and the plasma cholesterol concentration of the other increases to 7 mmol per litre. Which man is now at the greater risk from heart disease? Explain your answer. *(3 marks)*

(Total 15 marks)

AQA June 2001, HB (A) BYA3, No.8

3 Malaria is a disease caused by the parasite *Plasmodium*.

a Explain why *Plasmodium* is described as a parasite. *(2 marks)*

b Most of the life cycle of *Plasmodium* is spent inside a red blood cell of its main host. Suggest how each of the following is an adaptation to its way of life.

(i) *Plasmodium* has no locomotory structures. *(1 mark)*

(ii) *Plasmodium* has no mechanism for regulating its water content. *(1 mark)*

c *Plasmodium* survives for long periods in the blood of its main host. Explain how. *(1 mark)*

(Total 5 marks)

AQA Jan 2001, HB (A) BYA3, No.2

4 The table shows the results of some blood tests carried out on a patient admitted to hospital suffering from a suspected myocardial infarction (heart attack).

Substance	Concentration in patient's blood / arbitrary units	Range of concentration in blood of healthy individuals / arbitrary units
Urea	5.7	2.5 – 6.7
Cholesterol	8.2	3.6 – 6.7
Lactate dehydrogenase enzyme	2263	300 – 600
Potassium	4.3	3.4 – 5.2

a A myocardial infarction results in damage to the muscle of the heart.

(i) Explain how a blood clot may cause damage to the muscle of the heart. *(2 marks)*

(ii) Lactate dehydrogenase is an enzyme found inside healthy heart muscle cells. Suggest why the concentration of this enzyme in the blood can be used to confirm that this patient had suffered a myocardial infarction. *(2 marks)*

b Use the table to explain what is meant by a *risk factor*. *(2 marks)*

(Total 6 marks)

AQA (specimen) 1999, HB (A) BYA3, No.7

5 a (i) Name the organism that causes tuberculosis (TB). *(1 mark)*

(ii) Describe how tuberculosis is spread from infected to uninfected people. *(3 marks)*

The figure shows the number of cases of tuberculosis recorded by health authorities in England and Wales between 1913 and 1998.

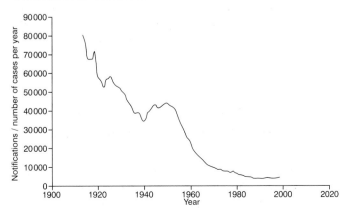

b With reference to the figure, describe the changes in the number of recorded cases of tuberculosis between 1913 and 1998 in England and Wales. *(4 marks)*

c Explain how **social** and **economic** factors have been important in reducing the number of cases of tuberculosis in developed countries such as England and Wales. *(4 marks)*

(Total 12 marks)

OCR 2802 Jan 2001, B(HHD), No.7

6 a Complete the table below to show which of the three statements about disease transmission apply to cholera, tuberculosis (TB) and HIV/AIDS. Put a tick (✔) to show if the statement applies.

Statement	Cholera	Tuberculosis (TB)	HIV/AIDS
causative organism is a bacterium			
transmission is via drinking water			
sexually transmitted			

(3 marks)

b Suggest **two** reasons why antibiotics are **not** suitable for treating all infectious diseases. *(2 marks)*

Tuberculosis is a disease considered to be endemic throughout the world.

c State the meaning of the term *endemic*. *(1 mark)*

The table below shows the estimated number of deaths in 1998 for five diseases in developing countries in South-East Asia and developed countries in Europe.

Disease	Number of deaths per 100 000	
	Developing countries in South-East Asia	Developed countries in Europe
tuberculosis	51.90	1.78
HIV/AIDS	23.66	3.05
diarrhoeal diseases, e.g. cholera	36.39	1.01
lung cancer	17.10	53.00
chronic obstructive pulmonary diseases e.g. chronic bronchitis and emphysema	11.93	35.67

d Explain the advantage of expressing numbers of deaths as *per 100 000 of the population*. *(2 marks)*

e With reference to the table, describe and explain the differences between standards of health in developing countries and developed countries.

(In this question, one mark is awarded for the quality of written communication). *(8 marks)*

(Total 16 marks)

OCR 2802 June 2001, B(HHD), No.3

247

Defence against disease

Diagnosis of disease

There is no doubt that prevention is better than cure, but sometimes prevention is not enough, and cure is the only answer. The earlier a disease is diagnosed, the greater the chances of cure. There are now a number of techniques for providing early warnings of disease. Especially useful is the screening of apparently healthy people, to detect disease even before there are any symptoms. As there are not the economic resources, let alone the know-how, to screen everyone for every disease, the process is targeted on the most vulnerable group for any particular disease, e.g. amniocentesis for pregnant women over 35 years of age.

14.1.1 DNA probes and the detection of disease

It requires the genetic material from only a few cells to enable detection of certain genetic diseases. In pregnant women, for example, a sample of amniotic fluid will contain enough foetal cells to reveal Down's syndrome. This is relatively easy, because the defect is due to an additional chromosome 21, which is easy to observe. Less obvious are the 4300 other genetic diseases caused by a single gene mutation. To diagnose those genetic disorders, we need to identify the particular gene amongst the total DNA of the relevant chromosome. This is achieved using **DNA probes**. A DNA probe is a piece of DNA with a sequence of bases that complements that of the possibly defective gene. The preparation of the material and the use of the DNA probe is the same as that for **genetic fingerprinting** (unit 12.8), which is summarised below:

- DNA is extracted from the cell sample
- the DNA is cut into fragments using **restriction endonucleases**
- the fragments are separated using gel **electrophoresis**
- the double strands of DNA are separated into single strands
- the single strands are transferred onto a nylon membrane by **Southern blotting**
- radioactive DNA probes are attached to the single fragments
- an X-ray film is put over the nylon membrane
- the film is 'fogged' (exposed) where there are radioactive DNA probes
- a series of bands is revealed (DNA fingerprint)
- the pattern of banding produced is compared with that of an individual who is normal for the defect being screened for
- specific variation of the banding in particular regions of the DNA fingerprint indicates the presence of a defective gene.

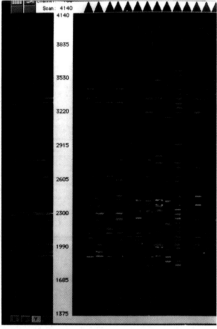

DNA fingerprint

14.1.2 Enzymes and diagnosis

Genetic diseases, and a number of non-genetic ones, cause changes in the concentration and distribution of enzymes. One such disease is **pancreatitis** – inflammation of the pancreas. The symptoms include abdominal pain, nausea and vomiting. The pancreas is the main organ producing digestive enzymes (section 6.8.2). These enzymes break down fats, carbohydrates and proteins and are produced by epithelial cells in the pancreas. The enzymes are in an inactive form at first, to prevent them from digesting the cells of the pancreas itself. They are then activated when they reach the intestine. However, in patients with pancreatitis these enzymes can become active in the pancreas. The two protein-digesting enzymes then break down the pancreatic cells, leading to the destruction of the whole pancreas. The causes of pancreatitis are not clear, although the blockage of the bile duct (part of which carries pancreatic secretions into the intestine) by gallstones is one known cause. Other contributing factors include:

- excessive alcohol consumption
- particular foods or diet
- volatile chemicals and solvents (e.g. those in paint or dry-cleaning fluids)
- certain rare inherited genes.

Whatever the cause, something appears to go wrong with the secretion of the pancreatic enzymes. This results in the amount of pancreatic enzymes

- decreasing in the intestines
- increasing in the blood.

It seems therefore that, instead of the secreted enzymes going down the pancreatic duct, they enter the blood stream. Diagnosis of pancreatitis is therefore made by testing for the presence of pancreatic enzymes in the blood. This is important, to distinguish the disease from other ailments such as food poisoning, which have the same symptoms, but not the presence of pancreatic enzymes in the blood.

The success of any treatment can be measured by a return to normal of the occurrence of pancreatic enzymes, i.e.:

- present in the faeces
- absent from the blood.

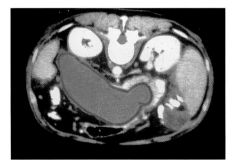

False-colour computed tomography scan through the upper abdomen showing inflammation of the pancreas (blue)

14.1.3 Enzymes as analytical agents

Enzymes are used extensively as biosensors – agents which allow a rapid and accurate measurement of the amount of a particular chemical. Enzymes are excellent for this purpose because they are:

- **highly specific** in the reactions they catalyse, and therefore may be used to identify a particular molecule precisely, amongst a mixture of many others
- **highly sensitive**, enabling them to detect molecules at very low concentrations.

One application of enzymes in this way is in the diagnosis of diabetes and to allow diabetic people to adjust their insulin dose at each injection, according to their blood sugar level at the time. The enzyme used is **glucose oxidase**, which catalyses the conversion of glucose to hydrogen peroxide, as follows:

$$\text{glucose} + \text{oxygen} \quad \xrightarrow{\text{glucose oxidase}} \quad \text{gluconic acid} + \text{hydrogen peroxide}$$

The hydrogen peroxide produced is then reacted with a colourless substance in the presence of the enzyme, peroxidase, causing the colourless substance to become coloured:

$$\text{colourless hydrogen donor} + \text{hydrogen peroxide} \quad \xrightarrow{\text{peroxidase}} \quad \text{water} + \text{coloured compound}$$

The exact colouration depends on the concentration of hydrogen peroxide present which, in turn, depends upon the concentration of glucose originally present. The test is carried out using 'Clinistix'. Clinistix are small sticks, with a small pad at one end containing glucose oxidase, peroxidase and the colourless hydrogen donor. When a small sample of blood or urine is added, the amount of glucose present in the sample causes the development of a particular colour. This colour is then compared against a chart that shows the amount of glucose corresponding to that colour (section 2.4.4).

SUMMARY TEST 14.1

Diagnosing diseases such as Down's Syndrome can be achieved by sampling the **(1)** in pregnancy. Other genetic diseases such as cystic fibrosis can be detected using **(2)**, which entails cutting sample DNA into fragments using **(3)**, separating them by **(4)** and transferring them to a nylon membrane by the technique of **(5)**. By use of X-ray film, a series of bands called the **(6)** is revealed that indicates if a defective gene is present. Enzymes can be used to detect diseases like **(7)** by looking for the presence of certain digestive enzymes in the **(8)** where they do not normally occur, or their absence in the **(9)** where they ordinarily are present. Enzymes may also be used as **(10)** to accurately measure levels of a particular chemical. One example is the use of the enzyme **(11)** in measuring the levels of glucose in blood in individuals suffering from **(12)**.

14.2

Defence against infection – blood clotting and phagocytosis

AQA.A (Human)

OCR

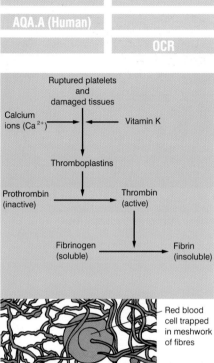

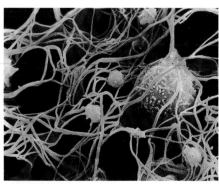

Fig 14.1 Summary of the clotting process

EM of the clotting process

There are two general mechanisms of defence against disease. First, any injury that causes the rupture of a blood vessel is likely to provide a point of entry for infection, and to lead to its being distributed around the body by the blood. Plugging any such opening by clotting the blood is therefore a first line of defence. Second, it is important to trap and destroy any foreign cells that do enter, before they can multiply. This is achieved by **phagocytosis**. Both forms of defence are non-specific, i.e. they occur in response to all potential types of infection.

14.2.1 Blood clotting

It is important to seal any rupture to a blood vessel quickly, for two reasons:
- to prevent excessive blood loss, which would make the blood pressure dangerously low
- to prevent the entry of foreign material.

The body therefore needs a mechanism that ensures the blood never clots under normal circumstances, but does so rapidly when there is damage to the blood vessel. The mechanism that has evolved has a number of stages, rather than a single one. This helps to ensure that clotting does not take place, except under very special circumstances. This avoids the possible lethal consequences of a clot occurring in the circulating blood. It is like making an important decision in a business – it has to go through many stages and involve many people, to ensure that the checks needed to avoid a disastrous error are made. Cell fragments, known as **platelets** or **thrombocytes**, are important in starting blood clotting, which occurs as follows:
- When a blood vessel is ruptured, **collagen** fibres in its wall are exposed.
- Platelets quickly attach themselves to these fibres, and release clotting factors called **thromboplastins**.
- A chain reaction takes place in which one factor is converted from an inactive form into an active one.
- This factor then activates the next one in the chain, in a cascade effect that starts as a tiny reaction but increases in size, to produce a major effect. The final clotting factor in the chain converts (in the presence of calcium ions and vitamin K) an inactive plasma protein called **prothrombin** into its active form, called **thrombin**.
- Thrombin is a plasma enzyme which converts another plasma protein, called **fibrinogen**, from its soluble form into its insoluble form, called **fibrin**.
- Fibrin forms a meshwork of threads, which trap red blood cells and so prevent them from escaping (Fig 14.1).
- The trapped cells dry and harden to form a protective scab, which both stops further blood loss and prevents the entry of foreign material.

14.2.2 Phagocytosis

Phagocytosis is the process by which large particles are taken up by cells, in the form of vesicles formed from the cell surface membrane (plasma membrane). In the blood, two types of white blood cells carry out phagocytosis – **monocytes** and **neutrophils** (section 8.1.3). These cells are known as **phagocytes** and are produced in the marrow of the limb bones. The process is illustrated in figure 14.2 and is summarised as follows:
- **Antibodies** attach themselves to **antigens** on the surface of the bacterium.
- Proteins, found in the plasma, attach themselves to the antibodies.

- As a result of a series of reactions, the surface of the bacterium becomes coated with proteins called **opsonins**. This process is called **opsonisation**.
- Complement proteins and any chemical products of the bacterium act as attractants, causing neutrophils to move towards the bacterium.
- Neutrophils attach themselves to the opsonins on the surface of the bacterium.
- Neutrophils engulf the bacterium to form a vesicle, known as a **phagosome**.
- Lysosomes (section 4.6.4) move towards the vesicle and fuse with it.
- The enzymes within the lysosomes break down the bacterium into smaller, soluble material.
- The soluble products from the breakdown of the bacterium are absorbed into the cytoplasm of neutrophil.

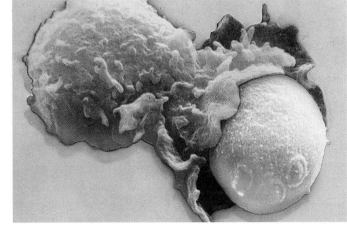

False-colour SEM of phagocytosis

Phagocytosis causes **inflammation** at the site of infection. This swollen area contains dead bacteria and phagocytes, which are known as **pus**. Inflammation is the result of the release of **histamine**, which causes dilation of blood vessels in order to speed up the delivery of antibodies and white blood cells to the site of infection.

1. The neutrophil is attracted to the bacterium by chemoattractants. It moves towards the bacterium along a concentration gradient

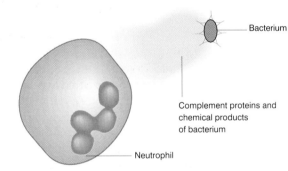

Bacterium

Complement proteins and chemical products of bacterium

Neutrophil

2. The neutrophil binds to the bacterium

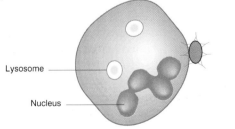

Lysosome

Nucleus

3. Lysosomes within the neutrophil migrate towards the phagosome formed by pseudopodia engulfing the bacterium

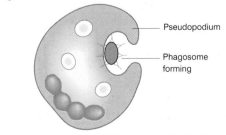

Pseudopodium

Phagosome forming

4. The lysosomes release their lytic enzymes into the phagosome, where they break down the bacterium

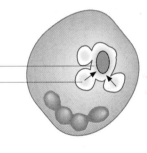

Phagosome

Lysosomes release lytic enzymes into phagosome

5. The breakdown products of the bacterium are absorbed by the neutrophil

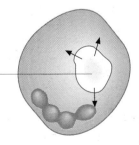

Breakdown debris of bacterium

Fig 14.2 *Summary of phagocytosis of a bacterium by a neutrophil*

Immunity is the ability of an organism to resist infection by protecting against disease-causing microorganisms that invade its body. It involves the recognition of foreign material, and production of chemicals which help destroy it.

14.3.1 Antigens

An **antigen** is any organism or substance that is recognised as non-self (foreign) by the immune system. Antigens are usually proteins that make up the cell surface membranes of invading cells such as microorganisms, or diseased cells such as cancer cells. The presence of an antigen triggers the production of an antibody as part of the body's defence system.

14.3.2 Antibodies

Antibodies are also known as **immunoglobulins**. They are proteins synthesised by cells in the blood called **B-lymphocytes** (section 8.1.3; unit 14.4). When the body is invaded by non-self (foreign) material, a B-lymphocyte produces antibodies, which react with antigens on the surface of the foreign material by binding to them precisely, in the way that a key fits a lock. They are therefore very specific, each antigen having its own separate antibody. Antibodies are made up of four polypeptide chains. One pair of chains is long and called **heavy chains**; the other pair is shorter and known as **light chains**. The chains are held together by **disulphide bridges** (section 1.6.3) but, to help the antibody fit around the antigens, it can change its shape by moving as if it had a hinge at the fork of the Y-shape. Antibodies have two sites called **binding sites**, which fit very precisely onto the antigen (Fig 14.3). The binding sites are different on different antibodies, and are therefore called the **variable region**. The rest of the antibody is the same in all antibodies, and is known as the constant region. There are a number of different antibodies, all of which function in one or more of the following ways:

- **Agglutination** – some antibodies have many binding sites, and so can attach to a number of different antigens at the same time. In this way, antibodies cause foreign material to clump together, which makes it more vulnerable to attack from phagocytes.
- **Precipitation** – some antigens are soluble, and so antibodies precipitate them out, so that they can be destroyed by phagocytes (section 14.2.2).
- **Neutralisation** – foreign cells often produce toxins, which are chemicals that cause many of the symptoms of a disease. Antibodies can bind to these toxic molecules, and so neutralise their harmful effects.
- **Lysis** – having attached themselves to antigens on foreign cells, antibodies then attract other compounds, which bind to them. These include enzymes that help to break down the foreign cells.

These methods of defending the body from infection are illustrated in figure 14.4 opposite.

14.3.3 The concept of self and non-self

To be able to defend the body from invasion by foreign material, B-lymphocytes must be able to distinguish the body's own cells and chemicals (self) from those that are foreign (non-self). If they could not, B-lymphocytes would produce antibodies that would destroy the organism's own tissues. How, then, do B-lymphocytes recognise their own cells?

- Each antibody is produced by a different lymphocyte.

The variable region differs with each antibody. It has a shape which exactly fits an antigen. Each antibody therefore can bind to two antigens.

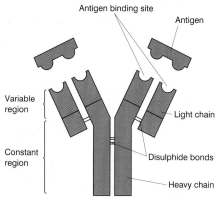

Fig 14.3 Structure of an antibody

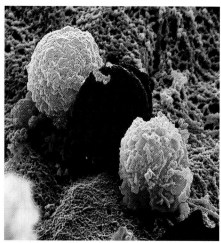

SEM of lymphocytes (orange)

Table 14.1 Antigens and antibodies of the ABO blood grouping system

Blood group	Antigen present	Antibodies present
A	A	anti-b
B	B	anti-a
AB	A and B	None
O	None	anti-a and anti-b

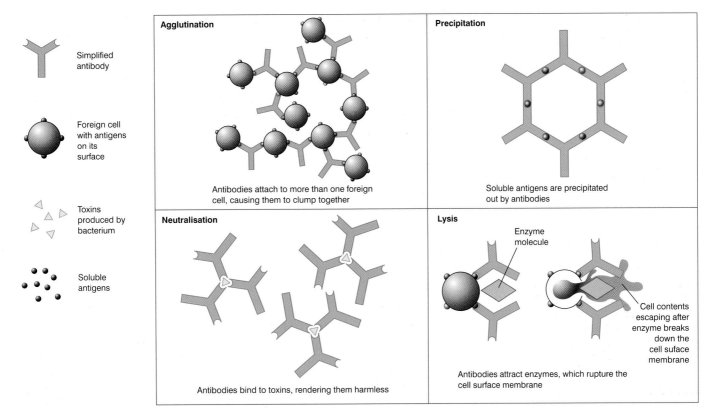

Fig 14.4 *Ways in which antibodies defend the body*

- There are more than 10 million different lymphocytes, each capable of recognising a different chemical shape.
- In the foetus, these lymphocytes are constantly colliding with other cells.
- Infection in the foetus is rare, as it is protected from the outside world by the mother and, in particular, the placenta (section 5.13.1).
- Lymphocytes will therefore collide almost exclusively with the body's own material (self).
- Some of the lymphocytes will have receptors which exactly fit those of the body's own cells.
- These lymphocytes either die, or are suppressed.
- The only remaining lymphocytes are those that fit foreign material (non-self), and therefore the only antibodies produced are those that attack foreign material.

14.3.4 Blood grouping

Blood groups are an example of an antigen–antibody reaction. The cell surface membrane of red blood cells possesses polysaccharides which act as antigens. These antigens may induce the production of antibodies if introduced into another individual. The best known type of blood grouping is the ABO system, in which red blood cells possess two possible antigens, A and B. If only antigen A is present, the blood group is A, whereas antigen B only gives blood group B. Both antigens make up group AB, and if neither is present, the group is O. For each antigen, there is a corresponding antibody, e.g. **antigen A** has a corresponding antibody – **anti-a**. If blood containing one antigen is mixed with its corresponding antibodies, these antibodies cause the

blood cells to clump together (**agglutinate**) (section 14.3.2). Ultimately, the red blood cells break down. For this reason, an individual does not produce antibodies corresponding to the antigens present on its red cells. It does, however, produce any other possible antibodies, as a matter of course (table 14.1). This reaction forms the basis of blood matching, which is carried out before a transfusion.

14.3.5 Allergies and the immune system

An allergy, such as asthma or hay-fever, is the result of an excessive reaction of the body's immune system to certain substances. Such substances, called **allergens**, include pollen, animal hairs, dust and dust-mite faeces. When an allergen is detected for the first time, e.g. when breathed into the lungs, it induces lymphocytes to produce antibodies. These antibodies bind to the surface of white blood cells, called **mast cells**, in the lungs. The person is now said to be **sensitised**. If they meet the allergen on a second occasion, the mast cells cause the release of a chemical called **histamine**. Histamine causes dilation of blood vessels, and swelling. In the case of hay-fever, it results in runny nose and eyes, and sore throat, as these are the areas affected. With asthma, the consequences are more severe, as the lungs are the affected region. Here, the air-ways become narrowed and there is difficulty in breathing. This disorder now affects 10% of the population across the world, and accounts for around 2000 deaths a year in the UK alone. Asthma can be triggered by many factors, including specific substances, environmental pollution, exercise, cold air, infection, anxiety and stress.

The immune response

Immune responses such as phagocytosis (section 14.2.2) are non-specific, and occur whatever the infection. The body also has specific responses, which fight individual forms of infection. These are slower in action at first, but they can provide long-term immunity. This specific immune response depends on a type of white blood cell called a **lymphocyte**. There are two types of lymphocyte, each with its own immune response:

- B-lymphocytes (B-cells) – humoral immunity (involves antibodies which are present in body fluids or 'humour')
- T-lymphocytes (T-cells) – cell-mediated immunity (involves cells).

Both types of lymphocyte are formed from **stem cells** found in the bone marrow. Their names, however, indicate where they develop and mature:

- B-lymphocytes mature in the **B**one marrow
- T-lymphocytes mature in the **T**hymus gland.

The maturation process takes place in the foetus, and in the case of B-lymphocytes it results in more than 10 million different types, each capable of responding to a different antigen.

14.4.1 B-lymphocytes and humoral immunity

B-lymphocytes produce antibodies (section 14.3.2). They do so in response to a specific antigen (section 14.3.1). When a B-lymphocyte recognises its specific antigen, it divides rapidly to form a clone of cells which produce the relevant antibody. As most foreign cells possess many antigens, many B-lymphocytes are stimulated to form clones in this way. This is known as **polyclonal activation**. The B-lymphocytes change into two types of cell:

- **Plasma cells**, which secrete antibodies. Plasma cells survive for only a few days, but each can make around 2000 antibodies every second during its brief life-span. They are responsible for the immediate defence against infection. This is known as the **primary immune** response.
- **Memory cells** live considerably longer, often for decades. Memory cells quickly develop into antibody-producing plasma cells when they encounter antigens for a second time. In this way, they provide long-term immunity against the original infection. This is known as the **secondary immune response**; it is both more rapid and of greater intensity than the primary response (Fig 14.5).

14.4.2 T-lymphocytes and cell-mediated immunity

While B-lymphocytes attack foreign material which is **outside** the body cells, e.g. in the blood plasma, T-lymphocytes attack the foreign material that is **within** body cells, e.g. cells infected by a virus. This is why they are said to provide **cell-mediated immunity**. T-lymphocytes respond to specific antigens, but only when they are attached to the surface of one of the body cells. For example, a cell infected by a virus will have viral molecules on its cell surface membrane. Once attached to their antigen on the surface of a cell, T-lymphocytes divide by **mitosis** to form a clone. The cells of this clone then change into three different cell types, each with a specific function:

- **T-helper cells** have a number of functions:
  - they activate phagocytic cells to engulf foreign cells and body cells infected with foreign antigens
  - they assist T-killer cells in the destruction of foreign cells
  - they activate B-lymphocytes to divide and differentiate into plasma cells.
  T-helper cells clearly have a crucial role in immunity, because the human

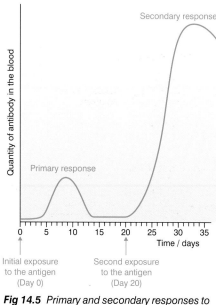

Fig 14.5 *Primary and secondary responses to an antigen*

immunodeficiency virus (HIV) prevents them from operating, with the consequent loss of an immune response.

- **T-killer cells (cytotoxic cells)** kill body cells which are infected by foreign cells. They make holes in the cell surface membrane, causing the cell to burst. As viruses need living cells to reproduce, this sacrifice of body cells effectively prevents viruses multiplying.
- **T-suppressor cells** – suppress the activities of lymphocytes by turning off their action once the foreign cells have been eliminated.

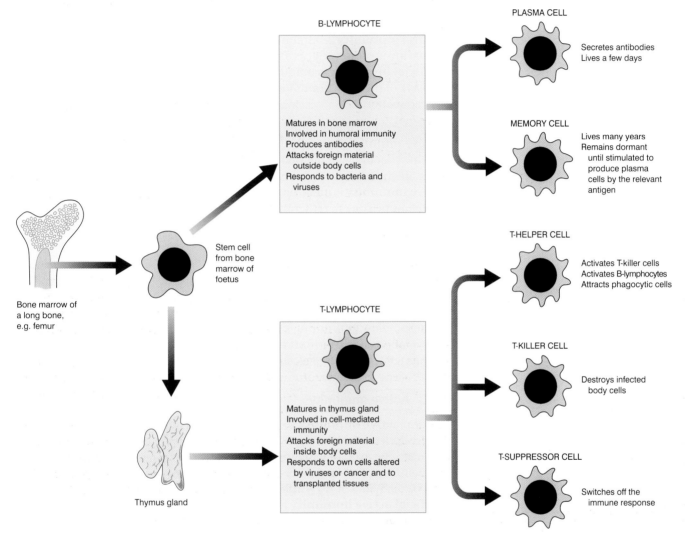

Fig 14.6 *The origin and roles of lymphocytes in immunity*

SUMMARY TEST 14.4

If an individual is infected by a foreign body such as a bacterium, cells called B-lymphocytes produce antibodies in response to specific **(1)** on the surface of the bacterium and at the same time divide themselves to form a **(2)** of identical cells. As many different B-lymphocytes produce groups of identical cells, the process is known as **(3)**. Each set of B-lymphocytes divides into two types. Those that produce the antibodies are called **(4)** and are the immediate defence against infection, known as the **(5)**. The second type called **(6)** remain in the blood system ready to divide into antibody-producing cells should the same infection occur again. This is known as the **(7)**.

Immunisation

Immunity is the ability of an organism to resist infection (unit 14.3). This immunity may be naturally acquired or artificially induced. The process of artificially inducing immunity is known as **immunisation**.

14.5.1 Types of immunity

- **Natural immunity** is immunity which is either inherited, or acquired as part of normal life processes, e.g. as a result of having had a disease.
- **Artificial immunity** is immunity acquired as a result of the deliberate exposure of the body to antibodies or antigens in non-natural circumstances, e.g. vaccination.

Both natural and artificial immunity may be passively or actively acquired.
- **Passive immunity** is immunity acquired from the introduction of antibodies from another individual, rather than from one's own immune system. It is generally short lived.
- **Active immunity** is immunity resulting from the activities of an individual's own immune system, rather than from an outside source. It is generally long lasting.

14.5.2 Passive immunity

- **Natural passive immunity** is gained when an individual receives antibodies from their mother via
  - the placenta as a foetus
  - the mother's milk during suckling.
- **Artificial passive immunity** occurs when antibodies from another individual are injected. This takes place in the treatment of diseases such as tetanus and diphtheria.

14.5.3 Active immunity

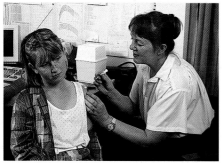

Vaccination in progress

- **Natural active immunity** results from an individual becoming infected with a disease under normal circumstances. The body produces its own antibodies, and may continue to do so for many years. It is for this reason that many people suffer diseases such as measles only once in a lifetime. The immunity results from the activities of **B-lymphocyte** memory cells (section 14.4.1).
- **Artificial active immunity** forms the basis of immunisation. It involves inducing an immune response in an individual, without them suffering the symptoms of the disease. This is achieved by introducing the appropriate disease antigens into the body, either by injection or by mouth. The process is called **vaccination**, and the material introduced is called **vaccine**. There are different forms of vaccine:
  - **Living attenuated microorganisms** are living microorganisms which have been treated, e.g. by heat, so that they do not cause symptoms, but still multiply. Although harmless, they stimulate the body's immune system. Measles, TB and poliomyelitis can be vaccinated against in this way.
  - **Dead microorganisms** have been killed by some means. Again, they are harmless, but induce immunity. Typhoid, cholera and whooping cough can be controlled by this means.
  - **Genetically engineered microorganisms** can be produced in which the genes for antigen production are transferred from a harmful organism to a harmless one. These are then grown in fermenters (unit 12.5), and the

extracted antigen is separated and purified before injection. Hepatitis B vaccine is of this type.

14.5.4 Control of disease by vaccination

The programme of vaccination against various diseases has had considerable success in controlling them – and, in cases such as smallpox, eliminating them altogether (section 14.5.5). To be effective, however, a programme of vaccination depends upon:
- a suitable vaccine being economically available in sufficient quantity
- few, if any, side effects from vaccination, to avoid discouraging the population from seeking vaccination
- the mechanisms to produce, store, transport and administer the vaccine properly at the appropriate time
- the ability to vaccinate the vast majority (all, if possible) of the vulnerable population.

Even where these criteria are met, it can still prove extremely difficult to eradicate a disease by vaccination, for the following reasons:
- Vaccination fails to induce immunity in certain individuals, e.g. ones with defective immune systems.
- Individuals may develop the disease immediately after vaccination, but before their immunity levels are high enough to fight it.
- The disease-causing agent (pathogen) may mutate frequently, so that its antigens change suddenly **(antigenic shift)**, rather than gradually **(antigenic drift)**. This means that vaccines suddenly become ineffective, as happens with influenza vaccines.
- There may be so many varieties of a particular pathogen that it is all but impossible to develop a vaccine which is effective against them all. The common cold has over 100 varieties, for example.

- There are, as yet, no effective vaccines against pathogens that are **eukaryotes** (unit 4.4). Diseases such as malaria and sleeping sickness cannot therefore be tackled in this way.
- Certain pathogens 'hide away' from the body's immune system, either by concealing themselves inside cells, or by living in places out of reach, such as within the intestines, e.g. the cholera pathogen.
- Some pathogens suppress the body's immune system, so stimulating it through vaccination is ineffective, e.g. human immunodeficiency virus.

14.5.5 Smallpox – the last Jenneration!

The first ever vaccinations were carried out by Edward Jenner in 1794, against smallpox, and this very same disease was the first to be completely eradicated from the world, by the same process, in 1977. Why have we been able to remove smallpox when many other diseases such as measles, tuberculosis, malaria and cholera, are still around? There are many reasons, which include:
- a simple, safe, easily stored vaccine against smallpox
- the vaccine was easily and economically produced and simply administered
- the smallpox virus was genetically stable, and so it did not develop resistance to the vaccine
- the lethal nature of the disease (up to 30% of the victims died) encouraged people to be involved in the vaccination programme
- the symptoms of the disease were easily recognised, and so infected patients could be isolated and treated before they spread the disease further
- the virus did not remain in the body after an infection
- there was a worldwide vaccination programme, coordinated by the World Health Organisation.

Table 14.2 *Summary of different types of immunity*

	Natural	Artificial
	Inherited or acquired naturally, not deliberately	Acquired deliberately by exposure to causative agent
Passive Results from the introduction of antibodies from another organism's immune system, rather than one's own Short lived	Antibodies pass from mother • to foetus via placenta • to baby during suckling	Antibodies from a different individual or organism are injected
Active Results from the activities of an individual's own immune system Long lasting	Antibodies acquired as a result of a previous infection producing B-lymphocyte memory cells, which are reactivated on the second infection	Antigens are injected or given by mouth as a vaccine. They induce the body to produce its own antibodies to the disease. Vaccine may contain • dead pathogen • attenuated pathogen • genetically engineered antigens

14.6

Drugs in the treatment of disease

A wide range of new drugs is now available to treat the huge number of diseases and disorders that affect humans. Some of the most important groups are **beta blockers**, **antibiotics** and **monoclonal antibodies**.

14.6.1 Beta blockers

Beta blockers are drugs which interrupt the transmission of stimuli through receptors in the body that are known as beta-receptors. They modify the stimulating effects of hormones such as adrenaline and noradrenaline by occupying the beta-receptors that these hormones normally attach to. Beta blockers therefore reduce the force and speed of the heart beat and prevent dilation of the blood vessels around the brain. Consequently, they are used to treat

- angina
- hypertension (high blood pressure)
- migraine headaches.

14.6.2 Antibiotics

First discovered by Alexander Fleming in 1928, but not produced commercially until 1941, antibiotics are now used extensively to treat infections. One in every six prescriptions issued by British doctors each year is for one antibiotic or another. The word 'antibiotic' derives from Greek words meaning 'against life'. In practice, these drugs are rather narrower in their operation, and are generally substances made by microorganisms which are either

- **biocidal (bacteriocidal)** – kill other microorganisms (bacteria)
- **biostatic (bacteriostatic)** – prevent the growth and multiplication of microorganisms (bacteria).

No antibiotic is effective against all forms of infection, although two groups are recognised:

- **Broad-spectrum antibiotics** are effective against a large range of bacteria
- **Narrow-spectrum antibiotics** are effective only against a selected type of bacteria.

Antibiotics do not function against viruses, although they may sometimes be prescribed to vulnerable people with viral diseases such as influenza, to prevent secondary bacterial infections like pneumonia. In general, antibiotics work in two ways – either by preventing bacteria from making normal cell walls, or by interfering with the internal biochemistry of the bacteria. More particularly, they operate as follows:

- **Preventing cell wall synthesis**. The cell walls of bacteria, like those of plant cells, are essential to prevent the cells from bursting when water enters them by **osmosis (osmotic lysis)**. Antibiotics inhibit the synthesis and assembly of important peptide cross-linkages in bacterial cell walls, weakening them and causing them to burst, thereby killing the bacterium. As these antibiotics inhibit the correct formation of cell walls, they are effective only when bacteria are actively growing. Penicillin is an antibiotic that works in this way.
- **Interfering with nucleic acid synthesis**. Some antibiotics, such as the anthracyclines, inhibit the synthesis of DNA (unit 3.3), while others like rifampicin inhibit the formation of messenger RNA and so prevent **transcription** (unit 3.5). In these cases, the bacterium is unable to grow or multiply.

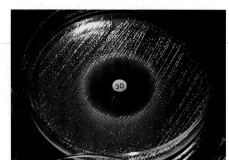

Action of antibiotic on a bacterial cell

- **Interfering with protein synthesis**. Antibiotics like streptomycin will bind to ribosomes of bacteria, but not those of mammalian cells. Other antibiotics, such as tetracycline, bind equally well to both mammalian and bacterial ribosomes, but the drug is taken up only by bacterial cells. In both cases, protein synthesis (units 3.5 and 3.6) is prevented, and the bacterial cells cannot grow and multiply.
- **Damaging cell membranes**. Amphotercin B is an antibiotic which distorts the lipid bilayer of the cell surface membrane of fungi (unit 4.8). The damaged membrane allows the cell contents to escape, killing the fungus.

Production of monoclonal antibodies

14.6.3 Monoclonal antibodies

A bacterium or other microorganism entering the body is likely to have on its surface many hundreds of different antigens, each of which will induce a different B-lymphocyte to multiply and **clone** itself (section 14.4.1). Each of these clones will produce a different antibody, collectively known as **polyclonal antibodies**. By fusing B-lymphocytes with cancer cells, it is now possible to get individual B-lymphocytes to divide and produce antibodies outside the body. These are called **monoclonal antibodies**, and have a number of useful functions in science and medicine:

- **Separation of a chemical from a mixture**. Monoclonal antibodies for the desired substance can be attached to resin beads in a column, and the mixture passed over them. The chemical (antigen) becomes attached to the antibodies. It can then be separated from the beads by washing them with a solution which causes the antibody to release the chemical.
- **Immunoassays**. These are a means of calculating the amount of a substance in a mixture. The monoclonal antibody is labelled in some way, e.g. using radioactivity or a fluorescent dye. These antibodies are then added to the mixture, where they attach only to the substance (antigen). The mixture is then washed in solutions that remove unattached antibodies, leaving only those bound to their antigen. The degree of radioactivity or fluorescence in the remaining mixture gives a measure of the amount of the substance. A similar technique, but using immobilised antibodies, is used in
  - pregnancy testing kits
  - testing for drugs in the urine of athletes
  - detecting the human immunodeficiency virus (AIDS test).
- **Cancer treatment**. Monoclonal antibodies can be made which have antigens found only on cancer cells. It is possible to attach enzymes to these monoclonal antibodies. The enzymes convert an inactive form of a cytotoxic drug (one that kills cells) into an active form. When introduced into the body, the monoclonal antibodies with the enzyme attach themselves only to the cancer cells. The inactive cytotoxic drug is then given to the patient. This drug will only be activated by cells with the enzyme attached – namely the cancer cells. The cancer cells will then be destroyed, with little if any damage to other cells. For obvious reasons, monoclonal antibodies used in this way are referred to as **magic bullets** and can be used in smaller doses, as they are targeted on specific sites.
- **Transplant surgery**. Even with close matching, a transplanted organ will normally suffer some rejection, because of the action of the **T-lymphocytes**. Monoclonal antibodies can be used to 'knock out' these specific T-lymphocytes. This method is preferable to using immunosuppressant drugs, which have a less specific effect and therefore leave the patient open to other forms of infection.

SUMMARY TEST 14.6

Antibiotics are substances made by **(1)** and they operate in two ways. Those called **(2)** prevent pathogens such as bacteria from multiplying and growing by interfering with the bacteria's ability to make either **(3)** or **(4)**. Those given the name **(5)** kill the pathogens by preventing them making **(6)**, with the result that the cell bursts due to the influx of water. This is called **(7)**. An example of an antibiotic that works in this way is **(8)**. Monoclonal antibodies are made by fusing **(9)** with cancer cells. They are useful in **(10)**, a technique for measuring the amount of a substance in a mixture, and as magic bullets in the treatment of **(11)**. Beta-blockers prevent the hormones **(12)** and **(13)** from operating normally and are useful in the treatment of **(14)** and **(15)**.

Balanced diet

Table 14.3 *Sources of some carbohydrates*

Carbohydrate	Sources
Glucose	Grapes, honey
Fructose	Honey, fruits
Sucrose	Sugar cane, sugar beet, and therefore jams, biscuits and chocolate
Starch	Potatoes, cereals, and therefore bread and pasta

In order to remain healthy, we need to eat and drink adequate amounts of nutrients:
- to provide energy
- for the growth and repair of cells
- for the functioning of all our vital organs.

We need relatively large amounts of the **macronutrients – carbohydrates, lipids** and **proteins** – and very much smaller amounts of the **micronutrients – vitamins** and **mineral ions**. We also need enough **water** to replace that lost each day, and **fibre (non-starch polysaccharide)** to help the movement of food through the digestive system.

14.7.1 The role of carbohydrates

Sugars and starch are carbohydrates whose main function is to **provide energy** – most carbohydrates yield approximately $16kJ\ g^{-1}$. Table 14.3 shows the sources of some carbohydrates. The level of carbohydrate, in the form of glucose, is kept more or less constant in the blood. Any excess is stored as glycogen, and later as fat. High levels of sucrose in the diet may lead to tooth decay.

14.7.2 The role of lipids

Lipids are required
- **as a source of energy** – lipid yields about $37kJ\ g^{-1}$
- **as components of some steroid hormones**, e.g. oestrogen (section 5.11.2).

High intakes of lipids are associated with obesity (section 14.9.5) and coronary heart disease (unit 13.4). It is therefore recommended that only one-third of our energy should be derived from lipids, and the rest from carbohydrates. Lipids are made up of fatty acids and glycerol (unit 1.4). Most of the fatty acids we require can be made within the body, but two cannot. These are **linoleic acid** and **linolenic acid**. They can only be made by plants, and have to be included as part of the diet in humans. They are known as **essential fatty acids**.

14.7.3 The role of proteins

The structure of proteins is discussed in unit 1.6. Although proteins may be broken down to release energy ($17kJ\ g^{-1}$), their main function is as a source of amino acids to synthesise new proteins for growth and repair of cells. Some amino acids **(non-essential amino acids)** can be made by human cells, such as those of muscle, liver, kidney and brain tissue. However, eight amino acids **(essential amino acids)** cannot be made in the body, and must be supplied in the diet. Table 14.4 lists the essential amino acids required by adults and children. Although histidine is made by adult humans, it has to be provided in the diet of young children, to ensure they gain in body weight. Tyrosine is an amino acid that can be made from phenylalanine. It only becomes essential if phenylalanine is absent from the diet. Similarly, cysteine only becomes essential if there is not enough methionine in the diet. Most animal protein, such as meat, eggs and milk, contains all the essential amino acids in the proportions required by an adult. However, many vegetable proteins can have a low level of one or more essential amino acids. This does not pose a problem if a variety of vegetable products is eaten. For example, beans (lacking methionine) may be eaten with bread (lacking lysine).

Table 14.4 *Essential amino acids required by adults and children*

For adults and children	Additional requirement of small children
Isoleucine	Histidine
Leucine	
Lysine	
Methionine	
Phenylalanine	
Threonine	
Tryptophan	
Valine	

14.7.4 The roles of vitamins

Vitamins are organic substances required by the body in very small amounts. Although vitamins D and K can be made in the body, all others are required in the diet. A summary of the roles of some important vitamins is shown in table 14.5. Vitamins A and D will be considered in little more detail.

- **Vitamin A (retinol)**
  This vitamin is found in
  - milk, eggs, liver and fish-liver oils
  - mango and papaya.

  It can also be formed in the body from the related compound, **β-carotene**, found in vegetables such as carrots and spinach. High concentrations of vitamin A in the blood are toxic, but the liver is able to store enough to last 1–2 years. Vitamin A is required by the body to make
  - **rhodopsin**, used by rod cells in the eye, to detect low light intensities
  - **retinoic acid**, used for the growth and maintenance of epithelial tissues.

  The problems associated with a lack of vitamin A are considered in section 14.9.3.
- **Vitamin D (calciferol)**
  This vitamin is found in
  - eggs and oily fish
  - meat and meat products
  - margarine and low-fat spreads (to which vitamin D has to be added, by law).

  It can also be made in the body, by the action of sunlight on the skin, although little is made in temperate climates by dark-skinned people. Vitamin D is required by the body
  - to stimulate epithelial cells in the intestine to absorb calcium
  - to act on bone cells to regulate the deposition of calcium.

  Vitamin D can be stored in muscles and fat. The problems associated with its deficiency are considered in section 14.9.4.

14.7.5 Mineral ions

Table 14.6 summarises the sources and functions of some mineral ions required by humans.

Table 14.5 *Some other vitamins required in the human diet*

Vitamin	Fat- or water-soluble	Major food sources	Functions
B_1 – thiamine	Water	Liver, legumes, yeast	Coenzyme needed for respiration
Folic acid	Water	Liver, vegetables	Making red blood cells
C – ascorbic acid	Water	Citrus fruits	Formation of collagen
E – tocopherol	Fat	Liver, green vegetables	Antioxidant

Table 14.6 *Sources and functions of some mineral ions required by humans*

Mineral ion	Major food source	Function
Calcium (Ca^{2+})	Dairy foods, eggs	Constituent of bones and teeth, needed for blood clotting
Phosphate (PO_4^{3-})	Dairy foods, eggs, meat	Constituent of nucleic acids, ATP, phospholipids, bones and teeth
Iron (Fe^{2+}/Fe^{3+})	Liver, green vegetables	Constituent of haemoglobin and myoglobin

---

SUMMARY TEST 14.7

A balanced diet requires carbohydrates to provide **(1)** and these occur in a number of forms. Starch is found in foods such as **(2)** and **(3)**. Honey contains the two monosaccharide sugars called **(4)** and **(5)**, while cane sugar is an almost pure source of **(6)**. Even more energy is available from the food type, **(7)**, which are made up of **(8)** and glycerol. Proteins, essential for **(9)** and **(10)**, are broken down into amino acids, of which a total of **(11)** cannot be synthesised by the body and must be supplied in the diet. These are called **(12)** amino acids. Vitamins play many roles. Vitamin **(13)** is essential for the functioning of rod cells in the eye, while vitamin **(14)** is essential for calcium absorption by the gut and can be manufactured by the **(15)** of humans. Minerals such as iron can be obtained by eating **(16)** or **(17)**; iron is an essential constituent of **(18)** and **(19)**. Two other essentials of a balanced diet are **(20)** and **(21)**.

Variations in dietary requirements

It has become increasingly apparent that diet plays an important role in the maintenance of health. Since the early 1940s, governments have commissioned studies to evaluate the effect our diet has on health. After each study, various recommendations are made.

14.8.1 Dietary reference values (DRVs)

DRVs were introduced as the result of a far-reaching and influential report published in 1991 by the Committee on Medical Aspects of Food Policy (COMA). DRVs do not attempt to give recommended daily intakes of particular nutrients for individuals. Instead, they refer to groups of people. Up to three Dietary Reference Values can be defined when substantial amounts of data have been collected. These are shown in figure 14.7 and described here:

- **Estimated average requirement (EAR)** – this is the average requirement of the population. As shown on the graph, 50% of the population will need less, and 50% more.
- **Lower reference nutrient intake (LRNI)** – most people will need more than this, but it is adequate for those with the lowest needs.
- **Reference nutrient intake (RNI)** – most people will need less than this, but it provides enough for those with the greatest needs.

Because energy requirements, even within a population, vary widely, only the EAR is given in tables of recommended energy intake. Sometimes there are not enough data for the construction of the normal distribution curve shown in figure 14.7. In these cases only, a second group of DRVs, known as **safe intakes**, are given. For example, low levels of vitamin A cause deficiency diseases (section 14.9.3), but a very high level can be toxic. The DRV is therefore set well below the dangerous level – in other words, a safe intake level is quoted.

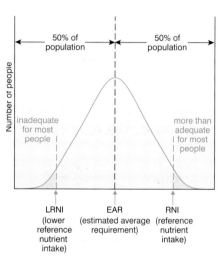

Fig 14.7 *Illustrating three dietary reference values*

14.8.2 Uses of dietary reference values

Although DRVs are not designed for individual use, it makes sense that, if you are eating a nutrient at around the EAR, it is unlikely to cause harm. If, however, you are eating more than the RNI for a particular nutrient, you may be having too much, as such a small percentage of the population has needs as high as this. DRVs are useful for dieticians and caterers, who need to design menus for those living in institutions such as prisons, schools or old people's homes. DRVs may need to be adapted when the needs of hospital patients are being considered. For example, someone who has trouble absorbing vitamin B may need to have more than the RNI in their diet.

14.8.3 Effect of sex

Some of the effects of sex on dietary requirements can be seen in tables 14.7 and 14.8. For example:
- infants have the same requirements, regardless of sex
- men generally have a larger body mass than women of the same age, and therefore have a higher requirement for protein
- during puberty, men require more calcium for additional bone growth, but after the age of 19 years males and females have the same requirement
- menstruating females have a higher iron requirement than men, to replace that lost in the haemoglobin of the menstrual blood
- men generally have higher energy requirements than women. This is because women have more **adipose tissue**, which uses energy only slowly.

Table 14.7 *Recommended daily intake of energy according to age, activity and sex*

Age / years	Average body weight / kg	Degree of activity, or circumstances	Energy requirements / kJ	
			Male	Female
1	7	Average	3200	3200
5	20	Average	7500	7500
10	30	Average	9500	9500
15	45	Average	11 500	11 500
		Sedentary	11 300	9000
25	65 (male)	Moderately active	12 500	9500
	55 (female)	Very active	15 000	10 500
		Sedentary	11 000	9000
50	65 (male)	Moderately active	12 000	9500
	55 (female)	Very active	15 000	10 500
75	63 (male)	Sedentary	9000	8000
	53 (female)			
Any	–	During pregnancy	–	10 000
Any	–	Breast-feeding	–	11 500

14.8.4 Effect of age

Tables 14.7 and 14.8 show the effect of age on dietary requirements; for example:
- protein requirements increase with age, especially around puberty
- calcium requirements peak with spurts in bone growth during puberty
- energy requirements generally increase with age, but fall in old age, as the mass of respiring cells reduces.

14.8.5 Effect of activity

As soon as a person becomes active, the energy requirement increases. As shown in table 14.7, the more strenuous the activity, the greater the amount of energy needed.

14.8.6 Effect of pregnancy

It is recommended that women planning to become pregnant should supplement their diet with folic acid tablets. Folic acid is known to protect against spina bifida – a defect in the formation of the nerve cord. Pregnant women are also advised to increase their intake of:
- energy (during the last 3 months)
- protein
- vitamins A, C and D.

These will support growth of the uterus, placenta and foetus.

14.8.7 Effect of lactation

Mothers who are breast-feeding not only need to maintain increased intakes of energy, protein and vitamins, as in pregnancy. They also need more calcium and zinc. The extra intake will supply the baby with the energy for growth and metabolism. Protein is especially needed for the growth of the young baby, and calcium is required for its bone formation.

Table 14.8 *Effect of age on dietary requirements for protein, calcium and iron*

Age / years	Protein / g day^{-1}		Calcium / mg day^{-1}		Iron / mg day^{-1}	
1–3	14.5		350		6.9	
7–10	28.3		550		8.7	
	Males	Females	Males	Females	Males	Females
11–14	42.1	41.2	1000	800	11.3	14.8
15–18	55.2	45.0	1000	800	11.3	14.8
19–50	55.5	45.0	700	700	8.7	14.8
50+	53.3	46.5	700	700	8.7	8.7

SUMMARY TEST 14.8

There are three important dietary reference values. The average amount of a particular nutrient needed by a population is called the **(1)**. The amount that is inadequate for all but a few people is called the **(2)** while the amount that satisfies the needs of almost all individuals in the population is called the **(3)**. Amounts of nutrient required in the diet vary. Men at puberty require more **(4)** while menstruating females need more **(5)**. During pregnancy vitamins **(6)**, **(7)** and **(8)** are essential, while breastfeeding mothers are recommended to increase their **(9)** intake to ensure proper bone development of their baby.

OCR

Diet and health

In order to remain healthy, we need to eat a range of nutrients in the amounts and proportions required by the body – in other words, a balanced diet (unit 14.7). A shortage, or even a complete lack, of food will lead to **starvation**. Taking in an excessive amount of food leads to **overnutrition** and **obesity**. Any imbalance of nutrients in the diet, whether it be too much of one thing or not enough, is referred to as **malnutrition**.

14.9.1 Protein–energy malnutrition

Protein–energy malnutrition is common in the poorer countries of the developing world. It is seen mainly in children when they are weaned from milk, to a diet which is often low in protein. These children grow slowly, and have high mortality rates. Two extreme cases of protein–energy malnutrition are known as **marasmus** and **kwashiorkhor**:

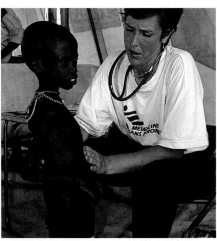

Child with marasmus

- **Marasmus** is often the result of early weaning to a diet low in both protein and energy. As can be seen in the photo, the children have the following characteristics:
  - extremely under-weight
  - muscle wasting
  - wrinkled skin, and an 'old' face.
- **Kwashiorkhor** is linked to a more severe protein shortage, and often appears in slightly older children. This is a difficult condition from which to recover because, not only is there liver damage, but the children become apathetic and lose their appetite. The child in the photo shows some of the following features of kwashiorkhor:
  - under-weight
  - oedema – this is swelling caused by the accumulation of fluid in the tissues (the fluid cannot return to the blood because the lack of plasma proteins in malnutrition upsets the osmotic balance (see section 8.4.2)
  - 'moon face', due to oedema
  - swollen abdomen, due to enlarged liver
  - dry, brittle hair, often reddish in colour.

14.9.2 Anorexia nervosa

Child with kwashiorkhor

This so-called 'slimmer's disease' is most common in western developed counties, and more often affects girls than boys. Underlying psychological causes lead to an obsessive desire to be thin. The weight loss can become so severe that

- muscles waste (including heart muscle)
- periods stop
- blood pressure falls
- hair becomes thin and sparse
- hands and feet are cold
- there is increased susceptibility to infection
- there are personality changes.

Anorexics often believe themselves to be fat, even when they are dangerously thin, and are slow to realise that they need help. Once this is recognised, prolonged psychotherapy is often needed to assist recovery.

Young woman with anorexia nervosa

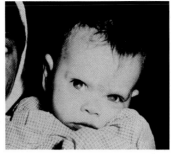

Face of someone with xerophthalmia

Rickets   *Obesity*

14.9.3 Vitamin A deficiency

The roles of vitamin A in a balanced diet are discussed in section 14.7.4. Vitamin A deficiency is often associated with poor diet in the developing world. The animal products which contain it are expensive, and the vegetables containing β-carotene (which can be converted to vitamin A) are not available. Diets are mainly based on cereals (maize and rice), which are low in vitamin A. There are a number of deficiency symptoms:

- **Dry, rough skin**.
- **Xerophthalmia**, which is a drying and scarring of the cornea. There is inflammation, and it may lead to blindness. Xerophthalmia is thought to be responsible for half a million cases of childhood blindness, worldwide (about 45% of the total).
- **Night blindness** is a reduced ability to see in a dim light; it can affect both children and adults. Vitamin A is needed to form rhodopsin. This is a pigment which is bleached when light enters the eye. A lack of vitamin A slows down the re-formation of this pigment.
- **Increased susceptibility to infections** because, without retinoic acid (made from vitamin A), the epithelial surfaces are not maintained properly. This allows pathogens to enter. The most likely infections are measles, and infections of the gut and lungs.

14.9.4 Vitamin D deficiency

The roles of vitamin D in a balanced diet are discussed in section 14.7.4. Vitamin D is rarely deficient in the diets of people in developed countries, because it is found in eggs, meat and oily fish and, in the UK, it is added to margarine and low-fat spreads. Vitamin D is also made in the skin. Problems may arise for dark-skinned people in temperate climates, or those whose religion requires them to wear clothes that do not expose the skin. Deficiency symptoms result from the link between vitamin D and calcium intake: vitamin D is required for the formation of a calcium–protein complex in the lining of the intestine. Without this, calcium absorption is poor.

- **Rickets** is the vitamin D deficiency disease in children. Their legs are bowed, because the bones lack calcium salts and are soft.
- **Osteomalacia** is a progressive softening of the bones in adults, which makes them more susceptible to fractures.

14.9.5 Obesity

Over-eating is a form of malnutrition. If people regularly take in more energy than they use (usually in the form of fats and carbohydrates), they will put on weight. The **body mass index** is used as a measure of whether a person is under- or over-weight. It should only be used to assess adults, because children store fat as part of the normal growth process.

$$\text{Body mass index (BMI)} = \frac{\text{body mass} / \text{kg}}{[\text{height} / \text{m}]^2}$$

Table 14.9 indicates which category a person comes into. About 60% of men and 40% of women in the UK are estimated to be either over-weight or obese. Obesity is an increasing problem in the whole of North America, Europe and Australasia. It is especially associated with high-fat diets and an inactive lifestyle. It is not only a problem in adults. Increasingly, obesity is seen in children: 10% of UK secondary school age children are over-weight. Obese people have a higher mortality rate than people of normal weight. People with a BMI of 35 or more have a mortality rate twice that of people of the same age who have a BMI between 20 and 25. Obesity is a factor associated with, or even causing, a whole range of medical conditions, including:

- **coronary heart disease** (unit 13.4), because of high blood pressure and raised blood cholesterol levels
- **mature-onset diabetes**, because obese people are unable to control the concentration of sugar in their blood
- **complications during surgery**, because the organs are surrounded by large masses of fat
- **cancers** (unit 13.9):
  - breast, cervical and uterine cancers in women
  - colon, rectum and prostate cancers in men
- **osteoarthritis** of knees, hips and spine, because supporting a large mass puts a strain on the skeleton.

Table 14.9 Body mass index

BMI / kg m⁻²	Description
Under 20	Under-weight
20–24.9	Normal
25–29.9	Over-weight
Over 30	Obese

1 MMR is the combined vaccine used against measles, mumps and rubella. It contains attenuated microorganisms.

  a What is an *attenuated* microorganism? *(1 mark)*

  b Vaccines protect against disease by stimulating the production of memory cells. Describe how memory cells protect the body from disease. *(3 marks)*

  c The graph shows the number of reported cases of whooping cough during the period 1950 to 1975.

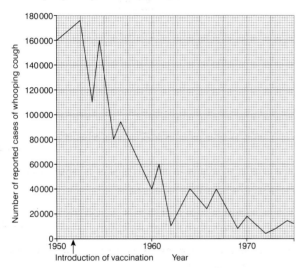

  Describe and explain what the graph shows about the number of reported cases of whooping cough during the period 1952 to 1960. *(2 marks)*

  d The number of reported cases of whooping cough increased during the 1980s.
  Suggest **one** reason why. *(1 mark)*
  (Total 7 marks)
  AQA Jan 2001, HB (A) BYA3, No.8

2 Plastic strips impregnated with certain chemicals may be dipped in urine. A change of colour indicates the presence of glucose. The test relies on the two chemical reactions shown by the equations.

Enzyme **A**
glucose + oxygen → gluconic acid + hydrogen peroxide

Enzyme **B**
blue dye + hydrogen peroxide → green-brown dye + water

  a Name enzyme **A**. *(1 mark)*

  b With which of the substances shown in the equations are the plastic strips impregnated? *(1 mark)*

  c Explain why:

  (i) the manufacturers recommend storing these strips in a cool place; *(1 mark)*

  (ii) the strips will only give a positive response to glucose. They will not give a reaction with other sugars. *(2 marks)*
  (Total 5 marks)
  AQA (specimen) 1999, HB (A) BYA3 No.5

3 The table shows some antibiotics and the way in which they work.

Antibiotic	Method of action
Penicillin	Prevents the formation of bacterial cell walls
Streptomycin	Distorts the shape of ribosomes in bacterial cells so that protein synthesis is stopped
Mitomycin C	Joins together the two polynucleotide chains which make a DNA molecule with strong chemical bonds so they cannot be separated

  a Explain why:

  (i) penicillin is effective against bacteria but not against fungi; *(1 mark)*
  (ii) streptomycin cannot be used to treat diseases caused by viruses. *(1 mark)*

  b (i) Explain why cells treated with mitomycin C cannot synthesise proteins. *(2 marks)*
  (ii) Suggest why it is thought that mitomycin C might be effective in treating cancer. *(2 marks)*
  (Total 6 marks)
  AQA (specimen) 1999, HB (A) BYA3, No.4

4 The pancreas secretes enzymes into the small intestine through the pancreatic duct. Pancreatitis is a condition in which the pancreas becomes inflamed. One type of pancreatitis is caused by a blocked pancreatic duct.
The diagram shows the positions of the pancreas and the small intestine.

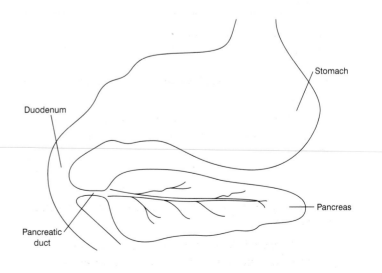

a To help diagnose pancreatitis, the contents of the small intestine can be tested for pancreatic enzymes.

   (i) Describe and explain the likely results of such a test in a patient suffering from this type of pancreatitis. *(2 marks)*

   (ii) Describe **one** other way in which any change in the level of pancreatic enzymes could be detected. Give the result you would expect in a case of pancreatitis like the one described. *(2 marks)*

b Two of the enzymes secreted by the pancreas digest proteins. They are normally secreted in an inactive form. In patients with pancreatitis, these enzymes can become active in the pancreas. Suggest how activation of these enzymes could affect the pancreas. *(2 marks)*

(Total 6 marks)

AQA June 2001, HB (A) BYA3, No.5

5 Proteins are digested into amino acids. Nutritionists consider some of these amino acids to be essential amino acids. A deficiency of protein in the diet can lead to significant damage to health.

a Explain why some amino acids are described as *essential* in the diet. *(2 marks)*

b Describe the consequences for a young child of a **very** restricted intake of protein in the diet. *(3 marks)*

The table shows the daily energy requirements and some of the daily nutrient requirements for a woman aged 25 and a lactating (breast feeding) woman of the same age and body mass.

energy or nutrients	Daily requirements	
	Woman aged 25	Lactating woman
energy / kJ kg^{-1}	145	185
protein / g kg^{-1}	0.75	0.93
vitamin A / μg	600	950
vitamin D / μg	0	10
calcium / mg	700	1250

c With reference to the table, explain why the energy and nutrient intakes are greater for the lactating woman. *(8 marks)*

(In this question one mark is awarded for the quality of written communication).

d State **three** further components of a balanced diet, other than digestible carbohydrate, that are not given in the table. *(3 marks)*

(Total 16 marks)

OCR 2802 Jan 2001, B(HHD), No.3

6 People may gain active and passive immunity to disease, such as measles.

a Describe **two** different ways, one active and one passive, in which babies may become immune to measles. *(4 marks)*

b State **two** reasons why measles has not been eradicated by vaccination. *(2 marks)*

(Total 6 marks)

OCR 2802 Jan 2001, B(HHD), No.5a, b

7 During an immune response to a bacterial infection many different antibody molecules may be produced. The figure below shows the structure of a typical antibody molecule.

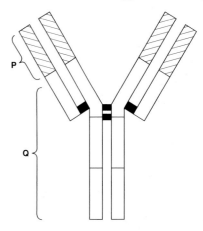

a Name the regions of the antibody molecule labelled **P** and **Q**. *(2 marks)*

Antibodies, like enzymes, are highly specific. This is because they are both proteins.

b Explain how the protein nature of antibodies allows the production of many different types. *(3 marks)*

During a bacterial infection, the number of phagocytes in the blood may increase.

c State where in the body phagocytes are made. *(1 mark)*

The figure below is a drawing made from an electron micrograph of a phagocyte engulfing some bacteria.

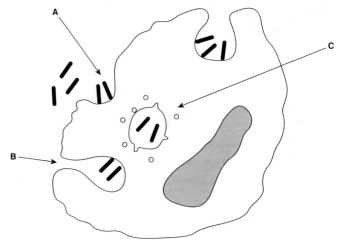

d With reference to this figure, describe the events that are occurring at **A**, **B** and **C**. *(6 marks)*

e Explain how antibodies, such as those shown in the first figure, may help phagocytes engulf bacteria. *(3 marks)*

(Total 15 marks)

OCR 2802 June 2001, B(HHD), No.4

Glossary

acetylcholine one of a group of chemicals, known as *neurotransmitters*, released by nerve cells. It diffuses across the gap between adjacent nerve cells (synapse) and so passes an impulse from one nerve cell to the next.

actin filamentous protein which is involved in contraction within cells, especially muscle cells. See also *myosin*.

active transport movement of a substance from a region where it is in a low concentration to a region where it is in a high concentration. The process requires the expenditure of energy.

adhesion attraction between the molecules of different types. See also *cohesion*.

adipose tissue a form of connective tissue which is made up of cells storing large amounts of fat.

aerobic connected with the presence of free oxygen. Aerobic respiration requires free oxygen to release energy from glucose and other foods.

aerotropism movement of part of an organism in response to the stimulus of air, e.g. pollen tubes grow away from air (i.e. are negatively aerotropic).

afforestation planting of trees in areas where they have not been grown in the recent past.

algal bloom rapid growth of algae in freshwater rivers and lakes, often as the result of an increase in nitrates and other nutrients due to the 'run-off' of artificial fertilisers from nearby agricultural land. See also *eutrophication*.

allele one form of a gene. For example, the gene for the shape of pea seeds has two alleles, one for 'round' and one for 'wrinkled'.

amnion membranous sac which encloses the embryo in reptiles, birds and mammals. It contains the amniotic fluid.

anaerobic connected with the absence of free oxygen. Anaerobic respiration releases energy from glucose or other foods without the presence of free oxygen.

anion negatively charged ion which is attracted to the anode during electrolysis. See also *cation*.

antibiotic resistance the development in microorganisms of mechanisms which prevent antibiotics from killing them.

antibody a protein produced by lymphocytes in response to the presence of the appropriate antigen.

antigen a molecule which triggers an immune response by lymphocytes.

antioxidant chemical which reduces or prevents oxidation. Often used as an additive to prolong the shelf-life of certain foods.

apoplast pathway route through the cells walls and intercellular spaces of plants by which water and dissolved substances are transported. See also *symplast pathway*.

atheroma fatty deposits on the walls of arteries, often associated with high cholesterol levels in the blood.

atherosclerosis narrowing of arteries due to thickening of the arterial wall caused by fat, fibrous tissue and salts being deposited on it. The condition is sometimes referred to as 'hardening of the arteries'.

ATP (adenosine triphosphate) nucleotide found in all plants and animals which is produced during respiration and is important in the transfer of energy.

biodiversity the range and variety of living organisms within a particular region.

biomass the total mass of living material, normally measured in a specific area for given amount of time.

blastocyst a stage in the early development of a mammal during which it implants in the uterus wall.

B-lymphocyte type of white blood cell that is produced and matures within the bone marrow. B-lymphocytes produce antibodies as part of their role in immunity. See also *T-lymphocyte*.

Calvin cycle a biochemical pathway which forms part of the light-independent stage of photosynthesis during which carbon dioxide is reduced to form carbohydrate.

cambium dividing layer of cells parallel to the surface of stems and roots which produces new cells, leading to an increase in their diameter.

carcinogen chemical, form of radiation or other agent which causes cancer.

cardiac output the total volume of blood which the heart can pump each minute. It is calculated as the volume of blood pumped at each beat (stroke volume) multiplied by the number of heart beats per minute (heart rate).

Casparian strip a distinctive band of *suberin* around the endodermal cells of a plant root which prevents water passing into xylem via the cell walls. The water is forced through the living part (protoplast) of the endodermal cells.

cathode ray oscilloscope an electronic instrument which measures variation in quantities, such as voltage or current, over time.

cation positively charged ion which is attracted to the cathode during electrolysis. See also *anion*.

chloride shift the movement of negatively charged chloride ions from the plasma into red blood cells to replace the loss of negatively charged hydrogen carbonate ions during the transport of carbon dioxide. In this way the overall electrochemical neutrality of the red blood cells is maintained.

chromatin the material which makes up chromosomes. It consists of DNA and proteins, especially *histones*.

chromatography technique by which substances in a mixture are separated according to their different solubilities in a solvent.

clone a group of genetically identical organisms formed from a single parent as the result of asexual reproduction or by artificial means.

cohesion attraction between molecules of the same type. *See also* adhesion.

collagen fibrous protein that is the main constituent of connective tissues such as *tendons*, cartilage, and bone.

collenchyma plant tissue which has cell walls thickened by cellulose, especially in the corners. It provides mechanical support, especially in young stems and leaves.

colloid substances made up of fine particles or large molecules which are evenly dispersed within a second substance. Examples include emulsions (liquid dispersed in a liquid) aerosols (solid/liquid dispersed in a gas) and foams (gas dispersed in a liquid/solid).

complementary DNA (cDNA) DNA which is made from *messenger RNA* using the enzyme *reverse transcriptase* in a process which is the reverse of normal *transcription*.

condensation reaction chemical process in which two molecules combine to form a more complex one with the elimination of a simple substance, usually water. Many biological polymers such as polysaccharides and polypeptides are formed by condensation reactions. See also *hydrolysis*.

consumer any organism that obtains energy by 'eating' another. Herbivores feed on plants and are known as primary consumers and carnivores feeding on herbivores are known as secondary consumers.

corpus luteum yellow tissue which forms in the *Graafian follicles* of the mammalian ovary after *ovulation*. If fertilisation takes place, the corpus luteum produces the hormone progesterone.

covalent bond type of chemical bond in which two atoms share a pair of electrons, one from each atom.

cytokinins group of plant growth substances which stimulate cell division.

cytoskeleton network of microfilaments and microtubules which give structural shape to *eukaryotic* cells.

deciduous term applied to plants which shed all their leaves together at one season.

decomposer a microorganism that breaks down the organic matter of dead organisms and waste products to form water, carbon dioxide and inorganic ions.

defoliation the removal of leaves (foliage) from a plant.

diastole the stage in the cardiac cycle when the heart muscle relaxes and ventricles fill with blood. See also *systole*.

dicotyledonous plant any member of the class of flowering plants called Dicotyledonae. Their features include: having two seed leaves (cotyledons), broad leaves, flower parts in whorls of 4 or 5 and vascular tissue arranged in a ring in stems. See also *monocotyledonous plants*.

diffusion the movement of molecules from a region where they are in high concentration to one where their concentration is lower.

diploid a term applied to cells in which the nucleus contains two sets of chromosomes. See also *haploid*.

DNA probe a single strand of DNA which has a base sequence complementary to a gene which it is being used to identify. The DNA probe is labelled radioactively to enable the gene to be located.

ecosystem all the living (biotic) and non-living (abiotic) components of a particular area.

electron transport system a chain of carrier molecules along which electrons pass, releasing energy in the form of *ATP* as they do so.

electrophoresis technique for separating a mixture of charged particles in a fluid by applying a voltage across the fluid. It is used in the analysis of mixtures of substances, especially proteins.

embryo sac large cell of the ovary of a flowering plant which contains a number of nuclei including the egg nucleus (female gamete).

endemic describes any disease that occurs regularly in a particular region or amongst a particular population.

endothermic an animal which uses physiological processes to maintain its body temperature at a more or less constant level. Birds and mammals are endotherms.

eugenics the science of changing the human species by selectively breeding individuals who possess desired genetic characteristics.

eukaryote an organism whose cells have a membrane-bounded nucleus which contains chromosomes. The cells also possess a variety of other membranous organelles such as mitochondria and endoplasmic reticulum. See also *prokaryote*.

eutrophication term applied to an increase in nutrients, especially nitrates and phosphates, in fresh-water lakes and rivers. This often leads to decrease in *biodiversity*. See also *algal blooms*.

fertilisation the fusion of the nuclei of male and female gametes to produce a zygote during the process of sexual reproduction.

gene technology general term that covers the processes by which genes are manipulated, altered or transferred from organism to organism. Also known as *genetic engineering*.

gene therapy a mechanism by which genetic diseases such as cystic fibrosis may be cured by replacing defective genes with healthy ones, or by masking the effect of the defective gene by another gene.

genetic engineering see *gene technology*.

genetic fingerprinting also known as DNA profiling, is a forensic technique for identifying individuals by means of the unique sequence of nucleotides in their DNA.

genotype the genetic composition of an organism.

global warming the recent increase in average temperatures at the Earth's surface thought to be the result of the increased production of greenhouse gases such as carbon dioxide and methane. These gases help to trap solar radiation at or near the Earth's surface.

glycoprotein substance made up of a carbohydrate molecule and a protein molecule. Part of cell surface membranes and certain hormones are glycoproteins.

goblet cell mucus-producing cell found in the epithelium of the intestines and bronchi, so called because its shape resembles a wine glass or goblet.

gonadotrophic hormones group of sex hormones produced by the anterior pituitary gland of both sexes. Examples include follicle stimulating hormone (FSH) and luteinising hormone (LH) in females and interstitial cell stimulating hormone (ICSH) in males. Gonadotrophic hormones stimulate the gonads (ovaries and testes) to produce their sex hormones.

Graafian follicle fluid-filled cavity in the ovaries of mammals which surrounds and protects a developing *oocyte*. Also called an ovarian follicle.

gross primary production the total energy production in the form of *biomass* made by *producers* (plants) during photosynthesis. It is normally expressed as the biomass per unit area in unit time. See also *net primary production*.

Haber process industrial process in which nitrogen and hydrogen are passed over a heated catalyst at a pressure of 1000 atmospheres (100 000 kPa), to produce ammonia.

habitat the place where an organism normally lives and which is characterised by physical conditions and the types of other organisms present.

haploid term referring to cells which contain only a single copy of each chromosome, e.g. the sex cells or gametes.

herbaceous term applied to non-woody plants.

histones proteins associated with the DNA in chromosomes. They function to condense *chromatin* and coil the chromosomes during cell division.

homeostasis the maintenance of a more or less constant internal environment.

homologous chromosomes a pair of chromosomes that have the same gene loci and therefore determine the same features. They are not necessarily identical, however, as individual *alleles* of the same gene may vary, e.g. one chromosome may carry the allele for blue eyes, the other the allele for brown eyes. Homologous chromosomes are capable of pairing during *meiosis* I.

human genome the entire complement of genetic material on the chromosomes of a single human cell.

hydrogen bond chemical bond formed between the positive charge on a hydrogen atom and the negative charge on another atom of an adjacent molecule, e.g. between the hydrogen atom of one water molecule and the oxygen atom of an adjacent water molecule.

hydrolysis the breaking down of large molecules into smaller ones by the addition of water molecules. See also *condensation reaction*.

hydrophyte plant that is adapted to live in water or wet habitats. See also *xerophyte*.

hydrostatic skeleton the supporting framework provided by the fluid-filled cavity of soft-bodied invertebrates such as an earthworm.

hydrotropism the movement of part of an organism in response to the stimulus of water, e.g. most plant roots grow towards water (i.e. they are positively hydrotropic).

introns portions of DNA within a gene which do not code for a polypeptide. The introns are removed from *messenger RNA* after *transcription*.

ion an atom or group of atoms which have lost or gained one or more electrons. Ions therefore have either a positive or negative charge. See also *anion* and *cation*.

ion channel a passage across a cell surface membrane made up of a protein which spans the membrane and opens and closes to allow *ions* to pass in and out of the cell.

isotonic solutions which possess the same concentration of solutes and therefore have the same *water potential*.

isotope variations of a chemical element which have the same number of protons and electrons but different numbers of neutrons. While their chemical properties are similar, they differ in mass. One example is carbon, which has a relative atomic mass of 12 and an isotope with a relative atomic mass of 14.

Krebs cycle series of biochemical reactions in most *eukaryotic* cells by which energy is obtained through the oxidation of acetyl coenzyme A produced from the breakdown of glucose.

leaching process in which chemicals are removed from soil by being dissolved in rainwater and washed away.

ligament tough fibrous connective tissue rich in *collagen* that joins bone to bone at joints. See also *tendon*.

lignin a complex, non-carbohydrate polymer associated with cellulose in plant cell walls. Lignin makes the cell walls stronger, allowing them to resist tension and compression. It also makes them more waterproof.

lumen the hollow cavity inside a structure such as the gut or a xylem vessel.

meiosis the type of nuclear division in which the number of chromosomes is halved.

menstrual cycle cycle in female humans, apes and monkeys of reproductive age during which the body is prepared for pregnancy. It differs from the oestrous cycle in other mammals in that, if *fertilisation* does not occur, the lining of uterus is shed along with a little blood, in a process called *menstruation*.

menstruation the shedding of the uterus lining along with a little blood during the menstrual cycle.

messenger RNA form of ribonucleic acid which carries the genetic code from the DNA of the nucleus to the ribosomes in the cytoplasm where it acts as a template on which polypeptides are assembled.

microflora the population of various types of microorganism in a particular place, e.g. in the gut.

mitosis the type of nuclear division in which the daughter cells have the same number of chromosomes as the parent cell.

monocotyledonous plant any member of the class of flowering plants called Monocotyledonae. Their features include having a single seed leaf (cotyledon) and leaves which are parallel veined. See also *dicotyledonous plants*.

monoculture the agricultural practice of growing a single crop over a wide area.

mutualism a form of *symbiosis* in which both species benefit.

myocardial infarction otherwise known as a heart attack, it results from the interruption of the blood supply to the heart muscle, causing damage to an area of the heart with consequent disruption to its function.

myosin the thick filamentous protein found in *striated muscle*. See also *actin*.

net primary production the rate at which material produced during photosynthesis is built up in a plant. Also known as the net assimilation rate, it is the *gross primary production* less the 20% or so used by the plant in processes such as respiration.

neurotransmitter one of a number of chemicals which are involved in communication between adjacent nerve cells or between nerve cells and muscles. Two important examples of neurotransmitters are *acetylcholine* and noradrenaline.

nitrogen-fixing bacteria group of microorganisms that incorporate atmospheric nitrogen into nitrogen-containing compounds, using the enzyme nitrogenase. They may be either free living or act in conjunction with leguminous plants.

nucleotides complex chemicals made up of an organic base, a sugar and a phosphate. They are the basic units of which the nucleic acids DNA and RNA are made.

nymph young stage of certain insects, e.g. dragonfly, which resembles the adult form.

oocyte cell which is undergoing *meiosis* during oogenesis. Primary oocytes undergo the first meiotic division; the secondary oocyte undergoes the second meiotic division, and is the female gamete in humans.

osmosis the passage of water from a region where it is highly concentrated (high *water potential*)to a region where its concentration is lower (low *water potential*), through a partially permeable membrane.

ovulation release of the secondary *oocyte* (female gamete) from the ovary in mammals.

parasite an organism which lives on or in a host organism. The parasite gains a nutritional advantage and the host is harmed in some way.

parasympathetic nervous system the part of the autonomic nervous system (controls involuntary actions) which prepares the body for a relaxed state. The *neurotransmitter* in the parasympathetic nervous system is *acetylcholine*. See also *sympathetic nervous system*.

parenchyma plant tissue made up of unspecialised cells with thin walls which are used to provide support through turgidity.

pectase enzyme which hydrolyses the polysaccharide pectin which is found in the cell walls and middle lamellae of plant tissues. It is used to clarify fruit juices.

peristalsis wave-like contractions of the muscles in the wall of the alimentary canal which move food along its length.

phagocytosis mechanism by which cells transport large particles across the cell surface membrane.

phospholipid lipid molecule in which one of the three fatty acid molecules is replaced by a phosphate molecule. Phospholipids are important in the structure and functioning of the cell surface membrane.

phytoplankton small, often microscopic, photosynthesising organisms which live suspended in large bodies of water such as oceans and lakes. See also *zooplankton*.

pinocytosis form of endocytosis by which cells take up liquids from their environment. Also known as 'cell drinking'.

plasmid a small circular piece of DNA found in bacterial cells and often used as a *vector* in *gene technology*.

plasmodesmata fine strands of cytoplasm that extend through pores in adjacent cell walls, connecting the cytoplasm of one cell with another.

platelets cell fragments found in blood which play an important role in blood clotting.

pollen grain spore containing the male gamete produced in the anthers of a flowering plant.

pollination the transfer of pollen from the anther to the stigma in flowering plants.

polymerase chain reaction process of making many copies of a specific sequence of DNA or part of a gene. It is used extensively in *gene technology* and *genetic fingerprinting*.

polymerases group of enzymes which catalyse the formation of long chain molecules (polymers) from similar basic units (monomers).

producer an organism which synthesises organic molecules from simple inorganic ones such as carbon dioxide and water (an autotrophic organism). Most producers are photosynthetic and form the first *trophic level* of a food chain. See also *consumer*.

prokaryote an organism belonging to the kingdom Prokaryotae which is characterised by having cells less than 5μm in diameter which lack a nucleus and membrane-bound organelles. Examples include bacteria and blue-green bacteria. See also *eukaryote*.

protoctist an organism belonging to the kingdom Protoctista which is made up of single-celled *eukaryotes*, such as certain algae and *protozoa*.

protoplast the living portion of a plant cell, i.e. the nucleus and cytoplasm along with the organelles it contains.

protozoa a sub-group of the kingdom *Protoctista* made up of single celled organisms such as *Amoeba*.

restriction endonucleases a group of enzymes that are able to cut DNA into shorter lengths at specific points. Found naturally in certain bacteria, they are important in *gene technology*.

reverse transcriptase enzyme capable of producing a DNA molecule from the corresponding *messenger RNA*. Produced naturally by many viruses, reverse transcriptase is used in *gene technology*.

root nodule swelling on the roots of certain plants such as legumes (bean family) which contain *nitrogen-fixing bacteria* which live mutualistically with the host plant.

ruminant a mammal which possess a modified region called the rumen at the front end of the alimentary canal. The rumen contains cellulose-digesting microorganisms which assist digestion. Ruminants 'chew the cud' and include sheep, cattle, goats, deer and antelope.

saprobiont also known as a saprophyte, a saprobiont is an organism that obtains its food from the dead or decaying remains of other organisms. Many bacteria and fungi are saprobionts.

sclerenchyma plant tissue in which the cells have become rigid due to the presence of cell walls thickened with *lignin*. The cells are dead and function to provide support to the plant.

sieve plate the perforated end wall of the phloem component called the *sieve tube* elements.

sieve tube phloem tissue made up of a series of sieve tube elements joined end to end to form a long column that carries organic material in the plant.

sinoatrial node an area of cardiac muscle in the right atrium that controls and coordinates the contraction of the heart.

smooth muscle also known as involuntary or unstriated muscle, smooth muscle is found in the alimentary canal and walls of blood vessels, its contraction is not under conscious control. See also *striated muscle*.

Southern blotting a technique for transferring DNA strands onto a nylon membrane during *genetic fingerprinting*.

spindle thread-like structure made up of microtubules which draws the chromosomes apart during *meiosis* and *mitosis*.

stem cells undifferentiated cells from which specialised cells arise during development.

stereoscopic vision the ability to see in three dimensions and thereby be able to judge distances.

suberin complex waxy waterproofing material made up of fatty acids which is found in some plant cell walls.

supernatant liquid the liquid portion of a mixture left at the top of the tube when suspended particles have been separated out at the bottom during centrifugation.

sympathetic nervous system the part of the autonomic nervous system (controls involuntary actions) which prepares the body for activity. The *neurotransmitter* in the sympathetic nervous system is noradrenaline. See also *parasympathetic nervous system*.

symplast pathway route through the cytoplasm and *plasmodesmata* of plant cells by which water and dissolved substances are transported. See also *apoplast pathway*.

systole the stage in the cardiac cycle in which the heart muscle contracts. It occurs in two stages: atrial systole when the atria contract and ventricular systole when the ventricles contract. See also *diastole*.

tendon tough, flexible connective tissue that joins muscle to bone. See also *ligament*.

T-lymphocyte type of white blood cell that is produced in the bone marrow but matures in the thymus gland. T-lymphocytes coordinate the immune response and kill infected cells. See also *B-lymphocyte*.

transcription the formation of *messenger RNA* molecules from the DNA which makes up a particular gene. It is the first part of protein synthesis.

transducer cells cells which convert a non-electrical signal, such as light or sound, into an electrical (nervous) signal, and *vice versa*.

transgenic animals or plants animals or plants which contain genes artificially transferred to them from another species.

transpiration evaporation of water from a plant.

trophic level the position of an organism in a food chain. See also *producer* and *consumer*.

ultrafiltration filtration under pressure. A term applied to the first stage of urine formation in the kidney.

vascular tissue any tissue that forms a network of vessels through which fluids are transported in organisms. Blood vessels are the usual vascular tissue of vertebrates, while in plants it is the phloem and xylem.

vasoconstriction narrowing of the internal diameter of blood vessels. See also *vasodilation*.

vasodilation widening of the internal diameter of blood vessels. See also *vasoconstriction*.

vector a carrier. The term may refer to something such as a *plasmid* which carries DNA into a cell, or to an organism which carries a *parasite* to its primary host.

vegetative propagation form of asexual reproduction in higher plants involving the separation of a piece of the original plant (root, stem or leaf) which then develops into a separate plant.

vital capacity the total volume of air that can be breathed in and out of the lungs during one deep breath.

water potential the measure of the extent to which a solution gives out water. The greater the number of water molecules present the higher (less negative) the water potential. Pure water has a water potential of zero.

xeromorphic possessing xerophytic features.

xerophyte a plant adapted to living in dry conditions. See also *hydrophyte*.

zooplankton small animals, often microscopic, which live suspended in large bodies of water such as oceans and lakes. See also *phytoplankton*.

Acknowledgements

The authors and publishers are grateful to the following awarding bodies for kind permission to reproduce examination questions:

AQA — Assessment and Qualifications Alliance
Edexcel — Edexcel Foundation
OCR — Oxford, Cambridge and RSA Exminations

Photograph acknowledgements

Axon Images p.224 (Forensic Science Service); Biophoto Associates pp.11, 49 (top), 61 (bottom), 71, 73 (top), 81 (top, bottom), 131 (bottom), 193 (bottom); Bruce Coleman pp.128 (Dr Frieder Sauer); 179 (bottom-John Murray), 181 (Dr Norman Myers), Clouds Hill Imaging Ltd pp. 84, 85 (left, right), 86, 89, 90, 93, 94 (bottom), 121, 131 (top), 132, 134, 160 (bottom), 192 (middle); Corbis pp.200, 265 (middle right); Corel (NT) pp.192 (top), 195 (bottom); Collections p.195 (middle-David M Hughes); Ecoscene p.190 (NASA); Erica Larkcom p.203; Forensic Science Service p.248; Gene Cox pp.52 (bottom), pp.195 (top); Geoscience Features Picture Library p.180 (D Hoffman); Getty Images p.126 (top-Andrew Syred, Stone); Griffin p.48; Holt Studios International p.188 (top); John Adds pp.140, 220 (top and bottom); Natural Visions p.192 (middle-Heather Angel); Oxford Scientific Films pp.163 (London Scientific Films), 188 (bottom-Kathie Atkinson); Panos Pictures p.179 (top-J Hartley); Robert Harding Picture Library p.199 (Paolo Koch); Roche Sight and Life p.265 (left, centre left-task force SIGHT AND LIFE); Sally and Richard Greenhill p.264 (bottom-Sally Greenhill); Science Photolibrary pp.30, 68, 73 (bottom), 82 (bottom), 91, 153, 212, 23, 127 (bottom-J C Revy), 29 (Saturn Stills), 49 (bottom-David Scharf), 52 (bottom), 106, 126 (bottom-Andrew Syred) 53 (Chris Priest), 55 (Juergen Berger, Max-Planck Institute), 56 (Biology Media), 58 (Secchi, lecaque, Roussel, UCLAF, CNRI), 59 (Professor P Motta, T Naguro), 61 (top-Quest), 74, 240, 250 (CNRI), 94 (top-Yorgos Nikas), 96, 160 (top), 233 (right) (Biophoto Associates), 104 (M I Walker), 105 (top), 166 (Claude Nuridsany, Marie Perrennou), 107 (Dr Jeremy Burgess), 117 (bottom), 230 (bottom-Manfred Kage), 117 (top), 230 (top-Astrid and Hans Frieder Michler), 120 (Professor P Motta, Department of Anatomy, University La Sapienza, Rome), 124 (John Mead), 151 (Professor P Motta, P Correr), 214, 256 (Simon fraser), 217 (Phillipe Plailly, Eurelios), 221 (James King-Holmes), 227 (top-Catherine Pouedras), (upper middle-Tim Beddows), 233 (left-Richard Kirby), 235 (Deep Light Productions), 238, 252 (Eye of Science), 239 (top-London School of Hygiene and Tropical Medicine) (middle BSIP-VEM) (bottom-Dr Tony Brain), 245 (Moredun Animal Health), 249 (BSIP-VEM), 251 (Biology Media), 258 (Charlotte Raymond), 259 (James Holmes, Cell Tech), 265 (right-De Planne, Jerrican); Still Pictures p.227 (lower middle-Mark Edwards), 264 (top-Harmut Schwarzbach) (middle-Carlos Guarita, Reportage); University of Strathclyde p.177 (Carl Schaschke, Department of Chemical and Process Engineering); Picture research by johnbailey@axonimages.com

Index

Bold page references refer to illustrations, figures or tables.